COMPUTATIONAL TOXICOLOGY

COMPUTATIONAL TOXICOLOGY
METHODS AND APPLICATIONS FOR RISK ASSESSMENT

Edited by

BRUCE A. FOWLER PH.D., A.T.S.
Senior Fellow
ICF International, Fairfax, VA, USA

AMSTERDAM • BOSTON • HEIDELBERG • LONDON
NEW YORK • OXFORD • PARIS • SAN DIEGO
SAN FRANCISCO • SINGAPORE • SYDNEY • TOKYO

Academic Press is an imprint of Elsevier

Academic Press is an imprint of Elsevier
32 Jamestown Road, London NW1 7BY, UK
225 Wyman Street, Waltham, MA 02451, USA
525 B Street, Suite 1800, San Diego, CA 92101-4495, USA

Copyright © 2013 Elsevier Inc. All rights reserved

All of the forms in the resources section only may be photocopied for individual use by therapists with patients. However, they may not be posted elsewhere, distributed to anyone other than an individual patient, or used as teaching material in courses without prior permission by Elsevier.

No other part of this publication may be reproduced, stored in a retrieval system or transmitted in any form or by any means electronic, mechanical, photocopying, recording or otherwise without the prior written permission of the publisher. Permissions may be sought directly from Elsevier's Science & Technology Rights Department in Oxford, UK: phone (+44) (0) 1865 843830; fax (+44) (0) 1865 853333; email: permissions@elsevier.com. Alternatively, visit the Science and Technology Books website at www.elsevierdirect.com/rights for further information

Notice
No responsibility is assumed by the publisher for any injury and/or damage to persons or property as a matter of products liability, negligence or otherwise, or from any use or operation of any methods, products, instructions or ideas contained in the material herein. Because of rapid advances in the medical sciences, in particular, independent verification of diagnoses and drug dosages should be made.

British Library Cataloguing-in-Publication Data
A catalogue record for this book is available from the British Library

Library of Congress Cataloging-in-Publication Data
A catalog record for this book is available from the Library of Congress

ISBN: 978-0-12-396461-8

For information on all Academic Press publications
visit our website at elsevierdirect.com

Typeset by MPS Limited, Chennai, India
www.adi-mps.com

Printed and bound in the United States of America

12 13 14 15 16   10 9 8 7 6 5 4 3 2 1

Contents

Foreword ix
List of Contributors xv

1. Introduction 1
BRUCE A. FOWLER

2. Quantitative Structure-Activity Relationship (QSAR) Models, Physiologically Based Pharmacokinetic (PBPK) Models, Biologically Based Dose Response (BBDR) and Toxicity Pathways 5
PATRICIA RUIZ, XIAOXIA YANG, ANNIE LUMEN, AND JEFF FISHER

Introduction 5
Application of Structure-Activity Relationship (SAR) and Quantitative Structure-Activity Relationship (QSAR) 6
Physiologically Based Pharmacokinetic (PBPK) Modeling Case Studies 7
VOC Models 8
Metals Models 10
References 16

3. Multiple Chemical Exposures and Risk Assessment 23
JOHN C. LIPSCOMB, NIKKI MAPLES-REYNOLDS, AND MOIZ MUMTAZ

Historical Perspective 23
Regulatory Perspective 24
Mixtures versus Components 26
Additivity Approaches 27

Future Directions 38
References 41

4. Modeling of Sensitive Subpopulations and Interindividual Variability in Pharmacokinetics for Health Risk Assessments 45
KANNAN KRISHNAN, BROOKS McPHAIL, WEIHSUEH CHIU, AND PAUL WHITE

Introduction 45
Physiological Differences and PBPK Modeling of Sensitive Human Subpopulations 47
Animal PBPK Models for Evaluating Sensitive Subpopulations 58
Concluding Remarks 61
Disclaimer 62
References 62

5. Integrated Systems Biology Approaches to Predicting Drug-Induced Liver Toxicity 67
KALYANASUNDARAM SUBRAMANIAN

Introduction 67
General Principles 68
Model Building 68
Energy Homeostasis 69
Glutathione Homeostasis 70
Fatty Acid Metabolism 71
Bile Salt Metabolism and Transport 73
Solving the Equation-Set 74
Model Validation and Predictions 74
Conclusions 79
References 80

6. Computational Translation and Integration of Test Data to Meet Risk Assessment Goals 85
LUIS G. VALERIO JR.

Introduction 85
Computational Analyses and Translational Research 86
Toxicology-Based (Q)SARs 87
Read-Across 97
Data Mining for Computational Translation and Integration of Test Data 98
High-Throughput Screening for Signal Detection in Risk Assessment 100
Integrating Computational Tools with Test Data for Risk Assessment 102
Disclaimer 107
References 108

7. Computational Translation of Nonmammalian Species Data to Mammalian Species to Meet REACH and Next Generation Risk Assessment Needs 113
EDWARD J. PERKINS AND NATÀLIA GARCIA-REYERO

A Changing Regulatory Environment 113
Nonmammalian Species Can Help to Reduce, Refine, and Replace Mammalian Animal Testing 115
Pathway-Based Hazard and Risk Assessment 115
Translating Effects on Nonmammalian Species to Mammalian Species 118
Translating Molecular Initiating Events: Gene/Protein Annotation and Mammalian Ortholog Identification 120
Annotation of Large Gene Sets 122
Pathway-Level Comparison/Translation 123
Pathway-Based Extrapolation to Mammals in Determining Chemical Mode of Action 124
Pathway-Based Dose-Response Relationships 125
Network Inference and Mapping 128
Cross-Species Analysis Using Networks 128
Translating Effects through Computational Modeling at the Systems Level 130
Future Efforts in Use of High-Throughput Screening and "Omics" Technology and Computational Tools in Translation of Nonmammalian Species to Mammalian Species to Meet REACH and Next Generation Risk Assessment Needs 130
References 131

8. Interpretation of Human Biological Monitoring Data Using a Newly Developed Generic Physiological-Based Toxicokinetic Model 137
FRANS JONGENEELEN, WIL TEN BERGE, AND PETER J. BOOGAARD

Introduction 137
The Generic PBTK Model IndusChemFate 139
Examples 141
Discussion 148
Supplementary Information 149
References 149

9. Uses of Publicly Available Data in Risk Assessment 151
ISAAC WARREN AND SORINA EFTIM

Introduction 151
Publicly Available Data Sets with Uses in Risk Assessment 152
Comparison of the NHANES IV and ToxCast™ Data Sets 159
Methods for Compiling Data from Multiple Sources for Risk Assessment 159
Designing Publicly Available Toxicological Data Sets 162
Analogies to the Human Genome Project in Computational Toxicology 163
Chemical Domain and Limitations to Data Analysis of Traditional and Computational Toxicology Data 164
Data Semantics and Limitations to Relating HTS Data to *In Vivo* Effects 165
Conclusions 166
References 166

10. Computational Toxicology Experience and Applications for Risk Assessment in the Pharmaceutical Industry 171

NIGEL GREENE AND MARK GOSINK

Background 171
Two Main Considerations 172
Summary 188
References 188

11. Omics Biomarkers in Risk Assessment 195

HONG FANG, HUIXIAO HONG, ZHICHAO LIU, ROGER PERKINS, REAGAN KELLY, JOHN BERESNEY, WEIDA TONG, AND BRUCE A. FOWLER

Abbreviations and Glossaries 195
Introduction 196
Biomarkers 198
Bioinformatics Approaches: Challenges and Solutions in Omics Biomarker Discovery 201
Decision Forest for Omics Biomarkers 206
Conclusion 210
Disclaimer 210
References 211

12. Translation of Computational Model Results for Risk Decisions 215

WILLIAM MENDEZ AND BRUCE A. FOWLER

Origins and Nature of the Computational Toxicology Applications in Risk Assessment 215
Drivers for the Application of Computational Toxicology to Risk Assessment 220
Translational Research 221
Computational Toxicology Applications in Risk- and Hazard-Based Screening 228
Current Status of Computational Toxicology in Quantitative Risk Assessment 236
Summary 238
References 242

13. Future Directions for Computational Toxicology for Risk Assessment 247

BRUCE A. FOWLER

Needed Essential Elements 247
Specific Elements in Computational Toxicology Needed for the Field to Move Forward 248

Index 251

Foreword

The roots of toxicology lie in antiquity with writings such as the Ebers Papyrus, from around the 1500 BC, describing the effects of hemlock, metals, and opium.[1] In fact, the very word "toxicology" comes from the Greek "toxicon" or "arrow poison." It arose as an observational science investigating the adverse effects of chemical, physical, and biological agents on humans and animals. However, it is only recently that toxicology has developed beyond an observational science. As the science advanced, safety concerns focused more on subtle effects of chemicals rather than gross adverse effects. In moving beyond its early roots, toxicology has taken advantage of the significant advances in technology and knowledge from fields as diverse as chemistry, medicine, computer sciences, mathematics, and statistics. Computational toxicology is the integration of advances in molecular biology and computational methods to improve our ability to estimate the adverse health effects of chemical and biological exposures. Computational toxicology has great promise to resolve seemingly intractable issues such as species extrapolation and population variability in addition to enhancing the scientific basis of risk assessment.

The first efforts in computational toxicology involved the development of physiologically based pharmacokinetic (PBPK) models. Classical pharmacokinetic modeling analyzed time course of tissue chemical concentrations using empirical approaches that estimate key kinetic parameters such as half-life and volume of distribution. Because these compartmental models do not directly represent the physiology or biology of the species under investigation, these models are best used for data interpolation rather than extrapolation.[2] In contrast, PBPK models include a mathematical description of actual tissues based on their volume, blood flow, and biochemical properties, making them readily useful for species extrapolation. When physiological parameters measured in an animal, such as blood flow and tissue volume, are replaced with human physiological parameters, a rodent model can be scaled to humans. Inclusion of metabolic parameters from human tissues, such as *in vitro* data using human hepatocytes, can dramatically improve the accuracy of the model.

The initial PBPK models used in toxicology, such as the methylene chloride models,[3] were relatively simple descriptions of the physiological and biochemical properties important in the kinetic behavior of chemicals. As our knowledge of toxicology and computational approaches matured, these models became more complicated and described gene expression changes[4,5] (and even endocrine homeostasis and disruption).[6] More recent advances in the field describe methods to incorporate human variability, due to life stage, genetic polymorphisms of metabolizing enzymes, and physiological differences, in PBPK models. Maturation of the field has led to the application of PBPK models in risk assessment as a more common practice.

PBPK modeling has developed to a point that it has been used in a predictive manner. There are several menu-driven PBPK models on the market (SimCyp, PK Sim, GastroPlus). Because these PBPK models already include physiological parameters for humans and experimental animals, the only information required is for chemical-specific parameters. The human physiological parameters in these models are based on databases containing physiological, genetic, and epidemiological data. In addition, these models have physiological descriptions of infants, neonates, and children. For example, using *in vitro* metabolism data, Gibson et al.[7] and Shaffer et al.[8] demonstrated that SimCyp accurately predicted the blood concentration–time course for pharmaceutical agents in preclinical trials. This *in vitro* to *in vivo* extrapolation approach is used to better design clinical trial studies of pharmaceutical agents. In addition, because these software applications incorporate population data on physiological parameters and population variability in metabolic capabilities, they can be used to assess population variance in pharmacokinetics prior to clinical trials.[8]

Experimental toxicology has often used animal or cell models to study mechanisms; however, the ultimate aim is to extrapolate the understandings gained to humans. Early mechanistic studies examined only single genes or enzyme activities, often only those that were already known. In some ways, this is the classic approach of "looking for keys under the lamppost." More often than not, these attempts proved unsatisfactory in describing detailed mechanisms of action for most chemicals. For example, 2,3,7,8-TCDD is one of the most well-studied chemicals, yet prior to the development of "omic" approaches, we understood only that the binding and activation of the Ah receptor is necessary but not sufficient for the toxicity of TCDD. However, despite the application of these newer technologies, our understanding of detailed mechanisms of action and their use in species extrapolation eludes us. For example, Perdew and colleagues compared the gene expression profile of mouse and human hepatocytes following exposure to TCDD.[9] Over 1,400 genes were altered by these treatments in both species; however, only 18% of the genes overlapped between species. Black et al.,[10] when comparing rat and human hepatocytes exposed to TCDD, observed similar results. Determining which of the 1,400 genes are involved in toxic responses becomes difficult. Therefore, while we understand that there are species differences in transcriptomic changes to dioxins, we are still uncertain if the overlapping responses are involved in the toxicity, or if there are true species differences in toxic response based on these data. These qualitative and correlative analyses provide some insight into species differences. However, the complexity of these data is not adequately addressed by these initial statistical approaches evaluating differences and similarities.

Our ability to collect data at multiple levels of biological organization has outstripped our ability to interpret intuitively these data. The development of high content technologies such as whole genome arrays, proteomics, and metabolomics allows us to query biological samples for thousands of changes in mRNAs, proteins, and small molecules. Each of these technologies requires bioinformatic or statistical approaches and database management tools. Even when these data are organized into pathways and networks, each network may consist of hundreds or thousands of gene, protein, and metabolite interactions. The complexity of these networks has necessitated the development and application of computational models to aid in data interpretation.

Computational toxicology has taken advantage of the advances in computational cell biology and has adopted many of its methods and approaches. Some of the approaches are more empirical in nature, such as KEGG pathway analysis (http://www.genome.jp/kegg/) or the Comparative Toxicogenomic Database (http://ctdbase.org/). Other approaches attempt to more quantitatively model organs or tissues. These attempts at virtual tissues take advantage of not only the high content "omic" data but of advances in multiresolution imaging and multiscale simulation to reconstruct tissues *in silico*.[11] The development of virtual organs may provide a platform for quantitative analysis of complex data sets collected using these high content technologies.

Another problem in toxicology is that there are too many chemicals and too many toxicities to test. In the past, we usually tested one chemical at a time using highly scripted guideline studies. The number of guideline studies, which began by examining overt toxicity and frank terata, has expanded to include a plethora of subtle endpoints and tests, resulting in a tremendous increase in the cost and time to get a chemical to market. There are possibly 80,000 chemicals in commerce.[12] The NAS reviewed toxicity testing for environmental assessment and concluded that the toxicity and exposure information available for chemicals is so incomplete that it does not support the review process.[13] Evaluating these chemicals using the traditional approaches is an intractable problem. We cannot traditionally test our way out of this problem, and alternative approaches to traditional toxicity testing are required.

The inability of traditional toxicity testing to address the data gaps for large numbers of chemicals was again recognized by the NAS in its report "Toxicity Testing in the 21st Century: A Vision and a Strategy."[14] In this paradigm-changing effort, the NAS proposed that toxicology take advantage of the high throughput screening technologies employed by pharmaceutical companies in drug discovery. These methods are capable of screening tens of thousands of chemicals using *in vitro* assays evaluating potential drug targets. Rather than screen for drug targets, the NAS proposed to screen for toxicity pathways. A toxicity pathway is a cellular response pathway that, when sufficiently perturbed, is expected to result in an adverse health effect. It is hoped that the high throughput *in vitro* screening data would enable the prioritization for targeted toxicity testing the large number of chemicals with inadequate toxicity data.

In recognition of this problem, the National Institute of Environmental Health Sciences/National Toxicology Program, the NIH National Chemical Genomics Center, and the Environmental Protection Agency's National Center for Computational Toxicology entered into a Memorandum of Understanding (http://ntp.niehs.nih.gov/?objectid=05F80E15-F1F6-975E-77DDEDBDF3B941CD) in 2008 on the research, development, validation, and translation of new and innovative *in vitro* and lower organism test methods that characterize key steps in toxicity pathways. This collaborative effort, known informally as Tox21, now includes the U.S. Food and Drug Administration. As part of theTox21 effort, these agencies have identified and/or supported the development of assays suitable for use in quantitative high throughput and high content screens (qHTS). This approach has resulted in data on thousands of chemicals for almost a hundred assays.[15–19] Once again, similar to our "omic" technologies, we have developed approaches that result in data sets that cannot be interpreted intuitively. Significant computational efforts resulted in the development of statistically based approaches for distinguishing between active, inactive, and inconclusive responses in these screens. These data are then used in the development of models that aid in identifying patterns predictive of toxicity. The hope is that these approaches can lead to avoidance of toxic chemicals and to safer products entering the market more quickly, while potentially toxic chemicals, already on the market, are identified sooner.

One of the first pilot projects using qHTS was initiated by the EPA in its ToxCast program. This program evaluated 309 chemicals on several hundred qHTS assays.[20–22] These data have been used to develop models that predict a variety of toxicities from cancer[20] to developmental toxicities.[22–24] These models bin chemicals into classes and do not provide dose response information. Rotroff et al.[25] demonstrated that by assuming media concentrations are equivalent to steady-state blood concentrations, one can estimate human daily intakes that result in equivalent blood concentrations to the media concentrations *in vitro* using the predictive pharmacokinetic models such as SimCyp. The lack of dose response information may limit the utility of using the *in vitro* data for risk assessment. However, it may be possible to use virtual tissue models to describe the dose response relationships.[26] One could envision a future where the *in vitro* qHTS data are input parameters for a virtual human model, which could then predict dose response relationships for adverse effects. How far off into the future these models are remains to be determined.

The approaches of computational toxicology are predicated on the concept of toxicity pathways. This concept is really an extension of cell biology and our understanding of metabolic processes of the cell. Similar to disease pathways, toxicity pathways are normal processes that, once perturbed, result in adverse response (toxicity pathways) or disease (disease pathways). In 1998, Daniel E. Koshland, Jr., Lasker Award winner and former editor of *Science,* described three eras in our knowledge of metabolic processes of cells: the Era of Pathway Identification, the Era of Pathway Regulation, and finally, the Era of Pathway Quantification.[27] In the 20th century, metabolic pathways were identified and their regulation was qualitatively described. Koshland proposed that the third and present era should focus on quantification of pathways. Only by understanding the quantitative relationships between pathways can we be sure that we have identified and described all the critical components in these metabolic pathways.

Koshland's ideas can be extended to toxicology. In its roots, the first era of toxicology was the Era of Hazard Identification. This first era of modern toxicology focused on indentifying toxicities and developing appropriate testing strategies to evaluate chemicals for these toxicities. Toxicology has now moved into an Era of Toxicity Pathway Identification and Quantification.[28] In order to truly identify a toxicity pathway, one must understand the quantitative relationship between altering the pathway and an adverse or toxic response. Toxicology must adapt the computational tools used in cell biology, physics, and engineering in order to build these relationships.

Over the past two decades, our knowledge of molecular biology and the tools available to query this knowledge have increased exponentially. We have gone from examining one or two proteins at a time to evaluating the expression of the entire genome and from testing single chemicals to thousands at a time. The advent of "omics" and qHTS technologies necessitates the expansion of computational toxicology. The simple statistical approaches of the past cannot be applied to these data. Translating these approaches to chemical risk assessment will not be trivial. While these approaches may replace our traditional risk assessment approaches, we need to be careful that we understand their uncertainty in their use. Research efforts should focus on not just developing these new technologies but also on their applications. These research efforts should be coordinated with regulators and the regulated community so that the end results of this work will provide a more scientific risk assessment process.

Linda S. Birnbaum, Ph.D., D.A.B.T., A.T.S.,
Director,
National Institute of Environmental Health Sciences and National Toxicology Program

Michael DeVito, Ph.D.
Senior Toxicologist,
National Institute of Environmental Health Sciences and National Toxicology Program

References

[1] Gallo, 2008.
[2] Andersen ME. Toxicokinetic modeling and its applications in chemical risk assessment. Toxicol Lett 2003;138(1–2):9–27.
[3] Andersen ME, Clewell III HJ, Gargas ML, Smith FA, Reitz RH. Physiologically based pharmacokinetics and the risk assessment process for methylene chloride. Toxicol Appl Pharmacol 1987;87(2):185–205.
[4] Emond C, Michalek JE, Birnbaum LS, DeVito MJ. Comparison of the use of a physiologically based pharmacokinetic model and a classical pharmacokinetic model for dioxin exposure assessments. Environ Health Perspect 2005;113(12):1666–8.
[5] Emond C, Birnbaum LS, DeVito MJ. Use of a physiologically based pharmacokinetic model for rats to study the influence of body fat mass and induction of CYP1A2 on the pharmacokinetics of TCDD. Environ Health Perspect 2006;114(9):1394–400.
[6] Fisher J, Lumen A, Latendresse J, Mattie D. Extrapolation of hypothalamic-pituitary-thyroid axis perturbations and associated toxicity in rodents to humans: case study with perchlorate. J Environ Sci Health C Environ Carcinog Ecotoxicol Rev 2012;30(1):81–105.
[7] Gibson CR, Bergman A, Lu P, Kesisoglou F, Denney WS, Mulrooney E. Prediction of Phase I single-dose pharmacokinetics using recombinant cytochromes P450 and physiologically based modelling. Xenobiotica 2009;39(9):637–48.
[8] Shaffer CL, Scialis RJ, Rong H, Obach RS. Using simcyp to project human oral pharmacokinetic variability in early drug research to mitigate mechanism-based adverse events. Biopharm Drug Dispos 2012;33(2):72–84.
[9] Flaveny CA, Murray IA, Perdew GH. Differential gene regulation by the human and mouse aryl hydrocarbon receptor. Toxicol Sci 2010;114(2):217–25.
[10] Black MB, Budinsky RA, Dombkowski A, Cukovic D, LeCluyse EL, Ferguson SS, et al. Cross-species comparisons of transcriptomic alterations in human and rat primary hepatocytes exposed to 2,3,7,8-tetrachlorodibenzo-p-dioxin. Toxicol Sci 2012;127(1):199–215.
[11] Wambaugh and Shah. 2010.
[12] Locke PA, Myers Jr. DB. A replacement-first approach to toxicity testing is necessary to successfully reauthorize TSCA. ALTEX 2011;28(4):266–72.
[13] NAS. 2006.
[14] NAS. 2007.
[15] Fox JT, Sakamuru S, Huang R, Teneva N, Simmons SO, Xia M, et al. High-throughput genotoxicity assay identifies antioxidants as inducers of DNA damage response and cell death. Proc Natl Acad Sci USA 2012;109(14):5423–8.
[16] Huang R, Xia M, Cho MH, Sakamuru S, Shinn P, Houck KA, et al. Chemical genomics profiling of environmental chemical modulation of human nuclear receptors. Environ Health Perspect 2011;119(8):1142–8.
[17] Lock EF, Abdo N, Huang R, Xia M, Kosyk O, O'Shea SH, et al. Quantitative high-throughput screening for chemical toxicity in a population-based in vitro model. Toxicol Sci 2012;126(2):578–88.
[18] Sakamuru S, Li X, Attene-Ramos MS, Huang R, Lu J, Shou L, et al. Application of a homogenous membrane potential assay to assess mitochondrial function. Physiol Genomics 2012;44(9):495–503.
[19] Shukla SJ, Huang R, Simmons SO, Tice RR, Witt KL, Vanleer D, et al. Profiling environmental chemicals for activity in the antioxidant response element signaling pathway using a high-throughput screening approach. Environ Health Perspect 2012 [Epub ahead of print] PubMed PMID: 22551509.

[20] Judson RS, Houck KA, Kavlock RJ, Knudsen TB, Martin MT, Mortensen HM, et al. In vitro screening of environmental chemicals for targeted testing prioritization: the ToxCast project. Environ Health Perspect 2010;118(4):485–92.

[21] Knudsen TB, Houck KA, Sipes NS, Singh AV, Judson RS, Martin MT, et al. Activity profiles of 309 ToxCast™ chemicals evaluated across 292 biochemical targets. Toxicology 2011;282 (1–2):1–15.

[22] Martin MT, Knudsen TB, Reif DM, Houck KA, Judson RS, Kavlock RJ, et al. Predictive model of rat reproductive toxicity from ToxCast high throughput screening. Biol Reprod 2011;85 (2):327–39.

[23] Reif DM, Martin MT, Tan SW, Houck KA, Judson RS, Richard AM, et al. Endocrine profiling and prioritization of environmental chemicals using ToxCast data. Environ Health Perspect 2010;118(12):1714–20.

[24] Kleinstreuer NC, Judson RS, Reif DM, Sipes NS, Singh AV, Chandler KJ, et al. Environmental impact on vascular development predicted by high-throughput screening. Environ Health Perspect 2011; 119(11):1596–603.

[25] Sipes NS, Martin MT, Reif DM, Kleinstreuer NC, Judson RS, Singh AV, et al. Predictive models of prenatal developmental toxicity from ToxCast high-throughput screening data. Toxicol Sci 2011;124(1):109–27.

[26] Rotroff DM, Wetmore BA, Dix DJ, Ferguson SS, Clewell HJ, Houck KA, et al. Incorporating human dosimetry and exposure into high-throughput in vitro toxicity screening. Toxicol Sci 2010;117(2):348–58.

[27] Shah I, Wambaugh J. Virtual tissues in toxicology. J Toxicol Environ Health B Crit Rev 2010;13 (2–4):314–28.

[28] Koshland Jr. DE. The era of pathway quantification. Science 1998;280(5365):852–3.

List of Contributors

John Beresney ICF International, Fairfax, VA, USA

Peter Boogaard Shell Health, Shell International, The Hague, The Netherlands

Weihsueh Chiu U.S. Environmental Protection Agency, National Center for Environmental Assessment, Washington, DC, USA

Sorina Eftim Department of Environmental and Occupational Health, George Washington University School of Public Health and Health Services, Washington, DC, USA

Hong Fang Office of Scientific Coordination, National Center for Toxicological Research, U.S. Food and Drug Administration, Jefferson, AR, USA

Jeff Fisher Food and Drug Administration, National Center for Toxicological Research, Jefferson, AR, USA

Bruce A. Fowler ICF International, Fairfax, VA, USA

Natàlia Garcia-Reyero Mississippi State University, Starkville, MS, USA

Mark Gosink Investigative Toxicology, Pfizer Inc., Groton, CT, USA

Nigel Greene Compound Safety Prediction, Pfizer Inc., Groton, CT, USA

Huixiao Hong Division of Bioinformatics and Biostatistics, National Center for Toxicological Research, U.S. Food and Drug Administration, Jefferson, AR, USA

Frans Jongeneelen IndusTox Consult, Universitair Bedrijven Centrum, Nijmegen, The Netherlands

Reagan Kelly Division of Bioinformatics and Biostatistics, National Center for Toxicological Research, U.S. Food and Drug Administration, Jefferson, AR, USA

Kannan Krishnan Université de Montréal, Montréal, QC, Canada; U.S. Environmental Protection Agency, National Center for Environmental Assessment, Washington, DC, USA

John C. Lipscomb U.S. Environmental Protection Agency, Office of Research and Development, National Center for Environmental Assessment, Cincinnati, OH, USA

Zhichao Liu Division of Bioinformatics and Biostatistics, National Center for Toxicological Research, U.S. Food and Drug Administration, Jefferson, AR, USA

Annie Lumen Food and Drug Administration, National Center for Toxicological Research, Jefferson, AR, USA

Nikki Maples-Reynolds ICF International, Fairfax, VA, USA

William Mendez ICF International, Fairfax, VA, USA

Brooks McPhail Agency of Toxic Substances and Disease Registry/Center for Disease Control and Prevention, Division of Toxicology and Human Health Sciences/Environmental Toxicology Branch, Atlanta, GA, USA

Moiz Mumtaz Centers for Disease Control and Prevention, Department of Toxicology and Environmental Medicine, Atlanta, GA, USA

Edward J. Perkins U.S. Army Engineer Research and Development Center, Vicksburg, MS, USA

Roger Perkins Division of Bioinformatics and Biostatistics, National Center for Toxicological Research, U.S. Food and Drug Administration, Jefferson, AR, USA

Patricia Ruiz Agency of Toxic Substances & Disease Registry (ATSDR), U.S. Centers for Disease Control, Chamblee, GA, USA

Kalyanasundaram Subramanian Strand Life Sciences P. Ltd., Bangalore, India

Wil ten Berge Santoxar, Westervoort, The Netherlands

Weida Tong Division of Bioinformatics and Biostatistics, National Center for Toxicological Research, U.S. Food and Drug Administration, Jefferson, AR, USA

Luis G. Valerio Jr. Science and Research Staff, Office of Pharmaceutical Science, Center for Drug Evaluation and Research, U.S. Food and Drug Administration, Silver Spring, MD, USA

Isaac Warren ICF International, Fairfax, VA, USA

Paul White U.S. Environmental Protection Agency, National Center for Environmental Assessment, Washington, DC, USA

Xiaoxia Yang Food and Drug Administration, National Center for Toxicological Research, Jefferson, AR, USA

CHAPTER 1

Introduction

Bruce A. Fowler
ICF International, Fairfax, VA, USA

The evolving field of computational toxicology includes a number of related subdisciplines, some of which are reviewed in this book by international class authorities. The purpose of this chapter is to provide an overview of why computational modeling has become so important to the field of toxicology in general and to the practical needs of risk assessment in particular.

First, it should be noted that computational modeling is accepted as a powerful risk assessment tool for a number of important aspects of contemporary life. Weather forecasting models are an outstanding example; the results save lives every year by predicting the paths of hurricanes and other storms so that needed actions may be initiated *before* the arrival of the adverse weather condition, and loss of life and property may be prevented. In the pharmaceutical industry, potentially useful drugs are screened for dangerous side effects early in development based on chemical structure analyses. Computer modeling of adverse side effects has been used with good results for decades for helping to interpret the results of *in vitro* studies, guide more costly *in vivo* animal studies, and ultimately provide credible information on likely human health effects for risk assessment. This general approach is depicted in Figure 1-1, which shows the forward extrapolation of basic scientific information through the basic levels of biological organization beginning with molecular reactions to cells to organs to whole organisms to human risk assessment. Aspects of this paradigm have been tried, tested, and vetted for a number of years; so there is circumspect reason to trust that information from these embedded components has proven useful in helping to protect the public's health. As these elements are better understood and their linkages more fully appreciated, results from computational toxicology will become more valuable. This book addresses merging information from different fields of science to provide a more coherent and richer view of mechanisms and risks.

More broadly, chemical modeling tools are being applied to the more than 80,000 chemicals in commercial use, plus 500–1,000 new ones each year. Both individual and mixtures' risks

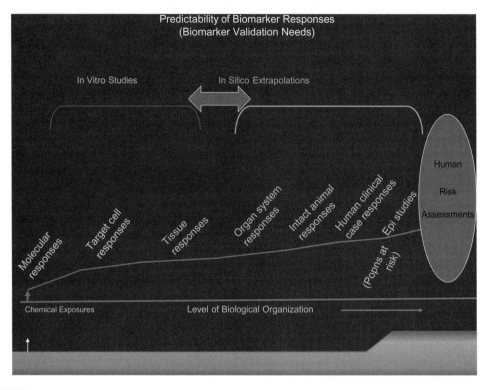

FIGURE 1-1 General diagram for utilizing computational toxicology methods to extrapolate basic molecular biomarker data from *in vitro* test systems to human health risk assessments.

are assessed to help prioritize regulatory or cleanup decisions. We live in a chemical-rich world, and there are quite simply not enough toxicologists or laboratory facilities capable of evaluating this large number of chemicals in a rapid and cost-effective manner. Yet society and societal decision makers must have guidance on chemical safety issues in a timely manner. Major chemical accidents, such as the Deepwater Horizon Gulf oil spill, are clear examples of this need and the effectiveness of computational modeling in providing needed answers with a short turnaround time so that important decisions could be undertaken.

The need for precision and expediency in risk assessment of chemicals exceeds the data available, and this is unlikely to change. To bridge this gap, creative computer modeling methods have been developed that consolidate and use all credible evidence. A strong underpinning of these approaches is the transparency and replication of each decision step. This book brings together some of the most promising methods that have been tested and successfully applied to real-world needs to meet the pressing challenge of assessing human risk from chemical exposures.

The dual overall goals of this book are to provide a summary of the state of the science of computational toxicology by presenting specific applications that have enhanced the response to a defined risk assessment challenge and to suggest future research needs based on a synthesis of the extant knowledge. Additionally, important areas, such as high-throughput screening of large numbers of chemicals, not addressed in this book, are making great advances and hold promise for improving risk assessment when applied to specific risk assessment situations in the future.

The applications of computational modeling presented cover a diverse range of exposures and needs for rapid risk assessment responses. They have been used to inform decision making in varying but challenging risk assessment situations confronting risk managers. These needs include risk of chemical mixtures encountered in the Deepwater Horizon Gulf oil spill; the identification of sensitive subpopulations as a function of age, gender, genetic inheritance, and diet; and the rapid development of preliminary health guidance values for emergency response situations. The computational methodologies have generated evidence-based, quantitative levels of risk. These methods include database mining; molecular pathway/network analyses; read-across matrices for REACH chemical registrations; alternatives to animal testing; and application of integrated QSAR, PBPK, and molecular docking approaches for predicting the toxicity of chemicals and their metabolites on an individual or mixture basis.

Optimism regarding the practical application of computational modeling stems not only from the examples discussed by experts in this book but also the 20+ years of successful and productive experience in the pharmaceutical industry in the design and evaluation of drugs. This clear track record of success will undoubtedly continue to expand. Computational toxicology is not a panacea that will resolve all current chemical risk assessment issues, but the judicious application of the available computational methods to specific problems **can** yield robust information to better inform wiser, cost-effective chemical risk decision making today. The examples provided should stimulate further advances in methods and importantly expand the number and types of risk assessment needs to which these methods may be credibly applied in a transparent manner.

A concerted effort has been made to provide international experts an opportunity to discuss applications in a readily understandable form so that persons with limited technical backgrounds can make optimal use of the information. Important practical advantages of computational toxicology for promoting chemical safety rest with initial screening of chemicals or drugs in order to focus limited laboratory resources on more precise and significantly important questions. A second important aspect from the perspective of risk assessment is the synthesis, analysis, and interpretation of data generated by laboratory studies. This bioinformatics aspect of computational methodologies is of ever-increasing importance, since modern molecular approaches to toxicology generate enormous quantities of complex and interrelated data sets. This voluminous amount of information must be analyzed, digested, and interpreted in order to be of practical use in risk assessment. A promising aspect for risk assessment in this area is the generation of molecular pathway analyses, which bring together several lines of information to gain insights into likely chemical modes of action. If the pathway analyses are sufficiently robust to predict cellular death or carcinogenic transformation, then the primary toxicity pathway or network of pathways may be designated as an adverse outcome pathway (AOP) and used for informing credible preliminary risk assessment decisions and further confirmatory laboratory research needs.

Specific chapters include PBPK, QSAR, and toxicity pathways for initial screening of chemicals; application of QSAR to chemical agents released into water environments; chemical mixtures; modeling of sensitive subpopulations for risk assessments; computational modeling of toxicogenomic data sets for risk assessment; and integrating systems biology approaches for predicting drug-induced liver toxicity. Other chapters focus on practical translation of computational methods for risk assessment; computational translation and integration of test data to meet risk assessment goals; computational translation of data from nonmammalian species to meet REACH informational strategies; development of *in silico* models for risk assessment; examples of simulations with a newly developed generic PBTK model for incorporating human biomonitoring data to meet REACH guidelines; use of public data sets for risk assessment, computational toxicology, and applications for risk assessment of pharmaceuticals; the decision forest—a novel pattern recognition method for *in silico* risk assessment; and translation of computational model results for risk assessment.

The value of these approaches has also been recognized by leading forward-thinking U.S. public health agencies, such as the NIEHS, FDA, EPA, and ATSDR, in fostering initiatives that utilize computational approaches alone or in combination with modern molecular toxicology biomarker tools. The excellent foreword to this book provides a prospective-looking overview of the interagency Tox21, and EPA ToxCast and NexGen programs, which are focused on moving the field of chemical risk assessment ahead by utilizing more modern 21st century tools and provides a rationale for why these approaches are important for risk assessment. It should be noted that the pharmaceutical and chemical industries have also been heavily committed to these approaches for many years and have made major contributions to thought leadership in this area.

It is hoped that the book will provide the reader with a good perspective of what is currently being done in computational toxicology and will stimulate insights that pinpoint where applications can better inform risk assessment decisions about chemicals and drugs.

CHAPTER 2

Quantitative Structure-Activity Relationship (QSAR) Models, Physiologically Based Pharmacokinetic (PBPK) Models, Biologically Based Dose Response (BBDR) and Toxicity Pathways
Computational Tools for Public Health

Patricia Ruiz[1], Xiaoxia Yang[2], Annie Lumen[2], and Jeff Fisher[2]

[1]Agency of Toxic Substances & Disease Registry (ATSDR), U.S. Centers for Disease Control, Chamblee, GA, USA, [2]Food & Drug Administration, National Center for Toxicological Research, Jefferson, AR, USA

INTRODUCTION

Human health risk assessment is "the process to estimate the nature and probability of adverse health effects in humans who may be exposed to chemicals in contaminated environmental media, now or in the future."[1] Currently, most data required for human risk assessment are derived from toxicological studies conducted in laboratory animals. The "Toxicology in the 21st Century" initiative[2] expands the toxicity testing tools to include the development of alternative toxicity testing methods that examine pathways of toxicity (on a large scale) and the employment of dose-response and extrapolation modeling tools. While the latter methodology is in its infancy, several methodologies for dose-response and extrapolation modeling are more mature. Over the last decade, physiologically based pharmacokinetic (PBPK) modeling has gained acceptance as a computational tool for use in public health assessments.[3,4] In this chapter, we present examples of quantitative structure-activity relationship (QSAR) models, physiologically based pharmacokinetic (PBPK) models, and biologically based dose response (BBDR) models that have been developed for use in public health assessments and advancing knowledge gained through *in silico* examinations of biological systems.

APPLICATION OF STRUCTURE-ACTIVITY RELATIONSHIP (SAR) AND QUANTITATIVE STRUCTURE-ACTIVITY RELATIONSHIP (QSAR)

Alternative methods, such as structure-activity relationship (SAR) and quantitative structure-activity relationship (QSAR), are invaluable tools for simulating necessary endpoints to correctly assess chemical hazards. SAR and QSAR models are mathematical relationships between the chemical's quantitative molecular descriptors and its toxicological endpoint.[5–10] SAR and QSAR are intended to complement, not replace, development and use of experimental, laboratory-based approaches. Some SAR and QSAR methods are based on a limited set of chemicals, whereas others are based on a vast or diverse database. Some are based on generic principles of toxicology; others use more specific mechanistic or mode of action information.[6,8,10–15]

A growing number of such methods and models are being developed for various specific toxicity endpoints, and are expected to play an increasingly important role in estimating and predicting toxicity for hazard and risk assessment. Many commercial and open source software packages, such as TOPKAT, CASE, and MultiCASE, have used QSAR models to predict human health effects and related toxicities.[16–25]

SAR/QSAR modeling plays an important and active role in the Agency of Toxic Substances and Disease Registry (ATSDR) programs in support of the agency mission to protect human populations from exposure to environmental contaminants.[26] It is used for cross-chemical extrapolation to complement the traditional toxicological approach when chemical-specific information is unavailable.[6,10,26–28] The key application of SAR/QSAR is filling chemical toxicity data gaps. QSAR methods are used to investigate adverse health effects and exposure levels, bioavailability, and pharmacokinetic properties of hazardous chemical compounds. QSAR analyses are incorporated into ATSDR documents (such as the Toxicological Profiles and chemical-specific health consultations) to support environmental health assessments and prioritization of environmental chemical hazards, and to improve study design.[10]

Two examples of the SAR/QSAR modeling were recently published.[28,29] In Ruiz et al.,[28] the Concise International Chemical Assessment Documents board requested quantitative structure toxicity relationship (QSTR) assessments of certain specific toxicity endpoints for ethylene glycol mono-*n*-alkyl ethers. These chemicals are a family of ethylene glycol ethers widely used as organic solvents and thinners for resins, paints, and dyes. QSAR models embedded in TOPKAT, a commercial toxicity prediction system, were applied to assess the toxicity potential of these compounds. Concordance between experimental and predicted mutagenicity data on many of these compounds confirmed the accuracy of QSAR model predictions. Although mutagenicity and carcinogenicity were not indicated for the majority of these molecules, predictions indicated their potential developmental toxicity. Applying unique features of the QSAR model helps to identify substructures and structural features that may lead to increased developmental toxicity potential and thus gain insight into the substructures that may alter potential for developmental toxicity.

In another example reported by Pohl et al.,[29] QSAR analysis was used to identify chemicals that affect joint toxicity of chemicals. Chemicals such as xylene and toluene have been shown to interact with ethoxyethanols at the enzymatic level and influence their toxicity outcome, particularly testicular toxicity. A unique feature of TOPKAT called "QSAR similarity analysis" was used to identify chemicals that could cause developmental toxicity and are structurally similar to xylene and toluene. Unlike other similarity measures, QSAR similarity search, expressed as the Euclidian distance computed from the values of model descriptors, is property sensitive because it reflects the similarity of descriptor values between two molecules with respect to a specific property or endpoint.[20] When implemented, this type of search mines the associated database of a QSAR model to look for the most similar compounds, and it assigns surrogate chemicals and prediction confidence based on the similarity distance and concordance between the experimental and predicted values of these similar compounds.[30,31] These two articles demonstrate the importance of an *in silico* method as a tool that can supplement traditional approaches to risk assessment of chemical mixtures.

SAR/QSAR tools play a role in pharmaceutical chemicals, especially when the only characteristic known is the structure of a chemical. The pharmaceutical industry has successfully used two- and three-dimensional QSAR approaches to study drug receptor interactions. Because these methods allow screening of large libraries of molecules for potential activity, such techniques could also be used for evaluating environmental chemicals. Additionally, this approach could eliminate the need for resource-intensive laboratory work for an increasing number of chemicals.

Practical use of such models can help estimate toxicity of chemicals that lack experimental data. These tools can also help prioritize chemicals for screening and subsequent toxicity testing while saving cost and time, minimizing experimental animal testing, and optimizing overall use of resources. Recent laws, such as the European Union's REACH regulation, are pushing acceptance of these methods by regulatory and public health communities in mitigating potential hazardous exposures. The versatility and utility of SAR/QSAR models could be further evaluated, demonstrated, and, if needed, refined through the *in vitro* data being generated through high-throughput and genomic studies. SAR/QSAR models are expected to play a critical role in toxicology and public health as the risk assessment process is modernized.

PHYSIOLOGICALLY BASED PHARMACOKINETIC (PBPK) MODELING CASE STUDIES

One major limitation to the PBPK models' acceptance was that they were developed using different simulation languages (e.g., MatLab™, Simusolve, AcslX™), thus making them difficult to use. Making the models more user-friendly will be an essential step to increase their use in the field by practitioners of risk and health assessments.

A General Approach to Developing a Human PBPK Toolkit

A human physiologically based pharmacokinetic (PBPK) model toolkit is being developed to assist with site-specific health assessments. The kit, designed to better serve ATSDR's health assessors and state partners, will consist of a series of published human PBPK models coded in a common simulation language (Berkeley Madonna).[4,32–36] The ultimate goal is to develop an online PBPK database, where health assessors and other related health workers can access PBPK models quickly and easily. This collection of models will be called the "ATSDR toolkit."[4,33]

For model selection criteria, information such as number of data sets used to calibrate and evaluate the model, model maturity (number of predecessor models from which the model was derived), and author experience was used. At present, the toolkit includes models of environmental contaminants, including volatile organic compounds (VOCs) and metals.[37–39]

VOC MODELS

A generic seven-compartment VOC model consisting of blood, fat, skin, kidney, and liver; rapidly and slowly perfused tissue compartments; and a gas exchange compartment was developed.[39] These compartments were included in the model based on their use in previously published PBPK models. The generic VOC PBPK model can be used for VOCs such as benzene (BEN), carbon tetrachloride (CCl_4), dichloromethane (DCM), perchloroethylene (PCE), trichloroethylene (TCE), and vinyl chloride (VC).[40–46] All compartments were described as well mixed and flow-limited. Published literature provided chemical-specific and biochemical parameters for the models.[47–54] This model code allowed simulation of three routes of individual or simultaneous exposure: inhalation, oral, and dermal. However, lack of available published human data sets prevented comparison of the generic model predictions for the dermal route. In the current version of the model, it did not include original model simulations for metabolites and metabolite data; however, a critical improvement for post-screening use in the future will be incorporating metabolites information in this model, particularly when toxicity is mediated by metabolite(s).

The applicability of the model was first assessed by comparing it to the published human kinetic data for each VOC and the corresponding published model predictions. To ensure further the reliability of the model, the area under the concentration curve (AUC) for blood or exhaled breath for each VOC using both the generic and original published model was calculated as shown in Table 2-1. The mean of the sum of the squared differences (MSSDs) between model prediction and observation for each kinetic time course data set was also calculated; MSSD was computed by squaring the difference between a measured data point and the value of the simulation at the corresponding time. These squares were summed and then the sum was divided by the number of data points. The MSSD was thus determined for both the published model and for the generic VOCs model.[39] The VOC PBPK model was used to estimate the blood concentrations for the available minimal risk levels (MRLs) values[55,56] of each of the specific VOCs for which

TABLE 2-1 Physiologically Based Pharmacokinetic (PBPK) Volatile Organic Compounds (VOCs) Model Comparison

VOCs	AUC$_r$		MSSD	
	Generic Model	Original Model	Generic Model	Original Model
BEN[a]	0.9	1.6	0.0008	0.0009
CCl$_4$[b]	2.5	1.9	0.4515	0.2344
DCM[c]	1.1	1.1	3.8214	1.1722
PCE[c]	0.6	0.8	0.0805	0.0164
TCE[c]	0.8	0.8	0.0095	0.0089
VC[b]	1.2	1.1	0.1875	0.1831

BEN, benzene; CCl$_4$, carbon tetrachloride; DCM, dichloromethane; PCE, perchloroethylene; TCE, trichloroethylene; VC, vinyl chloride.
[a]μM;
[b]ppm;
[c]mg/L.

TABLE 2-2 Comparison of Minimal Risk Level (MRL) Simulated Blood Concentration of Each Solvent, Assuming Simultaneous Inhalation (24 h/day) and Oral Ingestion (4 Drinking Bouts per Day) to the Measured Blood Concentration of Solvent Reported by National Health and Nutrition Examination Survey (NHANES) 2003−2004. The Simulated Solvent Exposure Is Set to the MRL for Inhalation of the Solvent in Air and Ingestion of the Solvent in Water

	BEN[+]	CCl$_4^+$	DCM[+]	PCE[+]	TCE[+]	VC[+]
MRL*	0.003/0.0005	0.03/0.007	0.6/0.2	0.3/0.06	0.2/0.05	2/0.2	none
Exposure Duration	Chronic	Intermediate	Acute	Chronic	Acute	Acute	—
PBPK MODEL	Blood Concentration (ng/mL)					
Predicted Peak	0.04	0.40	18.12	6.70	10.76	111.65	—
NHANES**	Blood Concentration (ng/mL)					
	0.260 (0.210−0.320)	<LOD	<LOD	0.140 (0.091−0.300)	<LOD	ND**
Limit of Detection (LOD)	0.024	0.005	0.07	0.048	0.012	ND

Ben[+], benzene; CCl$_4^+$, carbon tetrachloride; DCM[+], dichloromethane; PCE[+], perchloroethylene; TCE[+], trichloroethylene; VC[+], vinyl chloride.
*Inhalation concentration (ppm)/Oral ingestion rate (mg/kg-day);
**NHANES 2003−2004. 95th percentiles of blood concentration (in ng/mL) for U.S. population, ND = Not Done.

biomonitoring data on human blood levels were available from the National Health and Nutrition Examination Survey (NHANES).[57] Steady-state VOC concentrations in venous blood were then compared with NHANES data using these simplified assumptions about exposure frequency and duration (Table 2-2).[39] If the measured NHANES blood levels were below those estimated from the simulations, the exposures were regarded as "safe."

METALS MODELS

Published human metals PBPK models for arsenic, mercury, and cadmium were reviewed, using Berkeley Madonna to select and recode the best model available based on performance, accuracy, and reproducibility.[38] Human physiological and chemical-specific parameters describing the absorption, distribution, and blood and tissue partitioning of arsenic (As), mercury (Hg), and cadmium (Cd) were obtained from the literature.[1,58–62] The PBPK models allow individual and simultaneous simulation of different routes of exposure.

A published Cd toxicokinetic model describes aggregated lung, liver, kidney, blood, and other tissues.[60–61] Intake by oral and inhalation routes is transferred to an uptake pool that distributes to three blood compartments. The model predicted the urinary concentrations of Cd that are considered a surrogate for body burden in assessing health risk from exposure, including the sex- and age-stratified geometric urinary mean. This model was used to predict the creatinine-corrected urinary Cd concentrations among women and men from the *Fourth National Report on Human Exposure to Environmental Chemicals*,[37] as shown in Table 2-3.

The recoded human PBPK model for arsenic consists of interconnected submodels for inorganic arsenic and its metabolites, monomethyl arsenic (MMA), and dimethyl arsenic (DMA).[58] The model includes compartments for lung, liver, gastrointestinal (GI) tract, kidney, muscle, brain, skin, and heart. Simulated arsenic exposures are controlled by a series of parameters that allow single or continuous dietary exposures to inorganic arsenic in the +3 or +5 valence state or exposures via drinking water. The recoded model

TABLE 2-3 Dietary Cadmium Intake, Model Predictions, and Geometric Mean Urinary Cadmium Concentrations in Nonsmoking Male U.S. Population (National Health and Nutrition Examination Survey: NHANES 2003–2004)

	Men			Women		
	Urinary Cd (µg/g creatinine)*		Cd Intake	Urinary Cd (µg/g creatinine)*		Cd Intake
Age Group (Years)	Measured	Predicted	GM (µg/day)	Measured	Predicted	GM (µg/day)
6–11	0.088 (0.071 – 0.11)	0.101 (0.071 – 0.11)	15.0	0.088 (0.072 – 0.108)	0.172 (0.152 – 0.188)	13.5
12–19	0.074 (0.066 – 0.083)	0.087 (0.078 – 0.095)	19.7	0.103 (0.089 – 0.118)	0.163 (0.136 – 0.190)	15.1
20–39	0.125 (0.114 – 0.137)	0.137 (0.082 – 0.190)	22.4	0.179 (0.159 – 0.202)	0.285 (0.182 – 0.386)	16.2
40–59	0.208 (0.184 – 0.234)	0.214 (0.188 – 0.241)	22.1	0.342 (0.305 – 0.383)	0.427 (0.377 – 0.477)	16.5
≥60	0.366 (0.324 – 0.414)	0.226 (0.221 – 0.232)	17.6	0.507 (0.460 – 0.558)	0.453 (0.447 – 0.459)	14.4

*From [60], 200 GM = geometric mean.

adequately simulated experimental human data found in the published literature. A visual comparison showed that the model performance corresponded well with that of the original model.[38] Performance was also evaluated by calculating values for percent median absolute performance error (MAPE%), median performance error (MPE%), and root median square performance error (RMSPE%) based on estimates of performance error (PE).[38]

The recoded human toxicokinetic model for methylmercury is based on the model described by the Carrier et al. model,[59] which consists of a total body compartment. Methylmercury enters this compartment from the GI tract by a first-order process. The amount of methylmercury in blood is proportional to that in the total body compartment.[1,59] The recoded model reproduced all the simulations of the original model.[38] A visual comparison showed that the model performance corresponded well with that of the original model. The model performance was evaluated by calculating a value for MAPE%, MPE%, and RMSPE% based on PE. The model could simulate and accurately predict the available total body burden of mercury experimental data, and its predictions were similar to those observed experimentally and found in published literature.[38] Overall, the current model could integrate various experimental data that are critical determinants of methylmercury kinetics. It also duplicates the time courses of various tissue burdens for different dose regimens and exposure scenarios.

In this chapter, a review of the ATSDR's progress toward making PBPK models available to the scientific community and health assessors, by bridging the gap between development and use, was represented. This human PBPK toolkit was done by recoding the best available published multiple simulation languages' PBPK models into a single simple simulation language (Berkeley Madonna). When completed, the models will be packaged into a human PBPK toolkit. Currently, the recoded models include three high-ranking metals and some commonly encountered VOCs from ATSDR's priority list of environmental contaminants. It has demonstrated that the toolkit can be used in the assessment of biomonitoring results as a screening tool. One major advantage of the human PBPK toolkit being developed at ATSDR is that it can be applied in the field by practitioners of risk and health assessments.

In the future, model developers and model users should work together to encourage use and acceptance of computational tools such as the human PBPK toolkit in the decision-making process. Such information exchange and shared expertise will

- Lead to the tools being more often used in the field;
- Increase awareness of their advantages and limitations; and
- Promote integration of the toolkit into the options available for decision makers.

It has been shown that models available in multiple simulation languages can be recoded into one simulation language. Thus, the end user has to learn only one simple language, rather than a multitude of computer languages, to derive the predictions needed for risk assessments. Such efforts will facilitate the models' integration into risk assessment processes.

The growing field of computational toxicology will produce innovative, increasingly available tools for chemical risk assessment. High-throughput screening and *in vitro* testing could change toxicity testing strategies. Computational tools are a necessary component for the development of new toxicity testing methods.

Physiological Model for Bisphenol A

Bisphenol A (BPA), a high production volume chemical commonly used to harden polycarbonate plastics and epoxy resins, is present in a wide variety of consumer products, such as hard plastic products and the lining of metal food and beverage cans.[63] The Food and Drug Administration (FDA) estimates of average dietary exposure to BPA are 0.2–0.4 μg/kg body weight (bw)/day for infants and 0.1–0.2 μg/kg bw/day for children and adults.[64] As currently stated on the FDA's website, "scientific evidence at this time does not suggest that the very low levels of human exposure to BPA through the diet are unsafe"; however, the FDA also realizes that uncertainties exist with regards to the overall interpretation of the experimental data. Thus, the FDA is still pursuing more investigations to clarify uncertainties about the potential risk of BPA on human health.[65]

One area of BPA research at the National Center for Toxicological Research (NCTR) is pharmacokinetic studies on BPA, in which several species (rats, mice, monkeys) have been dosed with deuterated BPA, including at different reproductive states and ages.[66–71] Phase II conjugation (metabolism) of BPA is a detoxification pathway resulting in production of BPA-monoglucuronide (BPA-glu) conjugates and much smaller amounts of BPA-sulfate conjugates.[72] BPA is subject to substantial presystemic metabolism in the liver and small intestine, resulting in low oral bioavailability. Since oral ingestion of BPA is the primary route of exposure, a quantitative understanding of first pass metabolism is important for estimating the potential health risks from exposure to BPA. In monkeys and adult humans, BPA conjugates (BPA-c) are eliminated mainly by renal excretion. In rodents, BPA conjugates, mainly excreted into the feces (as the aglycone BPA) via the bile, are subject to enterohepatic circulation.[73] Small amounts of BPA-c are excreted in urine of rodents.

At NCTR, the PBPK models for BPA are under development for rodents, while an adult and infant monkey model for BPA was recently constructed and used to predict internal doses of BPA in human adults and infants.[74] These PBPK models, calibrated with deuterated BPA, will assist in dose-response analyses as new toxicological data become available and provide a foundation to help interpret reports of unusual concentrations of native BPA in rodents or humans. Contamination of biological samples with native BPA is an issue for any laboratory involved in measuring low levels of native aglycone BPA in blood or urine.[75] Several approaches have been used to minimize possible background contamination, e.g., use of stable isotope labeled BPA, and/or use of glass and polypropylene plastic products during the sampling and analytical process.[75,76]

The model development for BPA was based on previous BPA modeling work by Teeguarden et al.; the rodent and human PBPK model for BPA consists of four compartments for BPA and a volume of distribution for BPA-glu.[77] They investigated plasma protein binding of BPA by ultrafiltration of rat plasma for BPA concentrations ranging from 5833 to 877017 nM. Equations were incorporated in the PBPK model to account for nonspecific plasma protein-bound BPA and unconjugated BPA. BPA doses simulated by Teeguarden et al.[77] included oral administration of 100 mg/kg of ^{14}C-BPA to male and female Fischer 344 rats[72] and intravenous (IV) dosing of 10 mg/kg of BPA in female DA/Han rats.[78] In addition, the model was used to extrapolate from rats administered large doses of BPA to low doses in humans. The human BPA model,[77] as well as the human BPA model reported by Fisher et al.,[74] relied on pharmacokinetic data reported by Volkel et al.[79] In this human study,

5 mg/individual of d_{16}-BPA was given in hard gelatin capsules (54–90 μg/kg bw).[79] The conjugated aglycone BPA was below the limit of detection in the human plasma (10 nM) and urine samples (6 nM).

Later, Fisher et al.[74] constructed a PBPK model for BPA in adult and infant rhesus monkeys, administering IV or oral bolus doses of 100 μg/kg of d6-BPA.[70] A sensitive and selective LC-MS/MS method was employed to determine the concentrations of both unconjugated aglycone and total BPA after enzymatic digestion with β-glucuronidase and sulfatase.[75] D6-BPA was used to avoid confounding the samples by native BPA contamination. In this PBPK model, phase II metabolism of BPA in the small intestine enterocytes was proposed to simulate the reduced serum aglycone BPA concentrations following oral administration. Also, systemic uptake of BPA metabolites (BPA-c) from the intestinal enterocytes was proposed to describe the rapid appearance of high serum levels of BPA-c after oral administration of BPA. In contrast to rats, urinary excretion of BPA-c is the major pathway for the elimination of BPA in non-human primates. Using the adult monkey model parameters for BPA and BPA-c, simulations of human ingestion of d_{16}-BPA[79] were accomplished to predict the unconjugated BPA concentrations in serum (Figure 2-1).

Currently, a maturing and adult rat PBPK model for BPA and BPA-c is in development to describe the age-dependent dosimetry of unconjugated BPA in serum.[80] While other PBPK models for rodents have been developed for BPA,[77,81–83] none of the PBPK models describe BPA deposition in young rats. Additionally, presystemic metabolism of BPA in the GI tract has not been adequately characterized in the previous models, or enterohepatic recirculation of BPA-c excreted in bile. Collectively, this suite of BPA PBPK models developed at NCTR for rodents, monkeys, and humans will provide a quantitative tool to

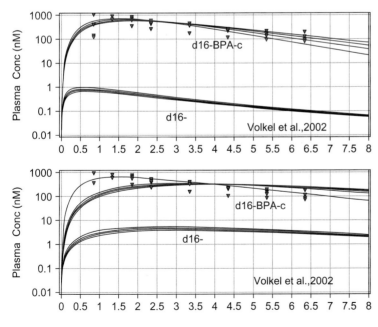

FIGURE 2-1 Simulations depicting plasma levels of the aglycone BPA and a metabolite, BPA-glucuronide (BPA-c),[74] from 4 adults who ingested deuterated BPA.[79] Only measurements of BPA-c in plasma were reported. The two panels represent model predictions in the 4 human volunteers based on *in vivo* nonhuman primate PBPK kinetics studies in which the monkeys were either dosed with deuterated BPA in an aqueous form (top panel) or with food (bottom panel).[74]

compare the dosimetry of BPA across species, life stages, and routes of administration and to conduct dose-response assessments.

BBDR HPT Axis Modeling

Pharmacokinetic analyses of the hypothalamic-pituitary-thyroid (HPT) axis several decades ago have helped us to understand complex relationships between thyroid hormone production, transport, distribution, and metabolism and the biological actions of thyroid hormones.[84,85] Recently, biologically based dose response (BBDR) models for the HPT axis were developed for adult rats[86–88] to better understand selected aspects of the HPT axis from a quantitative perspective. These models predicted changes in serum thyroid hormones in response to iodide deficiency or chemical insult[87] but did not describe the intracellular nature of thyroid hormone action (e.g., binding and transcription). One reason these BBDR models of the HPT axis were developed was to gain insights into the relationship between serum thyroid hormones and the status of the HPT axis, as measured by serum thyroid stimulating hormone concentrations, serum iodide concentrations, and thyroidal iodide stores (thyroidal stores of thyroid hormones). Linking changes in serum thyroid hormone levels with adverse outcomes mediated by the HPT status is difficult, particularly for the lower end of the dose-response spectrum. The endocrinology literature is replete with "high dose, sledge hammer" responses of the HPT axis to thyroid active compounds, and a dearth of carefully designed studies exists for examining the HPT axis at moderate or low doses of thyroid active chemicals.

The recent BBDR HPT axis models[86–88] for adult rats include submodels for dietary iodide, thyroid stimulating hormone (TSH), thyroxine (T4), and triiodothyronine (T3). These submodels are dependent on each other with interactions described using linear and nonlinear equations. Key biological processes described in the BBDR HPT axis models are the primary HPT axis negative feedback loop, TSH regulation of thyroidal gland by controlling the thyroid sodium iodide symporter (NIS) protein, the formation of thyroid hormones and the secretion of T4 and T3, and recycling of iodide liberated from deiodinase enzymes acting on thyroid hormones. TSH serum concentrations, under the control of serum levels of T4 (negative feedback loop), controlled the thyroid gland. Nested Michaelis–Menten equations were used to describe the control of TSH on thyroidal uptake of iodide by the NIS and the rate of formation of pools of bound precursor thyroid hormones in the thyroid gland, from which thyroid hormones were produced and secreted into systemic circulation. The rationale was to develop one set of model parameters that would describe, within reason, the data obtained from several laboratories. The BBDR HPT axis models evaluated iodide deficiency in albino rats using data published in the literature[86] in Long Evans rats using data collected as part of a larger research program,[89] and in euthyroid adult rats dosed with perchlorate, a drug and a food and environmental contaminant that blocks uptake of thyroidal iodide.[87] Significant strain differences in HPT axis responses to iodide deficiency were observed. Serum thyroid hormones and TSH changed less in the Long Evans rats compared to the Sprague Dawley rats on a comparable iodide deficient diet. The reason for this is unknown. Another discovery was that perchlorate may act on the thyroid gland by another mechanism, in addition to blocking uptake of iodide in the adult Sprague Dawley rats. The BBDR HPT axis model could not predict the rapid changes in serum TSH and serum thyroid hormones observed after perchlorate administration, based solely on depletion of thyroidal iodide caused by blocking its uptake.

BBDR HPT axis models for the fetal and young[88,90,91] rats and humans are in development. The goal is to predict changes in the HPT axis during the maturation phase *in utero* and in the young as a consequence of iodide deficiency and exposure to perchlorate. The fetus and young are thought to be sensitive to alternations in thyroid hormones, particularly the brain, during critical windows of development. Extending the BBDR HPT axis models to address these sensitive subpopulations is important to help protect the fetus and young from irreversible neurotoxicity by predicting perturbations in the serum thyroid hormone profiles that may be associated with adverse neurotoxicity.

In the lactating rat and postnatal day (PND) 14 nursing pup, an HPT axis model was recently calibrated for each dietary iodide intake, ranging from 39–6.4 µg/day for euthyroid lactating rats to 1.2–0.31 µg/day for iodide-deficient lactating rats.[92] In this BBDR HPT axis model, calibrations were carried out for each iodide diet. A single set of key model parameters described the euthyroid conditions, while key model parameters were changed in a dose-dependent manner to describe iodide deficient conditions. Serum T4 dosimetrics in the PND14 pup were calculated for the iodide-deficient conditions and compared to neurodevelopmental outcomes in adult offspring.[89,92]

Another effort ongoing in parallel with the lactating rat BBDR HPT axis model is the pregnant woman and fetus BBDR HPT axis model for moderate iodide deficiency and exposure to perchlorate.[91] Iodide deficiency in pregnant women, leading to hypothyroxinemia and hypothyroidism, has been shown to cause neurodevelopmental toxicity to the fetus.[93–95] A BBDR model for the HPT axis in the pregnant woman and the fetus was developed for near-term pregnancy (40 weeks) to quantitatively evaluate the effects of a range of iodide intake rates on serum-free thyroxine (fT4) levels, and the interactions of co-exposure to iodide and perchlorate on serum fT4 levels. The average intake of iodide for women of reproductive age (25–45 years) was estimated to be 145–197 µg/day by the FDA Total Diet Study (2003–2004), and the mean intake dose of perchlorate from food and water for females in age group 15–44 years was calculated to be 0.083 µg/kg bw/day.[96,97]

The BBDR HPT axis model is composed of four maternal and fetal submodels (iodide, perchlorate, T4, and T3). PBPK submodels for iodide and perchlorate were integrated based on the primary mode of action of perchlorate to inhibit competitively the sodium iodide symporter mediated uptake of iodide into the thyroid gland, and serum thyroid hormones were described using single compartments. Model parameterization for euthyroid conditions was carried out in a "piecewise" manner conforming to several quantitative thyroidal system constraints as determined from various literature sources.[98–104] The model was then extended to predict serum T4 and T3 levels in the mother and fetus for lower iodide intake conditions and for co-exposure to perchlorate. Maternal hypothyroxinemia is a condition in which the circulating free thyroxine (fT4) levels are in the lower regions of the normal euthyroid reference ranges with no expected change in the TSH levels beyond its normal reference range.[105] In marginal iodide deficiency conditions, evidence exists for thyroidal autoregulation prior to the engagement of the hypothalamus (TSH stimulation) to maintain thyroid hormone homeostasis.[106] Such a phenomenon was accounted for in the model by the adjustment of maternal parameters, like the renal clearance rate for iodide and the peripheral degradation rate for T3, as a function of iodide intake. The simulated maternal and fetal thyroid hormone levels, for perchlorate exposure, were in good agreement with the observed levels in the few epidemiological data available in the literature.[107] The dose-response model was thus successfully calibrated for euthyroid, mild to moderate iodide deficiency, and environmental and dietary perchlorate exposure conditions.[91]

This BBDR model for the HPT axis offers a valuable quantitative analysis, providing a better understanding for the conditions that may put the fetus at risk for perchlorate exposure and potential adverse neurodevelopmental outcomes as well as an assessment of the current exposure conditions for perchlorate. Expanding the deterministic BBDR HPT axis model for the pregnant mother and fetus to include distributions for model parameters (using Monte Carlo methods) will help provide model simulations that better represent a population of pregnant women. The significance of such computational tools lies in their ability to offer a fundamental framework for the biological systems and chemicals of interest by integrating the various conceptual, quantitative, and mechanistic segments of information available to better characterize the dose-response relationships and aid human health risk assessment measures to make informed regulatory decisions.

References

[1] Carrier G, Bouchard M, Brunet RC, Caza M. A toxicokinetic model for predicting the tissue distribution and elimination of organic and inorganic mercury following exposure to methyl mercury in animals and humans. II. Application and validation of the model in humans. Toxicol Appl Pharmacol 2001;171:50−60.
[2] Krewski D, Acosta Jr. D, Andersen M, Anderson H, Bailar III JC, Boekelheide K, et al. Toxicity testing in the 21st century: a vision and a strategy. J Toxicol Environ Health B Crit Rev 2010;13:51−138.
[3] Clewell RA, Clewell III HJ. Development and specification of physiologically based pharmacokinetic models for use in risk assessment. Regul Toxicol Pharmacol 2008;50:129−43.
[4] Mumtaz M, Fisher J, Blount B, Ruiz P. Application of physiologically based pharmacokinetic models in chemical risk assessment. J Toxicol 2012;2012:904603.
[5] Gombar VK. Quantitative structure-activity relationships in toxicology. In: Reiss C, et al., editors. Fundamentals to applications, advances in molecular toxicology. Utrecht, The Netherlands: VSP Publishers; 1998. p. 125−9.
[6] Ruiz P, Faroon O, Moudgal CJ, Hansen H, De Rosa CT, Mumtaz M. Prediction of the health effects of polychlorinated biphenyls (PCBs) and their metabolites using quantitative structure-activity relationship (QSAR). Toxicol Lett 2008;181:53−65.
[7] Cronin MT, Walker JD, Jaworska JS, Comber MH, Watts CD, Worth AP. Use of QSARs in international decision-making frameworks to predict ecologic effects and environmental fate of chemical substances. Environ Health Perspect 2003;111:1376−90.
[8] McKinney JD, Richard A, Waller C, Newman MC, Gerberick F. The practice of structure activity relationships (SAR) in toxicology. Toxicol Sci 2000;56:8−17.
[9] Jaworska JS, Comber M, Auer C, Van Leeuwen CJ. Summary of a workshop on regulatory acceptance of (Q) SARs for human health and environmental endpoints. Environ Health Perspect 2003;111:1358−60.
[10] Demchuk E, Ruiz P, Chou S, Fowler BA. SAR/QSAR methods in public health practice. Toxicol Appl Pharmacol 2011;254:192−7.
[11] Cronin MT, Dearden JC. QSAR in toxicology. 2. Prediction of acute mammalian toxicity and interspecies correlations. Quant Struct-Act Rel 1995;14:117−20.
[12] Zhu H, Martin TM, Ye L, Sedykh A, Young DM, Tropsha A. Quantitative structure-activity relationship modeling of rat acute toxicity by oral exposure. Chem Res Toxicol 2009;22:1913−21.

REFERENCES

[13] Papa E, Kovarich S, Gramatica P. On the use of local and global QSPRs for the prediction of physico-chemical properties of polybrominated diphenyl ethers. Mol Inform 2011;30:232–40.

[14] Li J, Gramatica P. Classification and virtual screening of androgen receptor antagonists. J Chem Inf Model 2010;50:861–74.

[15] Liu H, Papa E, Gramatica P. QSAR prediction of estrogen activity for a large set of diverse chemicals under the guidance of OECD principles. Chem Res Toxicol 2006;19:1540–8.

[16] Schultz TW, Seward JR. Health-effects related structure-toxicity relationships: a paradigm for the first decade of the new millennium. Sci Total Environ 2000;249:73–84.

[17] Simon-Hettich B, Rothfuss A, Steger-Hartmann T. Use of computer-assisted prediction of toxic effects of chemical substances. Toxicology 2006;224:156–62.

[18] Enslein K. QSTR applications in acute, chronic, and developmental toxicity, and carcinogenicity. In: Reiss C, et al., editors. Advances in molecular toxicology. Utrecht, The Netherlands: VSP Publishers; 1998. p. 141–64.

[19] Richard AM. Commercial toxicology prediction systems: a regulatory perspective. Toxicol Lett 1998;102–103:611–6.

[20] TOPKAT User Guide Version 6.2. San Diego, CA, USA: Accelrys; 2004.

[21] T.E.S.T. Tool, User's Guide for T.E.S.T., Version 4.0. U.S. EPA; 2011.

[22] Simulation Plus Inc. S. P. User Manual Version 5.5 ed., Lancaster, CA, USA. 2011.

[23] Tunkel J, Mayo K, Austin C, Hickerson A, Howard P. Practical considerations on the use of predictive models for regulatory purposes. Environ Sci Technol 2005;39:2188–99.

[24] Devillers J, Mombelli E. Evaluation of the OECD QSAR application toolbox and toxtree for estimating the mutagenicity of chemicals. Part 2. Alpha-beta unsaturated aliphatic aldehydes. SAR QSAR Environ Res 2010;21:771–83.

[25] Fjodorova N, Novich M, Vrachko M, Smirnov V, Kharchevnikova N, Zholdakova Z, et al. Directions in QSAR modeling for regulatory uses in OECD member countries, EU and in Russia. J Environ Sci Health C Environ Carcinog Ecotoxicol Rev 2008;26:201–36.

[26] Demchuk E, Ruiz P, Wilson JD, Scinicariello F, Pohl HR, Fay M, et al. Computational toxicology methods in public health practice. Toxicol Mech Methods 2008;18:119–35.

[27] Pohl HR, Chou CH, Ruiz P, Holler JS. Chemical risk assessment and uncertainty associated with extrapolation across exposure duration. Regul Toxicol Pharmacol 2010;57:18–23.

[28] Ruiz P, Mumtaz M, Gombar V. Assessing the toxic effects of ethylene glycol ethers using Quantitative Structure Toxicity Relationship models. Toxicol Appl Pharmacol 2011;254:198–205.

[29] Pohl HR, Ruiz P, Scinicariello F, Mumtaz MM. Joint toxicity of alkoxyethanol mixtures: Contribution of in silico applications. Regul Toxicol Pharmacol 2012;64:134–42.

[30] Moudgal CJ, Lipscomb JC, Bruce RM. Potential health effects of drinking water disinfection by-products using quantitative structure toxicity relationship. Toxicology 2000;147:109–31.

[31] Venkatapathy R, Wang CY, Bruce RM, Moudgal C. Development of quantitative structure-activity relationship (QSAR) models to predict the carcinogenic potency of chemicals I. Alternative toxicity measures as an estimator of carcinogenic potency. Toxicol Appl Pharmacol 2009;234:209–21.

[32] Macey RI, Oster GF. Berkeley Madonna, Version 8.3.9. Berkeley, CA, USA: University of California at Berkeley; 2006.

[33] Ruiz P, Ray M, Fisher J, Mumtaz M. Development of a human physiologically based pharmacokinetic (PBPK) toolkit for environmental pollutants. Int J Mol Sci 2011;12:7469–80.

[34] Reddy M, Yang RS, Andersen ME, Clewell HJI. Physiologically based pharmacokinetic modeling: science and applications. New York, NY, USA: John Wiley & Sons; 2005.

[35] Zhao P, Zhang L, Grillo JA, Liu Q, Bullock JM, Moon YJ, et al. Applications of physiologically based pharmacokinetic (PBPK) modeling and simulation during regulatory review. Clin Pharmacol Ther 2011;89:259–67.

[36] Thompson CM, Sonawane B, Barton HA, DeWoskin RS, Lipscomb JC, Schlosser P, et al. Approaches for applications of physiologically based pharmacokinetic models in risk assessment. J Toxicol Environ Health B Crit Rev 2008;11:519–47.

[37] Ruiz P, Mumtaz M, Osterloh J, Fisher J, Fowler BA. Interpreting NHANES biomonitoring data, cadmium. Toxicol Lett 2010;198:44–8.

[38] Ruiz P, Fowler BA, Osterloh JD, Fisher J, Mumtaz M. Physiologically based pharmacokinetic (PBPK) tool kit for environmental pollutants—metals. SAR QSAR Environ Res 2010;21:603–18.

[39] Mumtaz MM, Ray M, Crowell SR, Keys D, Fisher J, Ruiz P. Translational research to develop a human PBPK models tool kit—volatile organic compounds (VOCs). J Toxicol Environ Health A 2012;75:6–24.

[40] Thrall KD, Vucelick ME, Gies RA, Zangar RC, Weitz KK, Poet TS, et al. Comparative metabolism of carbon tetrachloride in rats, mice, and hamsters using gas uptake and PBPK modeling. J Toxicol Environ Health A 2000;60:531–48.

[41] David RM, Clewell HJ, Gentry PR, Covington TR, Morgott DA, Marino DJ. Revised assessment of cancer risk to dichloromethane II. Application of probabilistic methods to cancer risk determinations. Regul Toxicol Pharmacol 2006;45:55–65.

[42] Covington TR, Robinan Gentry P, Van Landingham CB, Andersen ME, Kester JE, Clewell HJ. The use of Markov chain Monte Carlo uncertainty analysis to support a Public Health Goal for perchloroethylene. Regul Toxicol Pharmacol 2007;47:1–18.

[43] Fisher JW, Mahle D, Abbas R. A human physiologically based pharmacokinetic model for trichloroethylene and its metabolites, trichloroacetic acid and free trichloroethanol. Toxicol Appl Pharmacol 1998;152:339–59.

[44] Clewell HJ, Gentry PR, Gearhart JM, Allen BC, Andersen ME. Comparison of cancer risk estimates for vinyl chloride using animal and human data with a PBPK model. Sci Total Environ 2001;274:37–66.

[45] Clewell III HJ, Gentry PR, Covington TR, Gearhart JM. Development of a physiologically based pharmacokinetic model of trichloroethylene and its metabolites for use in risk assessment. Environ Health Perspect 2000;108(Suppl. 2):283–305.

[46] Clewell HJ, Gentry PR, Kester JE, Andersen ME. Evaluation of physiologically based pharmacokinetic models in risk assessment: an example with perchloroethylene. Crit Rev Toxicol 2005;35:413–33.

[47] Brown RP, Delp MD, Lindstedt SL, Rhomberg LR, Beliles RP. Physiological parameter values for physiologically based pharmacokinetic models. Toxicol Ind Health 1997;13:407–84.

[48] Cowles AL, Borgstedt HH, Gillies AJ. Tissue weights and rates of blood flow in man for the prediction of anesthetic uptake and distribution. Anesthesiology 1971;35:523–6.

[49] Corley RA, Gordon SM, Wallace LA. Physiologically based pharmacokinetic modeling of the temperature-dependent dermal absorption of chloroform by humans following bath water exposures. Toxicol Sci 2000;53:13–23.

[50] Poet TS, Weitz KK, Gies RA, Edwards JA, Thrall KD, Corley RA, et al. PBPK modeling of the percutaneous absorption of perchloroethylene from a soil matrix in rats and humans. Toxicol Sci 2002;67:17–31.

[51] Poet TS, Corley RA, Thrall KD, Edwards JA, Tanojo H, Weitz KK, et al. Assessment of the percutaneous absorption of trichloroethylene in rats and humans using MS/MS real-time breath analysis and physiologically based pharmacokinetic modeling. Toxicol Sci 2000;56:61–72.

[52] Reitz RH, Gargas ML, Andersen ME, Provan WM, Green TL. Predicting cancer risk from vinyl chloride exposure with a physiologically based pharmacokinetic model. Toxicol Appl Pharmacol 1996;137:253–67.

[53] Fisher J, Mahle D, Bankston L, Greene R, Gearhart J. Lactational transfer of volatile chemicals in breast milk. Am Ind Hyg Assoc J 1997;58:425–31.

[54] Brown EA, Shelley ML, Fisher JW. A pharmacokinetic study of occupational and environmental benzene exposure with regard to gender. Risk Anal 1998;18:205–13.

[55] Chou CHS, Holler J, de Rosa CT. Minimal risk levels (MRLs) for hazardous substances. J Clean Technol Environ Toxicol Occup Med 1998;7:1–24.

[56] Agency for Toxic Substances and Disease Registry (ATSDR). Minimal risk levels for priority substances and guidance for derivation. Fed Regist 1996;61:33511–20.

[57] CDC. Fourth national report on human exposure to environmental chemicals. Available Online: <http://www.cdc.gov/exposurereport> [accessed 18.06.11].

[58] El-Masri HA, Kenyon EM. Development of a human physiologically based pharmacokinetic (PBPK) model for inorganic arsenic and its mono- and di-methylated metabolites. J Pharmacokinet Pharmacodyn 2008;35:31–68.

[59] Carrier G, Brunet RC, Caza M, Bouchard M. A toxicokinetic model for predicting the tissue distribution and elimination of organic and inorganic mercury following exposure to methyl mercury in animals and humans. I. Development and validation of the model using experimental data in rats. Toxicol Appl Pharmacol 2001;171:38–49.

REFERENCES

[60] Choudhury H, Harvey T, Thayer WC, Lockwood TF, Stiteler WM, Goodrum PE, et al. Urinary cadmium elimination as a biomarker of exposure for evaluating a cadmium dietary exposure–biokinetics model. J Toxicol Environ Health A 2001;63:321–50.

[61] Diamond GL, Thayer WC, Choudhury H. Pharmacokinetics/pharmacodynamics (PK/PD) modeling of risks of kidney toxicity from exposure to cadmium: estimates of dietary risks in the U.S. population. J Toxicol Environ Health A 2003;66:2141–64.

[62] Kjellstrom T, Nordberg GF. A kinetic model of cadmium metabolism in the human being. Environ Res 1978;16:248–69.

[63] Willhite CC, Ball GL, McLellan CJ. Derivation of a Bisphenol A oral reference dose (RfD) and drinking-water equivalent concentration. J Toxicol Environ Health B Crit Rev 2008;11:69–146.

[64] U.S. FDA Bisphenol A (BPA): Use in food contact application. <http://www.fda.gov/newsevents/publichealthfocus/ucm064437.htm>; [Last accessed July 2012].

[65] U.S. FDA Consumer Updates: FDA Continues to Study BPA. <http://www.fda.gov/ForConsumers/ConsumerUpdates/ucm297954.htm>; 2012 [Last accessed July 2012].

[66] Doerge DR, Fisher JW. Background Paper on Metabolism and Toxicokinetics of Bisphenol A. FAO/WHO Expert Meeting on Bisphenol A (BPA) Ottawa, Canada, 2–5 November 2010.

[67] Doerge DR, Twaddle NC, Vanlandingham M, Brown RP, Fisher JW. Distribution of Bisphenol A into tissues of adult, neonatal, and fetal Sprague-Dawley rats. Toxicol Appl Pharmacol 2011;255:261–70.

[68] Doerge DR, Twaddle NC, Vanlandingham M, Fisher JW. Pharmacokinetics of Bisphenol A in neonatal and adult Sprague-Dawley rats. Toxicol Appl Pharmacol 2010;247:158–65.

[69] Doerge DR, Twaddle NC, Vanlandingham M, Fisher JW. Pharmacokinetics of Bisphenol A in neonatal and adult CD-1 mice: inter-species comparisons with Sprague-Dawley rats and rhesus monkeys. Toxicol Lett 2011;207:298–305.

[70] Doerge DR, Twaddle NC, Woodling KA, Fisher JW. Pharmacokinetics of Bisphenol A in neonatal and adult rhesus monkeys. Toxicol Appl Pharmacol 2010;248:1–11.

[71] Doerge DR, Vanlandingham M, Twaddle NC, Delclos KB. Lactational transfer of Bisphenol A in Sprague-Dawley rats. Toxicol Lett 2010;199:372–6.

[72] Pottenger LH, Domoradzki JY, Markham DA, Hansen SC, Cagen SZ, Waechter Jr. JM. The relative bioavailability and metabolism of Bisphenol A in rats is dependent upon the route of administration. Toxicol Sci 2000;54:3–18.

[73] Inoue H, Yokota H, Makino T, Yuasa A, Kato S. Bisphenol A glucuronide, a major metabolite in rat bile after liver perfusion. Drug Metab Dispos 2001;29:1084–7.

[74] Fisher JW, Twaddle NC, Vanlandingham M, Doerge DR. Pharmacokinetic modeling: prediction and evaluation of route dependent dosimetry of Bisphenol A in monkeys with extrapolation to humans. Toxicol Appl Pharmacol 2011;257:122–36.

[75] Twaddle NC, Churchwell MI, Vanlandingham M, Doerge DR. Quantification of deuterated Bisphenol A in serum, tissues, and excreta from adult Sprague-Dawley rats using liquid chromatography with tandem mass spectrometry. Rapid Commun Mass Spectrom 2010;24:3011–20.

[76] Prins GS, Ye SH, Birch L, Ho SM, Kannan K. Serum Bisphenol A pharmacokinetics and prostate neoplastic responses following oral and subcutaneous exposures in neonatal Sprague-Dawley rats. Reprod Toxicol 2011;31:1–9.

[77] Teeguarden JG, Waechter Jr. JM, Clewell III HJ, Covington TR, Barton HA. Evaluation of oral and intravenous route pharmacokinetics, plasma protein binding, and uterine tissue dose metrics of Bisphenol A: a physiologically based pharmacokinetic approach. Toxicol Sci 2005;85:823–38.

[78] Upmeier A, Degen GH, Diel P, Michna H, Bolt HM. Toxicokinetics of Bisphenol A in female DA/Han rats after a single i.v. and oral administration. Arch Toxicol 2000;74:431–6.

[79] Volkel W, Colnot T, Csanady GA, Filser JG, Dekant W. Metabolism and kinetics of Bisphenol A in humans at low doses following oral administration. Chem Res Toxicol 2002;15:1281–7.

[80] Yang X, Doerge DR, Fisher JW. Prediction and evaluation of route dependent dosimetry of BPA in rats at different developmental stages using a physiologically based pharmacokinetic model (manuscript in preparation). 2012.

[81] Hutter JC, Luu HM, Kim CS. A dynamic simulation of Bisphenol A dosimetry in neuroendocrine organs. Toxicol Ind Health 2004;20:29–40.

[82] Shin BS, Kim CH, Jun YS, Kim DH, Lee BM, Yoon CH, et al. Physiologically based pharmacokinetics of Bisphenol A. J Toxicol Environ Health A 2004;67:1971–85.

[83] Kawamoto Y, Matsuyama W, Wada M, Hishikawa J, Chan MP, Nakayama A, et al. Development of a physiologically based pharmacokinetic model for Bisphenol A in pregnant mice. Toxicol Appl Pharmacol 2007;224:182–91.

[84] Oppenheimer JH. The nuclear receptor-triiodothyronine complex: relationship to thyroid hormone distribution, metabolism, and biological action. In: Oppenheimer JH, Samuels. HH, editors. Molecular basis of thyroid hormone action. New York: Academic Press; 1983. p. 1–34.

[85] DiStefano III JJ, Landaw EM. Multiexponential, multicompartmental, and noncompartmental modeling. I. Methodological limitations and physiological interpretations. Am J Physiol 1984;246:R651–64.

[86] McLanahan ED, Andersen ME, Fisher JW. A biologically based dose-response model for dietary iodide and the hypothalamic-pituitary-thyroid axis in the adult rat: evaluation of iodide deficiency. Toxicol Sci 2008;102:241–53.

[87] McLanahan ED, Andersen ME, Campbell JL, Fisher JW. Competitive inhibition of thyroidal uptake of dietary iodide by perchlorate does not describe perturbations in rat serum total T4 and TSH. Environ Health Perspect 2009;117:731–8.

[88] Gilbert ME, McLanahan ED, Hedge J, Crofton KM, Fisher JW, Valentin-Blasini L, et al. Marginal iodide deficiency and thyroid function: dose-response analysis for quantitative pharmacokinetic modeling. Toxicology 2011;283:41–8.

[89] Gilbert M, Hedge J, Zoeller R, Kannan K, Crofton K, Valentin-Blasini L, et al. Developmental iodide deficiency: reductions in thyroid hormones and impaired hippocampal transmission. #2608. The Toxicologist CD—An Official Journal of the Society of Toxicology, March 2011.

[90] Li S, Gilbert M, Zoeller T, Crofton K, McLanahan E, Mattie D, et al. A BBDR-HPT axis model for the lactating rat and nursing pup: evaluation of iodide deficiency. #894, The Toxicologist CD—An Official Journal of the Society of Toxicology, March 2010.

[91] Lumen A, Mattie DR, Fisher JW. A BBDR-HPT axis model for pregnant woman and fetus: evaluation of iodide deficiency, perchlorate exposure and their interactions #439. The Toxicologist CD—An Official Journal of the Society of Toxicology, March 2012.

[92] Fisher JG, Li S, Crofton K, Zoeller RT, McLanahan ED, de Lumen A, et al. Evaluation of iodide deficiency in the lactating rat and pup using a biologically based dose-response model. Toxicol Sci 2013;132:75–86.

[93] Haddow JE, Palomaki GE, Allan WC, Williams JR, Knight GJ, Gagnon J, et al. Maternal thyroid deficiency during pregnancy and subsequent neuropsychological development of the child. N Engl J Med 1999;341:549–55.

[94] Man EB, Brown JF, Serunian SA. Maternal hypothyroxinemia: psychoneurological deficits of progeny. Ann Clin Lab Sci 1991;21:227–39.

[95] Pop VJ, Kuijpens JL, van Baar AL, Verkerk G, van Son MM, de Vijlder JJ, et al. Low maternal free thyroxine concentrations during early pregnancy are associated with impaired psychomotor development in infancy. Clin Endocrinol (Oxf) 1999;50:149–55.

[96] Murray CW, Egan SK, Kim H, Beru N, Bolger PM. U.S. Food and Drug Administration's Total Diet Study: dietary intake of perchlorate and iodine. J Expo Sci Environ Epidemiol 2008;18:571–80.

[97] Huber DR, Blount BC, Mage DT, Letkiewicz FJ, Kumar A, Allen RH. Estimating perchlorate exposure from food and tap water based on U.S. biomonitoring and occurrence data. J Expo Sci Environ Epidemiol 2011;21:395–407.

[98] Oddie TH, Meade Jr. JH, Fisher DA. An analysis of published data on thyroxine turnover in human subjects. J Clin Endocrinol Metab 1966;26:425–36.

[99] Fisher DA, Odell WD, Hobel CJ, Garza R. Thyroid function in the term fetus. Pediatrics 1969;44:526–35.

[100] Fisher DA, Oddie TH, Thompson CS. Thyroidal thyronine and non-thyronine iodine secretion in euthyroid subjects. J Clin Endocrinol Metab 1971;33:647–52.

[101] Nicoloff JT, Low JC, Dussault JH, Fisher DA. Simultaneous measurement of thyroxine and triiodothyronine peripheral turnover kinetics in man. J Clin Invest 1972;51:473–83.

[102] Vulsma T, Gons MH, de Vijlder JJ. Maternal-fetal transfer of thyroxine in congenital hypothyroidism due to a total organification defect or thyroid agenesis. N Engl J Med 1989;321:13–6.

[103] Liberman CS, Pino SC, Fang SL, Braverman LE, Emerson CH. Circulating iodide concentrations during and after pregnancy. J Clin Endocrinol Metab 1998;83:3545–9.
[104] Braverman LE, Utiger RD, editors. Werner & Ingbar's the thyroid. A fundamental and clinical text. 9th ed. Philadelphia: Lippincott Williams & Wilkins; 2005.
[105] Moleti M, Trimarchi F, Vermiglio F. Doubts and concerns about isolated maternal hypothyroxinemia. J Thyroid Res 2011;2011:463029.
[106] Obregon MJ, Escobar del Rey F, Morreale de Escobar G. The effects of iodine deficiency on thyroid hormone deiodination. Thyroid 2005;15:917–29.
[107] Tellez Tellez R, Michaud Chacon P, Reyes Abarca C, Blount BC, Van Landingham CB, Crump KS, et al. Long-term environmental exposure to perchlorate through drinking water and thyroid function during pregnancy and the neonatal period. Thyroid 2005;15:963–75.

CHAPTER 3

Multiple Chemical Exposures and Risk Assessment

John C. Lipscomb[1], Nikki Maples-Reynolds[2], and Moiz Mumtaz, PhD[3]

[1]U.S. Environmental Protection Agency, Office of Research and Development, National Center for Environmental Assessment, Cincinnati, OH, USA, [2]ICF International, Fairfax, VA, USA, [3]Centers for Disease Control and Prevention, Department of Toxicology and Environmental Medicine, Atlanta, GA, USA

HISTORICAL PERSPECTIVE

Humans are exposed to chemicals through voluntary and involuntary actions; to natural and synthetic chemicals all day, every day. Single chemical risk assessments are complex in and of themselves, and the assessment of chemical mixtures exponentially increases the complexity for toxicologists, regulators, and the public. Chemicals produce effects in biological systems which may or may not be related to their toxicity; some effects may be adaptive or may not be a direct part of their mode or mechanism of toxic action. These terms are commonly used and may be distinguished based on the level of detail implied. Mode of action usually describes the effect of a toxicant at the cellular or organ level, while mechanism of action implies an understanding of the interaction of the toxicant at the molecular level. Chemicals can have the same mode of action, but act via different mechanisms. Components in a chemical mixture are characterized by mode and/or mechanism for the purpose of grouping, described later.

Mixture risk assessment (MRA) is an evolving discipline within toxicology. The history of toxicology has primarily been the pursuit of the characterization of single chemical hazards and rarely on chemical mixtures.[1] This is by and large due to the complexity and uncertainty of mixture hazard characterization. At the advent of modern toxicology, in the late 1950s and early 1960s, the primary focus of the discipline was chemical characterization and safety evaluation for occupational exposures and consumer use products (e.g., pesticides, drugs, or warfare agents), rather than mechanistic toxicology.[2] However, humans are never exposed to a single chemical whether it is in our diet, pharmaceuticals, drinking water, ambient air, or workplace.[1] Environmental media, namely, air, water, and soil monitored at waste sites, contain a variety of chemical mixtures.[3]

As early as 1939, Bliss[4] defined three main categories of joint chemical action that are still relevant today: (1) independent joint action, which refers to chemicals that act independently and have different modes of action and/or mechanisms; (2) similar joint action, which refers to chemicals that cause similar effects often through similar modes of action/mechanisms; and (3) synergistic action, which is defined as a departure from additivity.[5] The first two mixture categories can be predicted with knowledge of the individual chemical toxicity, and the third needs data on the mixture toxicity.

The basic toxicology database is inadequate for assessing risk for the vast majority of chemical mixtures. Toxicology studies addressing chemical mixtures or interactions mostly focus on sequential exposure to two different chemicals or exposure to binary mixtures.[6–8] A significant number of well-designed and decisive studies have been carried out that involve multicomponent mixtures,[9] and several are described in a review published by Kortenkamp et al.[10] In spite of these advances, for many chemical mixtures, the best advantage of these advances may be a logical framework by which to identify data gaps and prioritize research needs.

With advancements in the understanding of biochemistry, physiology, and metabolism of chemicals, the assessment of chemical mixtures has flourished in recent decades, particularly the advances in genomics, proteomics, and bioinformatics. Progress in MRA has advanced because of the transformative shift called for in the recent report on toxicology in the 21st century by the National Research Council.[11] Teuschler et al.[12] stated that new methods to advance mixture risk assessment can include (1) computational technology; (2) mathematical/statistical modeling; (3) mechanistically based, short-term toxicology studies; (4) the latest advances in cell and molecular biology methodologies; (5) new *in vitro* studies for screening mixtures toxicity; (6) approaches developed to understand and analyze data on genomics and proteomics (i.e., bioinformatics); and (7) technologies available in other disciplines beyond the normal toxicological boundaries such as engineering and computer science.

REGULATORY PERSPECTIVE

The science of toxicology and a brief overview of U.S. governmental regulatory actions are explored in this section and are treated in more detail in a recent review by Monosson.[1] The earliest U.S. regulation of chemicals was the Federal Food, Drug, and Cosmetic Act of 1938,[13] which authorized assessments of the safety of new drugs, food additives, and colors and specified tolerance levels for pesticides and other chemicals that may occur in foods. In 1958, the Food Additives Amendment[14] established the Delany Clause that prohibited the approval of carcinogenic food additives. With the advent of the Food Quality Protection Act (FQPA) of 1996,[15] the FFDCA was amended to eliminate the applicability of the Delany Clause to pesticides.[15,16] Concerns about multiple pesticide exposures from residues in foods, however, have resulted in amendments to the FFDCA that address pesticide mixtures in foods.[15]

The environmental concerns of the 1970s led to advances in regulatory authority and regulatory entities, including the establishment of the U.S. Environmental Protection Agency (EPA). The Clean Water Act of 1972[17] guided states, territories, and tribes to evaluate the impact of chemicals on water sources using the Total Maximal Daily Load (TMDL) approach, and the U.S. EPA now recommends calculation of the TMDL using methods that include a consideration of the cumulative effects.[18] The TMDL is a calculation of the maximum amount of a pollutant (e.g., pathogens, nutrients, sediment, mercury, metals, pesticides, organics) that a water source for a state, territory, or tribe can receive and still safely meet water quality standards on both a daily and non-daily time frame (e.g., weekly, monthly, yearly). The TMDL calculation is

$$TMDL = \Sigma WLA + \Sigma LA + MOS$$

where WLA is the sum of wasteload applications (point sources), LA is the sum of load allocations (nonpoint sources and background), and MOS is the margin of safety. TMDLs can be expressed in terms of mass per time, toxicity, or other appropriate measures that relate to a state's, territory's, or tribe's water quality standard.

Under this approach, a given effect that characterizes water quality should be examined to determine the total amount of all toxicants (combined) that may be allowed without detriment to the aquatic system. The approach may lead to the identification of the most problematic contaminants, which may not be those contaminants making up the largest share of the contaminant load.

Before the 1970s, the USDA's Federal Insecticide, Fungicide and Rodenticide Act (FIFRA) of 1947 was essentially geared toward protecting consumers from ineffective products.[19,20] The act was later amended several times, with the 1972 amendments providing the basis for current pesticide regulation. Jurisdiction of the act transferred from the USDA to the U.S. EPA in 1970. Recent amendments to both FIFRA and the FFDCA via the FQPA mandate the assessment of risks from mixtures of pesticides with common modes of action from any source. In response, the U.S. EPA has developed sophisticated guidelines to decide which pesticides qualify for inclusion in common mechanism groups.[21–23] The terms used to describe similarity have been technically presented as follows:

- *Mode of action (MOA)* is a series of key events and processes starting with interaction of an agent with a cell, and proceeding through operational and anatomical changes causing disease formation.
- *Mechanism of action* is a detailed understanding and description of the events leading to a health outcome, often at the molecular level.
- *Toxicologic similarity* represents a general knowledge about the action of a chemical or a mixture and can be expressed in broad terms such as at the target organ or tissue level in the body.

The U.S. EPA has acknowledged the weaknesses of this approach, which it identifies as omitting other chemicals that might also induce the effect of interest, although by different mechanisms (see reference 24). The approach presented here is that since chemicals cause several effects, since some of these effects may not be directly related to either mechanism of toxic action or mode of toxic action for that particular chemical, and since other components of the chemical mixture may have some biological effects in their mode or mechanism of action that may be affected by other components, there may be an advantage to extending the concept of similarity to include effects that are beyond the mode or mechanism of action for component chemicals.

For assessment purposes, the U.S. EPA defines chemical mixtures as either (1) simple mixtures containing "two or more identifiable components but few enough that the mixture toxicity can be adequately characterized by a combination of the components' toxicities and the components' interactions" or (2) complex mixtures containing "so many components that any estimation of its toxicity based on its components' toxicities contains too much uncertainty and error to be useful."[22] Current methodologies for human health risk assessment commonly treat mixtures as simple mixtures, deriving the combined toxicity of individual components primarily from single-chemical studies.

Several other EPA legislative mandates including the Comprehensive Environmental Response, Compensation and Liability Act (CERCLA),[25] the Superfund Amendments and Reauthorization Act,[26] the Safe Drinking Water Act,[27] and the Clean Air Act[28,29] also require the consideration of joint chemical exposures and of chemical mixture toxicity in regulatory decision making. The U.S. EPA Risk Assessment Guidance for Superfund acknowledges that "simultaneous subthreshold exposure to several chemicals could result in an adverse health effect" such that estimates based on single chemicals might underestimate the overall risk.[21]

Some of the first risk assessment techniques addressing chemical mixtures were developed for worker protection by the Occupational Safety and Health Administration (OSHA).[30,31] This was due in part in the mandate of OSHA to protect workers from exposure to harmful chemicals through establishment of standards and guidelines for individual chemicals. Later, the Agency for Toxic Substances and Disease Registry (ATSDR), the agency that conducts public health assessments at National Priorities List ("Superfund") sites under CERCLA,[21,32] also developed a dedicated chemical mixtures program[33] to provide assistance to communities potentially exposed to multiple chemicals.

MIXTURES VERSUS COMPONENTS

Information on toxicity or risk assessment is available for some but not all of the chemicals found in the environment. Thus, risk assessment of chemical mixtures can be a daunting task in the absence of critical information. Three basic types of approaches are often employed to make the most of available information and perform mixtures risk assessments that vary in accuracy and uncertainties: whole mixtures approaches, similar mixtures approaches, and component-based approaches.[34] While some dose-response information is usually required for MRA, one of the most frequently employed default assumptions regards the mode of action for a given chemical. This is important in that it drives the application of MRA methodologies and is discussed and demonstrated in the following sections. Fundamentally, two types of approaches may be broadly defined as whole mixture approaches and component-based approaches.

Whole Mixture Approaches

As implied, the whole mixture approaches apply to data generated through the experimental exposure to complete chemical mixtures. This falls into two categories: the mixture of concern or a sufficiently similar mixture. Having data on the mixture of concern is the simplest approach, and the approach embodying the least amount of uncertainty.[35] This approach has been applied to jet fuels, PCBs, and polybrominated biphenyls.[36,37] However, especially for fuels, the approach must be adjusted to account for blending issues, like those encountered in switching from summer blend to winter blend gasoline. In cases like that, the components of the two mixtures may be quite similar, but the ratios of the components may be different, or minor components may be added. Then the approach may involve treating the two mixtures as similar and making any necessary adjustments to account for key toxicities of the more potent components, whose blending ratios may change. Justifying a mixture as sufficiently similar to the mixture in question has been accomplished for coke oven emissions, diesel exhaust, drinking water disinfection byproducts, and woodstove emissions. However, with respect to diesel engine emissions, recent evidence has indicated that while adjustments may be made for the mass of particulate emitted, the toxicologic properties of the particulate matter, as well as the composition and toxicity of nonparticulate components of diesel exhaust, may differ between conventional and biodiesel.[38] This underlines the need to carefully compare mixtures and their components in determining an acceptable surrogate on the basis of similarity. A more complete treatment of the sufficient similarity of chemical mixtures is available in the current literature (e.g., see references 39 and 40).

Component-Based Approaches

When data are lacking for the mixture of concern and no toxicologically similar mixture can be justified as a surrogate, the approach is reduced to one involving the identification and toxicologic estimation of component chemicals (CC). Two types of additivity approaches, dose addition and response addition, are often used. Lacking full toxicologic characterization of CC (dose response and mode of action), the approach can be completed through the application of the assumption that component chemicals have the same mode or mechanism of action. This places them each into the same class for application of the dose addition model.

ADDITIVITY APPROACHES

Dose addition is frequently the default approach employed in the absence of key data. When available, data on toxic MOA are used to determine whether to apply dose additive or response additive models. Chemicals are subjected to dose additive models when they act through the same MOA, or for which no data can inform a decision about MOA, or at times when they affect the same organ or tissue. When chemicals are known or assumed to act through dissimilar MOA, response addition is applied.

Dose Addition (DA)

Several basic DA models are available, and the application of each will depend on the type of analysis, the sophistication of the analysis, and the type and extent of toxicity data available. For DA models, the chemicals are shown to act through a common or similar mode of action or mechanism to support grouping. Relative potency factor (RPF) models require a greater and more sophisticated level of data than hazard index (HI) types of models.

Relative Potency Factors

Relative potency factor models combine the doses of CC to estimate a total dose through four basic steps. First, an index chemical (IC) is identified and the potencies of the CC are scaled relative to that of the IC; these scaling factors are called relative potency factors (RPFs).[22] Second, exposures to CC are characterized. These exposures are multiplied by their respective RPFs to estimate an index chemical equivalent dose (ICED), and finally, the ICED values for each CC are summed and the response is determined from dose-response data for the IC.

These models develop RPFs for chemicals in the class. Under this approach, an IC is chosen for the class (with the components of the class defined by knowledge of the MOA or the default assumption to group CC) based on considerations that include the extent to which the dose-response function has been characterized, the richness of toxicologic data, and the extent to which the toxicity of the proposed IC represents the toxicity of the MOA class. Under DA models that use an IC, the point on the response scale for developing potency estimates is chosen, and doses producing that level of response are recorded for the IC and all CC and compared. Component chemicals may have RPF estimates higher or lower than the IC, which is assigned a potency value of 1. Figure 3-1 demonstrates the concept of dose addition. Using Figure 3-1 as the example, chemical 1 may be chosen as the IC based on the preceding considerations. Though RPF estimates are typically developed from point estimates of equipotent doses (Eq. 3-1), they may also be developed by comparing the slopes of dose-response functions.

$$\text{RPF} = \text{IC dose}/\text{CC dose} \tag{3-1}$$

If, at the chosen response level, chemical 2 has a dose of 3 mg/kg-day and the index chemical has a dose of 2 mg/kg-day, the potency of chemical 2 is 0.66 (2/3). If chemical

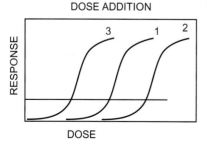

FIGURE 3-1   Dose addition.

3 has a dose of 1.5 mg/kg-day at the predefined response level, its potency would be 1.3 (2/1.5).

If the dose-response curves are not constantly proportional, the potency values will be a function of the response level chosen. When this is the case, it should be called out as an uncertainty when the results are communicated. RPF values can be developed for different portions of the dose-response function and can be limited by type of effect or by exposure route.[22]

Once potency estimates are developed, they are multiplied by the actual dose of the respective CC to develop an index chemical equivalent dose (ICED; Eq. 3-2).

$$ICED = RPF \times Dose \qquad (3\text{-}2)$$

ICED values for each CC are summed to estimate the total ICED, and response for the mixture is estimated from the dose-response function for the IC, using the total ICED to represent dose. If the actual doses of chemicals 1, 2, and 3 were 0.5, 0.25, and 1.5 mg/kg-day, respectively, the adjusted dose for chemical 1 would be 0.5 * 1 = 0.5 mg/kg-day; for chemical 2, it would be 0.25 * 0.66 = 0.165; and for chemical 3, it would be 1.5 * 1.3 = 1.95 mg/kg-day. The summed ICED would be 0.5 + 0.165 + 1.95 = 2.6 mg/kg-day. The anticipated mixture response for this group of three chemicals would be read from the dose-response function for the index chemical, given a dose of 2.6 mg/kg-day.

Two DA models have been frequently employed: the relative potency factor (RPF) approach and the hazard index (HI) approach. For the RPF approach, relative potency can be estimated as either the ratio of doses producing the same level or response, or as ratios of the slope of their respective dose-response functions. Regardless of the technical approach, RPF models should be employed in the low-dose region of the dose-response relationship. The shape of the dose-response curve may result in different potency estimates, so the decisions about where (what response level) along the dose-response curve a comparison is made should not be undertaken lightly and should be adequately justified.

Both the relative potency factor (RPF) approach and the HI approach (described later) rely on information describing both exposure (dose) and dose response. The RPF approach differs from the HI approach regarding the depth of knowledge required to group chemicals into classes. Whereas the HI approach uses *target organ* as the level at which chemicals are grouped, the RPF approach requires that similarity of toxic effect be defined at the level of the *type of effect* observed (e.g., cholinesterase inhibition). Once the chemicals are grouped, an index chemical is identified on the basis of the extent to which its toxicity represents the toxicity of the group and the degree of characterization of its dose-response relationship. It need not be the most potent chemical in its class. Data must be sufficient to defend the similarity of the shape of the dose-response relationship for other chemicals in the class—especially in the range of relevant exposures, doses, or concentrations.[22,41] Once estimated, ICED values for each chemical are summed to develop the total ICED for the mixture. The likelihood of a response is read from the DR relationship for the index chemical. This approach has been applied in risk assessments for organophosphate,[41,42] carbamate,[43] chloroacetanilide[44] and pyrethroid[45] insecticides, and polycyclic aromatic hydrocarbons.[46] The RPF approach extends dose addition through a consideration of similarity defined by the type of effect observed.

Toxicity Equivalence Factors (TEF)

The toxicity equivalence factors (TEF) approach is another dose addition model that is more restrictive than the RPF approach and has been applied only to the dioxin-like compounds. As in the RPF approach, chemicals are grouped according to similarity, but under the TEF approach, similarity is defined at level of *specific, molecular interactions* between the toxicant and biological target (i.e., AH receptor binding). This approach has been applied to dioxin-like compounds.[21,37,47–49] When the toxicity has been described at the level of the molecular interaction, TEFs developed can be applied to all health endpoints, routes, and time frames. In contrast, RPF values are generally constrained by route and health endpoint. TEF values, like RPF values, are derived as a ratio of doses or concentrations producing the same, predetermined level of the response. For TEF, the "response" is defined as interaction with the receptor, not biological (adverse health) outcome. Once TEF values are calculated for mixture components, their respective doses are scaled (through multiplication by TEF values) to develop toxic equivalence (TEQ values); TEQ values for all mixture components are summed. The anticipated mixture response is estimated from the DR relationship for the index chemical (TCDD) as a function of the sum of all TEQ values.

Hazard Index (HI) Approaches

The HI family of models requires that reference values be developed for the adverse health outcomes identified. Animal (or human) point of departure (POD) values may be fully translated to reference dose (RfD), reference concentration (RfC), or minimal risk level (MRL) values, or not. When dose-response data identify only POD values (NOAEL, LOAEL, BMDL), they must be extrapolated to RfC, RfD, or MRL values through the application of uncertainty factors (UF). While the values for UF may often be those developed on the basis of default approaches, processes to include the quantitative incorporation of toxicokinetic and toxicodynamic information from studies with test animals or humans through *in vitro* or *in vivo* study designs are available.[50] The composite uncertainty factor (CUF) is the product of UF values describing extrapolation from LOAEL to NOAEL, from animals to humans, human variability, from subchronic to chronic exposure durations and the fullness of the toxicologic database. Whether measures of acceptable exposure levels (AL) are available from peer-reviewed analysis or must be developed from POD values, they and their basis must be described in sufficient detail to include the species tested, the tissues or organs affected, and the type of effect (e.g., renal tubular hyperplasia, adenocarcinoma of the lung). Available MOA information should be briefly described. For some component-based approaches described later, it is important that dose-response information be presented for the critical effect as well as for secondary effects—those effects that are manifest at doses or exposure higher than those necessary to produce the critical effect. If AL values are not available, they can be developed according to Equation 3-3, applied to POD values for the given effects. It should be noted that secondary effects may be identified in the principal study or in other studies, and that UF values applied to the secondary effect(s) may differ from those applied to the critical effect.

$$RV = POD/CUF \qquad (3\text{-}3)$$

In this equation, RV (reference value) may be the value for the RfD, RfC, MRL, or other exposure value.

Several hazard index–based approaches are available to estimate the response from chemical mixtures. The decision among them may be based on programmatic need and/or the richness of available data; and decisions to move to more technical HI-based approaches may be based on results from initial (screening level) approaches which include health conservative assumptions. Other approaches are based on the identification of secondary effects, knowledge of the dose response for the critical effect as well as secondary effect(s), and information regarding MOA. Regardless of the approach, a hazard quotient (HQ) value is developed for each CC according to Equation 3-4:

$$HQ = E/AL \qquad (3\text{-}4)$$

In this equation, E is the exposure estimate in the same units as the AL value (e.g., mg/kg-day); hence, the HQ values are unitless. In each of the HI approaches, a hazard index for the mixture is calculated according to Equation 3-5:

$$HI = \sum HQ \qquad (3\text{-}5)$$

Mode of action information is used to place CC into groups acting through a common (or similar) MOA, and CC acting through different MOA. Within a chemical mixture, CC may be grouped into several MOA (sometimes called common mechanism groups). When the HI approach involves grouping into MOA classes, dose-response data for the MOA class are estimated and quantitatively employed. The default assumption is that CC act through a common MOA.

There are three basic approaches to HI: screening, HI, and Target Organ Toxicity Dose. The information for a hypothetical mixture in Table 3-1 is used to illustrate these approaches. The hypothetical mixture contains eight components, and among the mixture, seven different organs or tissues are affected. Organs or tissue are differentially sensitive to CC-induced insults, and not all organs or tissues are affected by every CC.

The simplest HI approach is commonly known as the *screening level approach*. It is assumed that all CC act through a common MOA, and information on dose response is evaluated only for the most sensitive organ or tissue a single HQ value is developed for each CC based on its *critical effect*. HQ values are summed for the CC, with the resulting value representing the HI for the mixture.

Using this hypothetical example, results in Table 3-2 indicate that concern for the mixture would be raised because the HI value exceeds 1. This approach guides an inspection more closely directed at the level of individual organs or tissues impacted through implementation of the hazard index approach. Here, the hazard quotients used in the screening approach are segregated according to impacted organ or tissue, and HI values are calculated for *each organ or tissue* (see Table 3-3). These results may the focus of the analysis on compounds producing developmental toxicity because that is the effect representing most of the concern for this mixture; none of the other effects demonstrate an HI value approaching 1.

To further refine the evaluation of mixtures risk through the HI approach, one may undertake a third type of evaluation. The Target Organ Toxicity Dose (TTD) method[22,33,51]

TABLE 3-1 Exposure, Effects, and RfDs Related to Chemical Components

Component	Exposure (E) (mg/kg-day)	Immunotoxicity	Cardiovascular	CNS	Testicular Toxicity	Liver	Developmental Toxicity	Kidney
1	0.15	0.51	0.22	1.42	2.1	0.75	0.1*	0.25
2	0.11	7.5	–	7.5	–	2.5	0.3*	1.5
3	0.025	–	0.25	0.1*	–	0.25	0.25	–
4	0.1	–	10	10	–	–	25	5*
5	0.12	–	0.257	1.429	2.1	–	0.096*	0.257
6	0.2	5	–	–	5	0.5*	5	–
7	0.02	0.1*	–	0.25	–	0.25	–	–
8	0.22	–	10	25	5*	–	25	25

Numerical values for effects are in units of mg/kg-day and represent RfD values for the critical effect and analogous values for secondary effects.
*Critical effect
– No effect noted

TABLE 3-2 Screening Level Hazard Index Calculation

Component	Exposure (E) (mg/kg-day)	RfD (AL) (mg/kg-day)	Critical Effect	Hazard Quotient
1	0.15	0.1	Developmental toxicity	1.5
2	0.11	0.3	Developmental toxicity	0.37
3	0.025	0.1	CNS toxicity	0.25
4	0.1	5.0	Kidney	0.02
5	0.12	0.096	Developmental toxicity	1.25
6	0.2	0.5	Liver toxicity	0.4
7	0.02	0.1	Immunotoxicity	0.2
8	0.22	5	Testicular toxicity	0.044
Hazard Index				3.67

is more robust than the previous HI models in that it takes into account *both the critical effects as well as secondary effects* for each chemical. Secondary effects are those that become evident as the dose increases above that necessary to produce the critical effect. Under this approach, multiple toxicities produced by a CC, rather than just the critical effect, are taken into account (see Table 3-4). For a given critical effect, the contributions of each CC to the toxicity are considered, whether they represent the critical effect or a secondary effect. HQ values for both the critical effect and secondary effects are calculated for each

TABLE 3-3 Hazard Index Calculation

Component	Exposure (E) (mg/kg-day)	AL (mg/kg-day) by Effect and HQ (unitless)						
		Immunotoxicity	Cardiovascular	CNS	Testicular toxicity	Liver	Developmental toxicity	Kidney
1	0.15					0.1		
	HQ					1.5		
2	0.11					0.3		
	HQ					0.367		
3	0.025			0.1				
	HQ			0.25				
4	0.1							5
	HQ							0.02
5	0.12						0.096	
	HQ						1.25	
6	0.2					0.5		
	HQ					0.4		
7	0.02	0.1						
	HQ	0.2						
8	0.22				5			
					0.44			
HI		0.2	—	0.25	0.44	0.4	3.11	0.02

CC and included in HI estimation for each effect the same way as HQ values for the critical effect (E/AL). Thus, the TTD approach allows a more comprehensive evaluation of the health effects of mixtures due to the inclusion of secondary effects of all CCs of the mixture.

Response Addition

Like dose additive approaches, response addition requires enough knowledge to assume chemicals can be grouped according to the CCs having independent MOAs. When the MOA for a given chemical component is toxicologically independent, the contribution to the mixture response is treated by response addition. Selecting an index chemical is obviated, and no estimates of scaled potency are carried out. The uniqueness of the MOA dictates that a single chemical be in each "class." The requirement for dose-response data is that they be sufficient to derive an estimate of the likelihood of a response at the anticipated dose (D) for each CC. For chemicals treated under the RA approach, the likelihood

TABLE 3-4 Target Organ Toxicity Dose Hazard Calculation

Component	Exposure (E) (mg/kg-day)	AL (mg/kg-day) by Effect and HQ (unitless)						
		Immunotoxicity	Cardiovascular	CNS	Testicular toxicity	Liver	Developmental toxicity	Kidney
1	0.15	0.51	0.22	0.42	2.1	0.75	0.1*	0.25
	HQ	0.294	0.682	0.106	0.071	0.2	1.5	0.6
2	0.11	7.5		7.5		2.5	0.3*	1.5
	HQ	0.015		0.015		0.044	0.367	0.073
3	0.025		0.25	0.1*		0.25	0.25	
	HQ		0.1	0.25		0.1	0.1	
4	0.1		10	10		25		5*
	HQ		0.01	0.01		0		0.02
5	0.12		0.257	1.429	2.1		0.096*	0.257
	HQ		0.467	0.084	0.057		1.25	0.467
6	0.2	5			5	0.5*	5	
	HQ	0.04			0.04	0.4	0.04	
7	0.02	0.1*		0.25		0.25		
	HQ	0.2		0.08		0.08		
8	0.22		10	25	5*		25	25
	HQ		0.022	0.009	0.044		0.009	0.009
TTD HI		0.549	1.28	0.553	0.213	0.824	3.269	1.169

*Critical effect.

of an adverse outcome is predicted from dose, and that likelihood (the single chemical response) is carried forward, to be combined with anticipated responses for other CC. This is depicted in Figure 3-2 in which the hypothetical mixture contains three chemicals, each with a different MOA. The responses (R) for each chemical are read from the respective dose-response function, and the likelihood of the response from each chemical is summed ($R_1 + R_2 + R_3$) to represent the likelihood of a response from the mixture (R_M). If D_1 results in a response of 0.05, if D_2 results in a response of 0.03, and if D_3 results in a response of 0.5, then $R_M = 0.58$.

Integrated Additivity

Mixtures may contain many chemicals, increasing the likelihood that more than a single MOA is represented among component chemicals. Such a condition requires a hybrid approach, that of integrated additivity,[41,52] that includes concepts from DA and

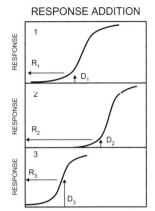

RESPONSE ADDITION   **FIGURE 3-2** Response addition.

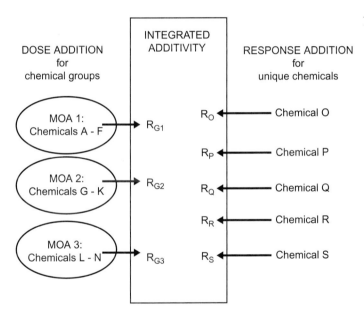

FIGURE 3-3 Integrated additivity.

RA approaches (see Figure 3-3). In this case, components are grouped according to MOA, recognizing that some MOA classes may contain only a single component chemical, and others may contain several component chemicals. Consider a hypothetical example where 19 chemicals (A through S) are grouped into 8 MOA classes with the number of CC in each class ranging from 1 to 6. In this example, chemicals A–F share a common MOA; Chemicals G–K share a (different) common MOA; and chemicals L–N share a common, third MOA. Chemicals O–S are unique, having modes of action distinct from one another and from that used as the basis for chemical grouping. The likelihood of a response is generated for all MOA classes using DA for classes (MOA groups 1–3) that contain several chemicals, and the likelihood of a response is also generated for each unique individual chemical (chemicals O–S). At the level of the complete mixture, the anticipated response from each MOA class is treated as if it represented the response for a single chemical with a unique MOA, and RA is employed to sum the anticipated responses for all 8 mixture components (MOA groups and single chemicals).

Weight of Evidence for Chemical Interactions

Since all chemicals use the same common physiological processes of absorption, distribution, metabolism, and excretion, they may influence each other's behavior. Consider the example of an HI in Table 3-4 for liver toxicity. The HQ for each of the eight chemicals reflects that chemical's individual contribution to hepatic toxicity. Consideration should be given to two sources of contributions to hepatic toxicity resulting from a single chemical by itself, indicated by the value of HQ, and the influence of all the other chemicals' interactions affecting the liver. In many cases, direct measurement of changes in liver toxicity will not be available. General changes affecting internal dose, such as the bioavailability or pharmacokinetics of the chemical, can then be substituted.[53,54] The need to focus on a single chemical's toxicity is illustrated by studies showing asymmetric interactions that depend on the sequence of exposure. For example, the influence of chemical A on chemical B's toxicity may be synergistic, whereas the influence of B on A's toxicity may only be dose additive. These differences should be incorporated into the risk assessment of mixtures. Component exposure levels also can affect the nature and magnitude of the interaction. The high-to-low-dose extrapolation is particularly problematic for mixtures. Many dramatic interactions occur at high exposure levels (e.g., the substantial synergism between tobacco smoking and radon exposure[55]). Several publications note the expectation that most high-dose interactions will be minimal at very low doses. There is a general acceptance that interactions between chemicals are possible, and when evident are most often a function of dose. Examples that include the dose dependence of the interaction, however, are sparse. Feron et al.[56] discussed some examples where interactions occur at exposures near individual minimal-observed-effect levels while only dose addition is apparent near individual no-effect levels; they do not present a quantitative relation between interaction and dose. The influence of the relative proportions is also of concern. For example, with respect to the loss of righting reflex in mice,[57] the ED50 results for the interaction between ethanol and chloral hydrate shows synergism at low ethanol levels, but concentration (dose) additivity at higher ethanol levels. One suggestion is that the interaction should become less important as one chemical begins to dominate the mixture toxicity.[22]

Interaction-Based Hazard Index

Several examples of such change in behavior and interactions have been documented.[6] The central issue that needs to be addressed is whether they are significant biologically, statistically, or both. Most often, toxicologic interactions have been studied with binary mixtures, very few of which quantify interaction, or describe interactions as a function of dose. Thus, realistic risk assessments cannot be conducted without adequately representing and integrating the interactions that represent departures from additivity outcomes (see Table 3-5).[58]

One way to integrate this information is to begin with the dose/response-additive HI, and then modify its calculation to reflect the interaction information, using plausible assumptions to fill the data gaps. One approach to integrate the role of interactions in the overall expression of toxicity is to modify the HI in light of the knowledge of these binary interactions.[58] This weight-of-evidence procedure allows integration of all the available interaction information.[58,59] Briefly, a review is conducted of all the relevant binary component interactions of a mixture that have been studied. Due consideration and weight are given to several factors, including the degree of understanding of the interaction, its relevance to toxicity, and the extent of extrapolation to the exposure conditions of interest (e.g., route and species conversions). The strength and consistency of this evidence could then be assigned binary weight-of-evidence (BINWOE) nomenclature that qualitatively describes the nature or the toxicological interactions based on available data.[33,58]

TABLE 3-5 Weight of Evidence Scheme for the Qualitative Assessment of Chemical Interactions

Determine if the interaction of the mixture is additive (=), greater than additive (>), or less than additive (<).
CLASSIFICATION OF MECHANISTIC UNDERSTANDING
I. Direct and Unambiguous Mechanistic Data:
The mechanism(s) by which the interactions could occur has been well characterized and leads to an unambiguous interpretation of the direction of the interaction.
II. Mechanistic Data on Related Compounds:
The mechanism(s) by which the interactions could occur is not well characterized for the compounds of concern, but structure/activity relationships, either quantitative or informal, can be used to infer the likely mechanisms and the direction of the interaction.
III. Inadequate or Ambiguous Mechanistic Data:
The mechanism(s) by which the interactions could occur has not been well characterized or information on the mechanism(s) does not clearly indicate the direction that the interaction will have.

CLASSIFICATION OF TOXICOLOGIC SIGNIFICANCE
A. The toxicologic significance of the interaction has been directly demonstrated.
B. The toxicologic significance of the interaction can be inferred or has been demonstrated in related compounds.
C. The toxicologic significance of the interaction is unclear.

MODIFIERS:
1. Anticipated exposure duration and sequence
2. A different exposure duration or sequence.
    a. *In vivo* data
    b. *In vitro* data
        i. The anticipated route of exposure
        ii. A different route of exposure

(Modified from reference 58.)

TABLE 3-6 BIMWOE Determinations for a Four-Component Mixture of Metals

Effect of	On toxicity of			
	Lead	Manganese	Zinc	Copper
Lead		= IIIC n	= IIB h	= IIIC p
Manganese	>IC n >IIBh		? h	? p
Zinc	<IB n <IA h	? n		<IB p
Copper	<IC n <IB h	? n	<IIA h	

n, neurological; h, hematological; p, hepatic. Direction of interaction: =, additive; >, greater than additive; <, less than additive; ?, indeterminate. Mechanistic understanding: I, clear; II, inferred; III, unclear. Toxicologic significance: A, clear; B, inferred; C, unclear.

The BINWOE is then scaled to reflect the relative importance of the component exposure levels. A main property of the Mumtaz and Durkin approach is that the scaled BINWOE decreases with decreasing exposure levels, reflecting a common observation that the significance of interactions in a mixture decreases as the exposure and likelihood of response decreases. In this procedure, individual hazard quotients (HQs) are summed to obtain the HI for the entire mixture and then, when needed, this scaled BINWOE (i.e., a single composite interaction factor) is used to obtain the modified HI. This procedure outlined by Mumtaz and Durkin[58] has been a major advancement in the risk assessment of chemical mixtures. The approach is quite feasible: it uses available information along with toxicological judgment and reflects many general concepts regarding toxicologic interactions (see Table 3-6). The approach has been tested for consistency of application in individuals and groups.[60] Based on this approach, ATSDR developed a series of documents for certain mixtures of chemicals found at hazardous waste sites.[61] An alternative approach that recommends first the modification of each hazard quotient (HQ) according to the interaction and summation of modified HQs to obtain the modified HI was proposed.[7,62,63]

FUTURE DIRECTIONS

The field of mixtures risk assessment has matured over the past few decades and has provided initial tools for evaluation of environmental chemicals. These tools have allowed categorizing exposures, particularly at waste sites, into two groups, namely, significant or insignificant. The former are those sites that need further analysis. Fundamentally, the risk assessment process uses all the current information that is available at the time. The advancement of the following issues is poised to further improve the certainty and accuracy of such estimates. Among them are physiologically based pharmacokinetic (PBPK) modeling, evaluations of alterations beyond those in the mode of action, use of biomonitoring data, improved experimental design, and refined whole mixtures approaches by identifying sufficiently similar mixtures.

The use of PBPK modeling: PBPK modeling is more frequently being cited as a tool to refine dose-response approaches and intra- and interspecies extrapolation in health risk assessments. While largely developed for application to single-chemical risk assessment, these models have been used to determine the likelihood of chemical interactions that are based on competition for metabolism. Inasmuch as toxicity is mediated by the interaction between the toxicologically active chemical species (parent or a metabolite) and the biological receptor, this tool offers valuable insight, since it provides the ability to predict time- and dose-dependent changes in tissue concentrations of toxicants and metabolites in experimental animals and humans. PBPK models have been developed for several environmentally relevant chemical mixtures such as gasoline, chlorinated solvents, pesticides, and trihalomethanes.[64–70]

Despite providing a framework through which to integrate biological and biochemical understanding of chemical metabolism and disposition, often attempts to apply these models to situations beyond their capacity or intended purpose results in skepticism. Recently, the WHO's International Program on Chemical Safety embarked on a project to document the best practices for PBPK modeling as applied to health risk assessment.[71] By addressing the toxicity, target organ/tissue, toxic chemical species and knowledge about mode of action early in the process, the developed PBPK model will be more purpose oriented and readily accepted. With the development and inclusion of an uncertainty analysis, identifying key determinants of dosimetry and documenting confidence in measured or estimated values for those parameters, an increased level of confidence in model application will follow.

Evaluations of alterations beyond those in the mode of action: Effects might be broadened to include those events that are not part of the mode of action. Chemicals are not neat in the adverse effects that they induce; rather, they produce an array of alterations, some of which contribute to the mode of action for the observed toxicity. The development of an adverse health effect is a function of the biochemical and cellular interactions events that comprise the mode of (toxic) action. The organization of these events into a mode of action serves as the basis for deciding whether chemicals act through common or dissimilar modes of action. While chemicals with similar modes of action are subjected to risk analysis via dose additive models, those with independent modes of action are relegated to analysis via response additivity. This choice is based on comparison of events within the mode of action for toxicity, and at present only seldom includes consideration of alterations beyond key steps in the mode of action, ignoring the potential that such alterations may be shared with the key steps in the mode of action for other mixture components.[24] Increased awareness of alterations of biochemical and physiological processes that are not part of the toxic mode of action is necessary to more fully understand the likelihood of chemical interactions. Certain alterations may not be part of the mode of action for one chemical but may be key processes in the mode of action for another component of the same mixture.

Use of biomonitoring data: Increasing numbers of biomarkers are being measured in human tissues and specimens in a growing number of population-based biomonitoring surveys.[72-75] There is growing interest to use such data to identify acceptable levels of individual chemicals and their combinations. If mixtures and the proportions of the CCs can be identified using data of such surveys, studies can be conducted to identify interactive toxicity and health effects of such combinations as commonly found.[61]

Advanced experimental design: An increased awareness of the dose-response relationship of environmental contaminants would benefit health risk assessment in general and mixtures and cumulative risk assessments in particular. Many of these agents have effects or mediators of dosimetry in common. Often, constraints on experimental design and/or analytical sensitivity force exposures of experimental animals to doses of chemicals that far exceed the anticipated human exposure. With an increasing dose, three events occur: an increase in the frequency of responding individuals, an increase in the severity of the response, and (importantly for chemical mixtures) an increase in the likelihood of chemical interactions. The first condition is often used as the basis for higher doses on the grounds that it may overcome analytical insensitivity. Often, the second condition is underappreciated or overlooked altogether. Shifts in toxicity and mode of action with changes in dose may result in shifts in the pattern of interactions predicted. An increased attention to low-dose responses would reduce the uncertainty in extrapolating high-dose events to lower doses.

Use of surrogate mixtures: Identification of surrogate mixtures has been recommended, since toxicity data on whole mixtures are limited.[22,33] Two criteria have been established to guide the identification of toxicity data from a mixture that may be sufficiently similar to serve as a surrogate for the mixture in question: (1) the mixture components must be qualitatively similar, and (2) the mixture components must occur in roughly the same proportions.[22] Often, large fractions of complex mixtures may be unidentified; in these cases, additional data such as the types of toxicities observed for each mixture can be informative. The continued advancement of this area by developing case studies where surrogate mixtures are identified and their dose-response data are applied to mixtures and cumulative risk assessments will be valuable. A valuable case exercise has been recently completed for drinking water disinfection byproducts.[76-81]

In conclusion, during the past few decades, reasonable progress has been made in methods development for risk assessment of chemical mixtures. With the increase in awareness of complex exposures, there will be an equally increased demand by the communities and stakeholders to make risk assessment inclusive of mixtures of chemicals and mixtures of chemical and nonchemical stressors. Thus, the risk assessment methods process will have to continue to evolve. The interdisciplinary approaches used in MRA force the scientists whose focus may be scientifically quite narrow to work together. It remains important to foster such cross-disciplinary discussions that may result in additional, fruitful risk assessment methods development. This will also ensure that discipline-specific advances can be brought to bear on these efforts. Incorporation of some of these advances will further bolster confidence in and acceptance of mixtures risk assessments.

References

[1] Monosson E. Chemical mixtures: considering the evolution of toxicology and chemical assessment. Environ Health Perspect 2005;113:383–90.
[2] Doull J. Toxicology comes of age. Annu Rev Pharmacol Toxicol 2001;41:1–21.

REFERENCES

[3] Fay RM, Mumtaz MM. Development of a priority list of chemical mixtures occurring at 1188 hazardous waste sites, using the HazDat database. Food Chem Toxicol 1996;34:1163–5.

[4] Bliss CI. The toxicity of poisons applied jointly. Ann Appl Biol 1939;26:585–615.

[5] Boobis A, Budinsky R, Collie S, Crofton K, Embry M, Felter S, et al. Critical analysis of literature on low-dose synergy for use in screening chemical mixtures for risk assessment. Crit Rev Toxicol 2011;41(5):369–83.

[6] Mumtaz MM, Hertzberg RC. The status of interactions data in risk asssessment of chemical mixtures. In: Saxena J, editor. Hazard assessment of chemicals, vol. 8. Washington, DC: Taylor & Francis; 1993.

[7] Hertzberg RC, Teuschler LK. Evaluating quantitative formulas for dose-response assessment of chemical mixtures. Environ Health Perspect 2002;110(Suppl. 6):965–70.

[8] Yang R. Introduction to the toxicology of chemical mixtures. In: Yang R, editor. Toxicology of chemical mixtures. New York: Academic Press; 1994. p. 1–10.

[9] Mileson BE, Chambers JE, Chen WL, Dettbarn W, Ehrich M, Eldefrawi AT, et al. Common mechanism of toxicity: a case study of organophosphorus pesticides. Toxicol Sci 1998;41:8–20.

[10] Kortenkamp A, Backhaus T, Faust M. State of the art review of mixture toxicity. Brussells, Belgium: Commission of the European Union; 2009.

[11] NRC. Toxicity testing in the 21st century: a vision and a strategy. Washington, DC: The National Academies Press; 2007.

[12] Teuschler LK, Klaunig J, Carney E, Chambers JE, Gennings CR, Giesy J, et al. Support of science-based decisions concerning the evaluation of the toxicology of mixtures: a new beginning. Regul Toxicol Pharmacol 2002;36:34–9.

[13] FFDCA (Federal Food, Drug, and Cosmetic Act). 1938. In Public Law 75-717, vol. 52 Stat. 1040, United States; 1938.

[14] FFDCA. Food Additive Amendments of 1958. In Public Law 85-929, vol. 72 Stat. 1784, United States; 1958.

[15] FQPA (Food Quality Protection Act). 1996. In Public Law 104-170, vol. 110 Stat. 1489, United States; 1996.

[16] U.S. FDA (Food and Drug Administration). Milestones in U.S. Food and drug law history. FDA Backgrounder. Washington, DC: U.S. Food and Drug Administration. Available: <http://www.fda.gov/opacom/backgrounders/miles.html>; 2002 [accessed 03.10.2012].

[17] CWA. (Clean Water Act) 1972. Public Law 1972;92–500.

[18] U.S. EPA. Guidance for water quality-based decisions: the TMDL process. Office of Water, Washington, DC; 1991.

[19] FIFRA (Federal Insecticide Fungicide and Rodenticide Act). 1972. In Public Law 92-516, vol. 86 Stat. 973, United States; 1972.

[20] Worobec M. Toxic substances controls primer: federal regulations of chemicals in the environment. 2nd ed. Washington, DC: Bureau of National Affairs, Inc; 1986.

[21] U.S. EPA. Risk assessment guidance for superfund, vol. 1. Human Health Evaluation Manual (Part A). Office of Emergency and Remdial Response, Washington, DC; 1989.

[22] U.S. EPA. Supplementary guidance for conducting health risk assessment of chemical mixtures. Washington, DC: Risk Assessment Forum; 2000.

[23] U.S. EPA. 2002. Guidance on cumulative risk assessment of pesticide chemicals that have a common mechanism of toxicity. Office of Pesticides, Washington, DC.

[24] Lambert JC, Lipscomb JC. Mode of action as a determining factor in human health risk assessment of chemical mixtures. Regul Toxicol Pharmacol 2007;49:183–94.

[25] CERCLA (Comprehensive Environmental Response, Compensation, and Liability Act). 1980. Comprehensive Environmental Response, Compensation, and Liability Act of 1980. Public Law 96-510.

[26] SARA (Superfund Amendments and Reauthorization Act). 1987. Revised Comprehensive Environmental Response, Compensation, and Liability Act (CERCLA).

[27] SDWA (Safe Drinking Water Act). 1996. Safe Drinking Water Act Amendment. Public Law of 1996. Public Law 93-523.

[28] CAA (Clean Air Act) 1955. Public Law 1955;91–604. Available: <http://epw.senate.gov/envlaws/cleanair.pdf> [accessed 03.10.2012].

[29] CAA (Clean Air Act). Clean Air Act Amendment. Public Law 1990;88–206.

[30] OSHA (Occupational Safety and Health Act). 1970. In Public Law 91-596, vol. 84 Stat. 1590, United States; 1970.

[31] OSHA (Occupational Safety and Health Act). Standards: Permissible Exposure Limits for Air Contaminants, vol. 29 CRF 19101000, United States; 1971.
[32] ATSDR. Public health assessment guidance manual (2005 update). Atlanta, GA: U.S. Department of Health and Human Services, Public Health Service; 2005.
[33] ATSDR. Guidance manual for the assessment of joint toxic action of chemical mixtures. Atlanta, GA: U.S. Department of Health and Human Services, Public Health Service; 2004.
[34] U.S. EPA. The risk assessment guideline of 1986. Washington, DC: Office of Health and Environmental Assessment; 1986.
[35] Mumtaz MM, Sipes IG, Clewell HJ, Yang RS. Risk assessment of chemical mixtures: biologic and toxicologic issues. Fundam Appl Toxicol 1993;21:258–69.
[36] Lambert JC, Lipscomb JC. Mode of Action as a Determining Factor in Human Health Risk Assessment of Chemical Mixtures. Regul Toxicol Pharmacol 2007;49:183–94.
[37] U.S. EPA. Framework for cumulative risk assessment. Washington, DC: Risk Assessment Forum; 2003.
[38] Bünger J, Krahl J, Schröder O, Schmidt L, Westphal GA. Potential hazards associated with combustion of bio-derived versus petroleum-derived diesel fuel. Crit Rev Toxicol 2012;42(9):732–50.
[39] Marshall S, Gennings C, Teuschler LK, Stork LG, Tornero-Velez R, Crofton KM, et al. An empirical approach to sufficient similarity: combining exposure data and mixtures toxicology data. Risk Anal 2013 Feb 11;10.1111/risa.12015 [Epub ahead of print].
[40] Stork LG, Gennings C, Carchman RA, Carter Jr WH, Pounds J, Mumtaz M. Testing for additivity at select mixture groups of interest based on statistical equivalence testing methods. Risk Anal 2006;26(6):1601–12.
[41] U.S. EPA. The feasibility of performing cumulative risk assessments for mixtures of disinfection by-products in drinking water. Cincinnati, OH: National Center for Environmental Assessment; 2003.
[42] U.S. EPA. Organophosphorus cumulative risk assessment 2006 update. U.S. Environmental Protection Agency, Office of Pesticide Programs, Washington, DC. Available at <http://www.epa.gov/oppsrrd1/cumulative/2006-op/op_cra_main.pdf>; 2006a.
[43] U.S. EPA. Revised N-Methyl carbamate cumulative risk assessment. U.S. Environmental Protection Agency, Office of Pesticide Programs, Washington, DC. Available at <http://www.epa.gov/oppsrrd1/REDs/nmc_revised_cra.pdf>; 2007.
[44] U.S. EPA. Cumulative risk from chloroacetanalide pesticides. U.S. Environmental Protection Agency, Office of Pesticide Programs, Washington, DC. Available at <http://www.epa.gov/pesticides/cumulative/chloro_cumulative_risk.pdf>; 2006b.
[45] U.S. EPA. Pyrethrins/Pyrethroid cumulative risk assessment. U.S. Environmental Protection Agency, Office of Pesticide Programs, Washington, DC. Available at <http://www.regulations.gov/#!documentDetail;D=EPA-HQ-OPP-2011-0746-0003>; 2011.
[46] Schoeny RS, Margosches E. Evaluating comparative potencies: developing approaches to risk assessment of chemical mixtures. Toxicol Ind Health 1989;5:825–37.
[47] Van den Berg M, Birnbaum LS, Denison M, et al. Toxicity equivalence factors (TEFs) for PCBs, PCDDs, PCDFs for humans and wildlife. Environ Health Perspect 1988;106:775–92.
[48] Van den Berg M, Birnbaum LS, Denison M, et al. The 2005 World Health Organization reevaluation of human and mammalian toxic equivalency factors for dioxins and dioxin-like compounds. Toxicol Sci 2006;93:223–41.
[49] U.S. EPA. Recommended Toxicity Equivalence Factors (TEFs) for Human Health Risk Assessments of 2,3,7,8Tetrachlorodibenzo-p-dioxin and Dioxin-Like Compounds. U.S. Environmental Protection Agency, Washington, DC, EPA/100/R 10/005; December 2010.
[50] IPCS World Health Organization, International Programme on Chemical Safety. Guidance document for the use of data in development of Chemical-Specific Adjustment Factors (CSAFs) for interspecies differences and human variability in dose/concentration–response assessment. Geneva: World Health Organization; 2005. Available at <http://www.who.int/ipcs/publications/methods/harmonization/en/csafs_guidance_doc.pdf>.
[51] Mumtaz MM, Poirier KA, Colman JT. Risk assessment of chemical mixtures: fine-tuning the hazard index approach. J Clean Technol Environ Toxicol Occup Med 1997;6(2):189–204.
[52] Teuschler LK, Rice GE, Wilkes CR, Lipscomb JC, Power FW. A feasibility study of cumulative risk assessment methods for drinking water disinfection by-product mixtures. J Toxicol Environ Health A 2004;67:755–77.

[53] Krishnan K, Andersen ME, Clewell HJ, Yang RSH. Physiologically based pharmacokinetic modeling of chemical mixtures. In: Yang RSH, editor. Toxicology of chemical mixtures: case studies, mechanisms, and novel approaches. San Diego, CA: Academic Press; 1994. p. 399−437.

[54] Krishnan K, Clewell HJ, Andersen ME. Physiologically based pharmacokinetic analysis of simple mixtures. Environ Health Perspect 1994;102:151−5.

[55] ATSDR Case Studies in Environmental Medicine: Radon Toxicity. <http://www.atsdr.cdc.gov/csem/radon/radon.pdf>; 2012.

[56] Feron VJ, Groten JP, Jonker D, Cassee FR, van Bladeren PJ. Toxicology of chemical mixtures: challenges for today and the future. Toxicology 1995;105:415−27.

[57] Gessner PK. Isobolographic analysis of interactions: an update on applications and utility. Toxicology 1995;105:161−79.

[58] Mumtaz MM, Durkin PR. A weight-of-evidence approach for assessing interactions in chemical mixtures. Toxicol Ind Health 1992;8:377−406.

[59] Mumtaz MM, DeRosa CT, Groten J, Feron VJ, Hansen H, Durkin PR. Estimation of toxicity of chemical mixtures through modeling of chemical interactions. Environ Health Perspect 1998;106:1353−60.

[60] Mumtaz MM, Durkin PR, Diamond G, Hertzbrg RC. Exercises in the use of weight of evidence approach for chemical mixtures interactions. The proceedings of the international congress on the health effects of hazardous waste. Princeton, NJ: Princeton Scientific Publishing Co.; 1994. p. 637−42.

[61] ATSDR. Interaction profiles for toxic substances. Atlanta, GA: U.S. Department of Health and Human Services, Public Health Service; 2013. <http://www.atsdr.cdc.gov/interactionprofiles/index.asp>.

[62] Hertzberg RC, Rice G, Teuschler LK. Methods for health risk assessment of combustion mixtures. In: Roberts SM, Teaf CM, Bean JA, editors. Hazardous waste incineration: evaluating the human health and environmental risks. Boca Raton, FL: CRC Press; 1999. p. 105−48.

[63] U.S. EPA. Supplementary Guidance for Health Risk Assessment of Chemical Mixtures. ORD/NCEA. Washington, DC. EPA/630/R-00/002. Online. <http://cfpub.epa.gov/ncea/cfm/recordisplay.cfm?deid=20533>; 2000 [accessed 03.10.2012].

[64] Dobrev ID, Andersen ME, Yang RS. Assessing interaction thresholds for trichloroethylene in combination with tetrachloroethylene and 1,1,1-trichloroethane using gas uptake studies and PBPK modeling. Arch Toxicol 2001;75:134−44.

[65] Dennison JE, Bigelow PL, Mumtaz MM, Andersen ME, Dobrev ID, Yang RS. Evaluation of potential toxicity from co-exposure to three CNS depressants (toluene, ethylbenzene, and xylene) under resting and working conditions using PBPK modeling. J Occup Environ Hyg 2005;2:127−35.

[66] Yang RS, Dennison JE. Initial analyses of the relationship between "thresholds" of toxicity for individual chemicals and "interaction thresholds" for chemical mixtures. Toxicol Appl Pharmacol 2007;223:133−8.

[67] Tan YM, Clewell H, Campbell J, Andersen M. Evaluating pharmacokinetic and pharmacodynamic interactions with computational models in supporting cumulative risk assessment. Int J Environ Res Public Health 2011;8:1613−30.

[68] Mumtaz M, Fisher J, Blount B, Ruiz P. Application of physiologically based pharmacokinetic models in chemical risk assessment. J Toxicol 2012;2012:904603. doi:10.1155/2012/904603. Epub 2012 Mar 19.

[69] U.S. EPA. Exposures and internal doses of trihalomethanes in humans: multi-route contributions from drinking water. Cincinnati, OH: Office of Research and Development, National Center for Environmental Assessment; 2006EPA/R-06/087.

[70] Timchalk C, Poet TS. Development of a physiologically based pharmacokinetic and pharmacodynamic model to determine dosimetry and cholinesterase inhibition for a binary mixture of chlorpyrifos and diazinon in the rat. Neurotoxicology 2008;29:428−43.

[71] WHO/IPCS. Characterization and application of physiologically based pharmacokinetic models in risk assessment. Geneva, Switzerland: World Health Organization; 2010. <http://www.who.int/ipcs/methods/harmonization/areas/pbpk_models.pdf>.

[72] NHANES (National Health and Nutritional Examination Survey), Centers for Disease Control and Prevention, Atlanta, GA; 2012.

[73] Tan YM, Sobus J, Chang D, Tornero-Velez R, Goldsmith M, Pleil J, et al. Reconstructing human exposures using biomarkers and other "clues." J Toxicol Environ Health B Crit Rev 2012;15:22−38.

[74] Hays SM, Aylward LL, LaKind JS, Bartels MJ, Barton HA, Boogaard PJ, et al. Guidelines for the derivation of biomonitoring equivalents: report from the biomonitoring equivalents expert workshop. Regul Toxicol Pharmacol 2008;51:S4–15.
[75] LaKind JS, Aylward LL, Brunk C, DiZio S, Dourson M, Goldstein DA, et al. Guidelines for the communication of biomonitoring equivalents: report from the biomonitoring equivalents expert workshop. Regul Toxicol Pharmacol 2008;51:S16–26.
[76] Rice GE, Teuschler LK, Bull RJ, Simmons JE, Feder PI. Evaluating the similarity of complex drinking-water disinfection by-product mixtures: overview of the issues. J Toxicol Environ Health A 2009;72:429–36.
[77] Bull RJ, Rice G, Teuschler LK. Determinants of whether or not mixtures of disinfection by-products are similar. J Toxicol Environ Health A 2009;72:437–60.
[78] Schenck KM, Sivaganesan M, Rice GE. Correlations of water quality parameters with mutagenicity of chlorinated drinking water samples. J Toxicol Environ Health A 2009;72:461–767.
[79] Feder PI, Ma ZJ, Bull RJ, Teuschler LK, Schenck KM, Simmons JE, et al. Evaluating sufficient similarity for disinfection by-product (DBP) mixtures: multivariate statistical procedures. J Toxicol Environ Health A 2009;72:468–81.
[80] Bull RJ, Rice G, Teuschler L, Feder P. Chemical measures of similarity among disinfection by-product mixtures. J Toxicol Environ Health A 2009;72:482–93.
[81] Feder PI, Ma ZJ, Bull RJ, Teuschler LK, Rice G. Evaluating sufficient similarity for drinking-water disinfection by-product (DBP) mixtures with bootstrap hypothesis test procedures. J Toxicol Environ Health A 2009;72:494–504.

CHAPTER

4

Modeling of Sensitive Subpopulations and Interindividual Variability in Pharmacokinetics for Health Risk Assessments

Kannan Krishnan[1,3], Brooks McPhail[2], Weihsueh Chiu[3], and Paul White[3]

[1]Université de Montréal, Montréal, QC, Canada, [2]ATSDR/CDC, DTHHS/ETB, Atlanta, GA, USA, [3]U.S. Environmental Protection Agency, National Center for Environmental Assessment, Washington, DC, USA

INTRODUCTION

Many health assessments of chemicals develop exposure limits intended to protect the general population, including sensitive subpopulations, from toxicity. The U.S. Environmental Protection Agency (EPA)[1] defined the sensitive subpopulations as groups of individuals (e.g., elderly, pregnant women, children) who respond biologically at lower levels of exposure to a contaminant or who have more serious health consequences than the general population for a given level of exposure. The biological factors, characteristic of each of the subgroups of population, make them more or less susceptible than typical members of the general population to the effects of chemicals.[2] These factors then lead to interindividual variability in toxicokinetics and/or toxicodynamics at the population level.

A human interindividual uncertainty factor (UF_H; also referred to as intraspecies uncertainty factor, interindividual variability factor, or human variability factor) is used in the health risk assessment process to account for the variability in the toxicokinetics and toxicodynamics of chemicals in a population. The historical use of UF_H in human health risk assessment has been reviewed by Dourson et al.,[3] Burin and Saunders,[4] and Price et al.[5]

Price et al.[5] suggest that the rationale underlying UF_H can be thought of in terms of two conceptual models: a sensitive population model (SPM) and finite sample model (FSM). The SPM views UF_H as correcting for the failure or inability of a study to include distinct sensitive subpopulations and to account for the possibly lower toxic doses in such sensitive subpopulations. An example where SPM considerations support a UF_H is a point of departure (POD) such as the no observed adverse effect level (NOAEL) determined using healthy worker data, i.e., a study population that does not include infants, pregnant women, the elderly, or individuals exhibiting enzyme polymorphism.[5] On the other hand, the finite sample model provides support for the use of UF_H based on the inference that POD obtained from any given study may fail to identify the true threshold in the population principally because the size of the study population is small or finite.[5] According to the FSM, then, there is always a chance of failing to include the most sensitive individuals in sufficient numbers because of small sample size of the study population relative to the target population for a risk assessment, thus supporting the need to use UF_H. A related question is whether and how the statistics developed in analysis of population data portray the variability in that population. As a simple example, a central measure (e.g., especially a median or geometric mean) can portray a shift in response for the population as a group, but not provide information on the variability of responses within the population. Not that the considerations behind the SPM and FSM approaches are not necessarily mutually exclusive, and both rationales may apply to a greater or lesser extent depending on the data set and the context in which it is used.

The use of UF_H represents a pragmatic way of accounting for one or both of the following scenarios/assumptions: (i) variability in toxicological responses between individuals occurs (but there is uncertainty about its magnitude), and (ii) some subgroups are more sensitive or more susceptible than the typical individual in terms of the toxicokinetics/dynamics of a particular chemical,[3] or data to support group-specific analysis are lacking. In order to account for these sources of uncertainty/variability, the NOAEL/POD dose for the typical human is divided by UF_H to derive what the U.S. EPA defines as a reference dose, i.e., an estimate of daily oral exposure to the human population (including sensitive subgroups) that is likely to be without an appreciable risk of deleterious effects during a lifetime (http://www.epa.gov/risk_assessment/glossary.htm#r).

UF_H has typically ranged from 1 to 10, with 1 being used in cases in which a POD is determined in the most sensitive subpopulation and 10 being used in the absence of any other information.[3,6] UF_H then accounts for the presumed but unknown extent of population variability in toxicokinetic and toxicodynamic processes, with a value of 10 implying that the potential dose associated with a certain level of response (or the no observable adverse effect level) may vary between the typical (e.g., median) individual and a sensitive subpopulation by up to an order of magnitude. The UF_H can also be thought of as accounting for interindividual differences as represented by the ratio of 50th percentile to an upper percentile (e.g., 95th or 99th) individual. The World Health Organization (WHO), U.S. EPA, and Health Canada have adopted a subdivision of UF_H into toxicokinetic and toxicodynamic components, with default values of 3.16 each.[7,8] The default UF_H is generally applied for all chemicals, regardless of mode of action, nature of the toxic moiety, and chemical-specific mechanisms of toxicokinetics and toxicodynamics.

Therefore, the default value of UF_H might be overly conservative in some cases, whereas it may not be adequately protective for others.[9-13] If data allow, chemical-specific adjustment factors can be computed on the basis of experimental data or using model simulations of human variability.[7-13] Toxicological studies in neonate, infant, and juvenile animals have also been conducted to determine the sensitivity relative to adult animals.[14] In such cases, the knowledge of toxicokinetics or serum levels of chemicals would be useful in determining the magnitude and basis of age-related difference in sensitivity.

This chapter focuses on the simulation of toxicokinetic variability and reviews the state of the art of computational approaches using PBPK models to simulate dose metrics in sensitive human subpopulations. The use of these models in addressing the issue of interindividual variability in risk assessments as well as interpreting animal studies to support toxicological evaluations focused on sensitive subpopulations is also discussed.

PHYSIOLOGICAL DIFFERENCES AND PBPK MODELING OF SENSITIVE HUMAN SUBPOPULATIONS

Physiologically based pharmacokinetic (PBPK) modeling approaches are effective tools for evaluating variability in target tissue dose of chemicals by taking into account the quantitative differences in physiological and biochemical determinants of toxicokinetics. Given adequate supporting data, PBPK models have the unique advantage of being able to account for physiological and metabolic changes as a function of age, gender, etc., in order to simulate the internal dose in selected subpopulations for a variety of exposure routes, scenarios, and doses. A specific subpopulation of interest (e.g., neonates, pregnant mothers, elderly) may exhibit certain characteristics that might make them less or more susceptible toxicokinetically compared to typical adults. For example, neonates have proportionally fewer alveoli and greater breathing rates compared to adults, such that the initial uptake rate of atmospheric contaminants is greater in neonates than adults.[15] Also, the heart rate and cardiac output in newborns are proportionally greater than in older children and adults.[16,17] Body composition (e.g., bone, fat, and muscle content) also changes markedly with growth and development. Moreover, the rate of blood flow to organs changes with age, and it is not always proportional to changes in organ weights.[15] Metabolism and elimination rates are generally lower in neonates compared to adults.[18,19] Renal clearance has been reported to be lower in neonates than older children and adults, for all classes of chemicals—lipophilic, hydrophilic, as well as organic ions.[19,20]

Pregnancy represents a phase during which many physiological changes occur in the maternal organ system as a consequence of fetal growth and support.[21-23] Cardiac output, respiratory volume, and peripheral blood flows are known to increase during pregnancy. The renal blood flow and glomerular filtration rate are increased during pregnancy, whereas the albumin concentration in plasma decreases to about two-thirds of the normal level. These changes are further compounded by changes in certain metabolizing enzymes during pregnancy,[24,25] leading to varying levels of influence on the uptake, absorption, distribution, metabolism, and excretion of xenobiotics not only in the mother but also in the developing fetus.[21-23]

Similarly, the elderly represent a subpopulation with toxicokinetic characteristics that influence their susceptibility.[26-37] A number of studies have shown that decreases in cardiac output and liver weight occur with the advancement of age.[26–28] It is generally thought that the metabolic activity in the elderly might decline compared to normal adults, due to decrease in blood flow to the liver and hepatic mass rather than due to change in protein concentration or enzyme activity.[29–36]

The preceding changes in the physiological and metabolic parameters, i.e., mechanistic determinants, can be integrated within PBPK models to simulate the internal dose in various subgroups of populations. Such an approach is scientifically sound and chemical-specific, and can provide a marked advance over the generic use of body surface area or allometric adjustment, to derive the toxicologically equivalent doses. The generic dose-scaling adjustment by definition does not utilize data on chemical-specific properties or dose metrics of a chemical and the associated mechanistic determinants in all subpopulations are not always simply related to body surface area or allometric differences. PBPK models, however, facilitate the consideration of typical or distributions of the values of physiological, metabolic, and physicochemical parameters in the population to simulate and characterize the interindividual differences in internal dose. Depending on the availability of data on input parameters, one of the following computational methods has been used for evaluating interindividual variability of internal dose with PBPK models:

- Individual- and subpopulation-based modeling
- P-bounds modeling
- Monte Carlo simulation
- Markov Chain Monte Carlo simulation

Individual- and Subpopulation-Based PBPK Modeling

The individual-based PBPK modeling involves the use of subject-specific parameters regarding the physiology and metabolic capacity (e.g., enzyme levels) to simulate the tissue dose in each individual of the population (see Figure 4-1). In this approach, PBPK models are developed for each individual, and the internal dose simulations are performed for a select group of individuals. This approach has been implemented either

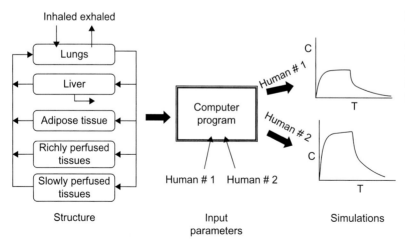

FIGURE 4-1 Illustration of the development and application of subject-specific PBPK models to simulate the kinetics of an inhaled chemical in each individual.

by constructing PBPK models for a typical individual belonging to a specified subgroup (e.g., adult women, pregnant women, lactating women, children) or by constructing models to represent each individual of a population of interest.[38–43]

The U.S. EPA,[44] in its evaluation of perchlorate, focused on the exploration of the sensitivity of various life stages on the basis of PBPK model simulations of percent inhibition of radioactive iodide uptake (RAIU). The approach involved the use of a PBPK model to evaluate the effect on a dose equal to the point of departure (POD) on RAIU in a number of different life stages. The model predictions of the percent RAIU inhibition were approximately similar in the 1–2-year-old child and adults but it was approximately 1–3-fold higher in breast-fed and bottle-fed infants, and 6.7-fold higher in the fetus (at gestational week 40) compared to the average adult. The PBPK modeling results obtained by the U.S. EPA,[44] consistent with the previous modeling analyses of Clewell et al.,[45] suggested that the near-term fetus was the most sensitive subgroup with respect to percent RAIU inhibition by perchlorate, at a maternal dose equal to POD. However, for an evaluation with a fixed drinking water concentration, the mg/kg/d exposures to perchlorate differ by subgroup and bottle-fed and nursing infants are predicted to have higher levels of RAIU inhibition.[46]

A number of PBPK modeling studies have focused on the simulation of internal dose in particular subgroups of interest to a risk assessment, by specifying physiological, biochemical, and physicochemical parameters (e.g., see References 15, 47, 48). In this regard, Price et al.[15] constructed PBPK models for a typical individual belonging to age groups of 6 years, 10 years, 14 years, and >18 years. The PBPK model simulations indicated that the blood concentrations of inhaled furan would be greater in children than in adults (see Figure 4-2), but the net amount metabolized, on the contrary, was predicted to be greater in adults than in children at all times. This study suggested that a typical value of the

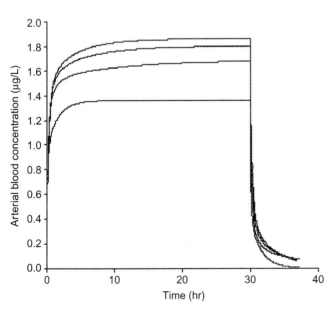

FIGURE 4-2 PBPK model simulations of the arterial blood concentration of furan in adults and children exposed to 1 μg/L of this chemical by inhalation during 30 hours. *Reproduced with permission from Price et al.*[15]

adult-children ratio, representing the differences in internal dose, corresponded to 1.5 for furan concentration in blood and 1.25 for its metabolite concentration in liver.[15] This modeling study on furan and other PBPK analyses suggested an adult-children dose ratio of <2 for inhaled lipophilic, low molecular weight VOCs that are substrates of CYEP2E1.[15,41] Note, however, that these analyses only address variability between "typical" members of each subpopulation, and not potential variability *within* a subpopulation.

Alternatively, when the "subject-specific" information can be obtained for *all* members of a subgroup, PBPK models can be constructed for each of the subjects with the ultimate goal of simulating the dose metric *distribution* in a population. This approach was implemented by Nong et al.[49] by incorporating subject-specific data on enzyme (CYP2E1) levels and physiological parameters within PBPK models. Essentially, this computational approach focused on constructing each individual of a given population and then simulating the distribution of internal dose in these individuals. Specifically, the data on body weight and liver volume for each child were obtained from autopsy record[50] and supplemented with information on quantitative relationships among body weight, age, and physiological parameters (i.e., alveolar ventilation rate, cardiac output, and blood flow rate to tissues).[51] Whereas the blood:air and tissue:blood partition coefficients were considered age-invariant (since the available data did not indicate significant variation in the lipid and water contents of most tissues and blood as a function of age), the intrinsic clearance in each individual child ($CL_{int-child}$) was computed as follows:

$$CL_{int-child} = CL_{int-CYP2E1} \times [CYP2E1]_{child} \times V_{liver-child} \qquad (4-1)$$

where $[CYP2E1]_{child}$ = subject-specific hepatic CYP2E1 protein content (pmol/mg protein) and $V_{liver-child}$ = subject-specific liver volume (L).

This modeling approach allowed the simulation of the blood-concentration profile of toluene in each individual child on the basis of subject-specific information on CYP2E1 content as well as physiological data (see Figure 4-3).[49] Even though the distributions of AUC in each age group could be computed on the basis of the totality of individual simulations of PK curves (see Table 4-1), such a computational analysis assumes that the finite sample of subject-specific information has sufficient coverage for the entire distribution, and that no other sources of variability (for which subject-specific data are not available) influence the outcome. In order to address these issues, probability-bounds (P-bounds) or Monte Carlo simulation approaches can be applied.[52]

Probability-Bounds Modeling

The probability-bounds (P-bounds) approach combines the probability theory and the interval theory, to handle variability and uncertainty, particularly when the knowledge about the statistical distribution of input parameters is not definitive or sufficient. In data-poor situations then, the P-bounds approach is useful to estimate the upper bound and lower bound of a probability distribution.[53,54] The P-bound method computes an area or envelope that is defined by the upper and lower bounds at each probability level. The envelope of P-bounds then represents all plausible cumulative probability associated with the output, regardless of the extent of knowledge about the distributions of input

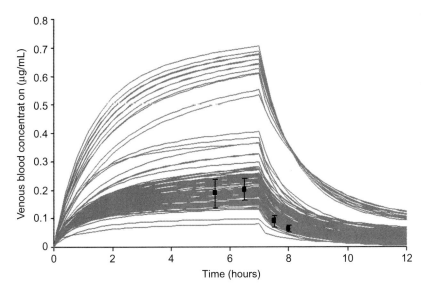

FIGURE 4-3 PBPK model simulations of venous blood concentrations of toluene in children ($n = 116$; from birth to 17 years old) exposed for 7 hours to 17 ppm of toluene. Experimental data (symbols) were obtained in adult volunteers exposed according to the same scenario. *Reproduced with permission from Nong et al.[49]*

TABLE 4-1 Intragroup and Adult-Child Factor Derived from Area under the Venous Blood Concentration Versus Time Curves (AUC) of Toluene in Children and Adults[a]

Age Group	Percentile Values of AUC (μg/ml × hr)			Intragroup Factor[b]	Adult-Child Factor[c]
	5%	50%	95%		
Adult	0.17	0.26	0.38	1.48	
Adolescent (12–17 years)	0.22	0.25	0.35	1.38	1.35
Child (1–11 years)	0.22	0.28	0.38	1.36	1.49
Infant (1 month–1 year)	0.24	0.31	0.40	1.29	1.57
Neonate (<1 month)					
low metabolizing	0.77	0.93	0.99	1.07	3.88
high metabolizing	0.31	0.48	0.63	1.31	2.46

[a] AUCs were calculated for a 24-hour period (exposure duration: 24 hours, exposure concentration: 1 ppm).
[b] The intragroup variability factor was calculated as the ratio of 95th percentile value over the 50th percentile value for the same age group.
[c] The adult-child variability factor was calculated as the ratio of the child 95th percentile value over the adult 50th percentile value.
(Reproduced with permission from Nong et al.[49])

parameters (see Figure 4-4). This approach has been applied to compute the interindividual variability in internal dose during chronic exposures.[55]

When the focus of a risk assessment is on the chronic exposure scenario (e.g., guideline development) and the chemical attains a steady state during that condition, then it is relevant to use steady-state algorithms instead of full-fledged PBPK models.[56–59] Accordingly,

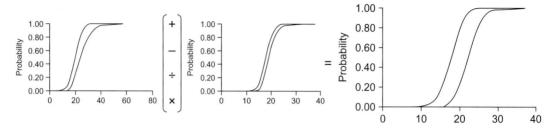

FIGURE 4-4 Illustration of the generation of the probability bounds (P-bounds) of an output based on P-bounds (i.e., lower and upper bounds) for two input parameters.

in case of inhalation exposures, the steady-state arterial blood concentration (Ca) and the rate of amount metabolized (RAM) for a given exposure concentration (Cinh) can be computed as follows[55–59]:

$$Ca = \frac{Cinh}{\left(\frac{1}{Pb}\right) + \frac{1}{Qp}\left(\frac{1}{\frac{1}{Ql}+\frac{1}{Cl_{int}}}\right)} \quad (4\text{-}2)$$

$$RAM = \frac{Cinh}{\frac{1}{Pb}\left(\frac{1}{\frac{1}{Ql}+\frac{1}{Cl_{int}}}\right) + \frac{1}{Qp}} \quad (4\text{-}3)$$

Qp, Pb, Ql, and Cl_{int} refer to alveolar ventilation rate, blood:air partition coefficient, hepatic blood flow, and intrinsic clearance, respectively.

Nong and Krishnan[55] performed the characterization of the P-bounds of the blood concentrations of benzene, carbon tetrachloride, chloroform, and methyl chloroform using RAMAS Risk Calc® software. The computer program uses semi-analytic algorithms instead of random sampling to determine the upper and lower bounds of probability distributions. The software provides the solutions in the form of the most likely bounds. Envelopes of P-bounds of blood concentration and amount metabolized during chronic exposures in humans were obtained by defining each input parameter based on its mean and standard deviation, or minimum, maximum, and mean values. Since the results of the P-bounds correspond to probability intervals (with minimum and maximum bounds), the magnitude of the human variability in kinetics (i.e., HKAF) was computed as the maximal value of the 95th percentile over the minimal value of the 50th percentile P-bound, as follows:

$$HKAF = \frac{[maximal \cdot 95\% \cdot Ca]}{[minimal \cdot 50\% \cdot Ca]} \quad (4\text{-}4)$$

Figures 4-5A and 4-5B depict the results of the P-bound analysis conducted for the amount metabolized and arterial blood concentration of chloroform, following continuous exposure to 1 ppm in the atmosphere. In this analysis, specifying the mean, standard deviation, and shape of the statistical distribution for all input parameters yielded tighter bounds (Figure 4-5A), whereas using distribution-free definitions for certain input parameters

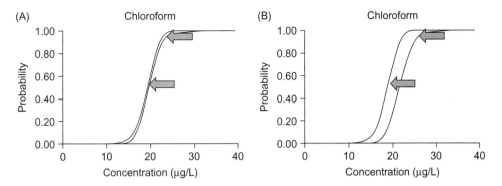

FIGURE 4-5 P-bounds of steady-state arterial blood concentrations (Ca, µg/L) in an adult population exposed to 1 ppm chloroform using input parameters characterized with (A) mean, standard deviation, and shape of the distribution; or (B) alternatively using the minimum, maximum, and mean values for the least sensitive parameter. *Reproduced with permission from Nong et al.*[55]

(i.e., least sensitive parameter) led to wider P-bounds (Figure 4-5B). The P-bounds method then allows the use of a range of specifications of input parameters, depending on data availability and purpose of the analysis. For example, when the distribution characteristics of an input parameter are not definitely known, then the mean, maximal, and minimal values based on limited available data can be used as the input to generate P-bounds of internal dose metrics for a defined subpopulation. The envelopes of P-bounds resulting from such an analysis will encompass the results of the Monte Carlo simulation method based on defined statistical knowledge of all input parameters (*vide infra*).

Monte Carlo Simulations and PBPK Modeling

The impact of interindividual variability in input parameters on the model output (i.e., internal dose measures), can be effectively evaluated using Monte Carlo methods, when there is sufficient information on the statistical distributions and interdependence of parameters. Using point estimates for input parameters or using intervals (i.e., P-bounds), does not permit the characterization of the probability or likelihood with which a particular value of internal dose may be observed in a population. When the variability in input parameters is characterized in terms of statistical distributions, the propagation of the variability leading to model output is facilitated by the Monte Carlo simulation method.[52,60,61] This method used for PBPK modeling consists of (i) specifying a probability distribution for each input parameter of the PBPK model, (ii) random sampling of each input parameter from its specified statistical distribution in consideration of possible interdependence or covariation with other input parameters, (iii) solving the PBPK model equations using the sampled set of input parameter values during each iteration, and (iv) computing the distributions of the various dose metrics of interest for a risk assessment based on the output of several thousand iterations (see Figure 4-6). In addition to specifying independent distributions for parameters, consideration of the specification of a joint probability distribution on the basis of correlation among certain parameters may be required. During

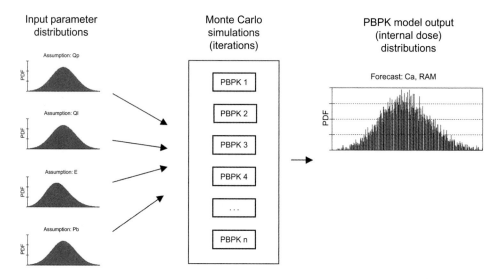

FIGURE 4-6 Illustration of the use of input parameter distributions for facilitating Monte Carlo simulations of the population distributions of the internal dose using PBPK models

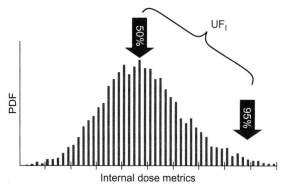

FIGURE 4-7 Computation of the magnitude of the human kinetic interindividual variability as the ratio of an upper percentile value (e.g., 95th) to the median value (50th) of the simulated internal dose metrics.

Monte Carlo simulations of internal dose using PBPK models, it is important to ensure during all iterations that

- The numerical values of physiological parameters are within known, plausible limits.
- The sum of compartment volumes is equal to or lower than the body weight.
- The sum of tissue blood flows is equal to cardiac output.
- The mass balance is respected (chemical absorbed = chemical in body + chemical eliminated).
- The covariant nature of the parameters is appropriately respected (e.g., the person with lowest breathing rate cannot be the one receiving the highest cardiac output).

TABLE 4-2 Reconstruction of the Hypothetical Populations of 100,000 People with the Canadian Demographic Profile

Subpopulation (age range)	Canadian Population in 2009 Median Age = 39.7 years	
	Population Size (%)	Corresponding Reconstructed Population Size and Number (n) of Monte Carlo Iterations
Adults (18–64 years)	21,685,253 (63.92)	63,923
Neonates (0–30 days)	31,303 (0.09)	93
Infants (1–12 months)	344,329 (1.02)	1,015
Toddlers (1–3 years)	1,126,896 (3.32)	3,322
Children and adolescents (4–17 years)	5,382,420 (15.87)	15,866
Elderly (65–90 years)	4,634,673 (13.66)	13,662
Pregnant women (15–44 years)	718,950 (2.12)	2,119
TOTAL	**33,923,824 (100)**	**100,000**

(Reproduced from Valcke and Krishnan[65])

The Monte Carlo sampling method implemented in PBPK modeling allows the simulation of the population distribution of internal dose based on specifications of input parameters.[52,62] The resulting distributions of internal dose can then be used to calculate the magnitude of UF_H (e.g., dividing a 95th or 99th percentile value of the internal dose by the 50th percentile value in the population; see Figure 4-7). Even though the distributions used in such an approach may appropriately represent the variability within adult populations, it is unclear as to the extent to which it would also be representative of the various sensitive subpopulations. It is for that reason that the "distinct subpopulation" approach is implemented, i.e., specifying the parameter distributions in presumed or selected sensitive groups (neonates, pregnant women, elderly, polymorphic individuals ...) and simulating distributions of internal dose in each subgroup. The PBPK model simulations of an upper percentile value of internal dose in the sensitive subgroup(s) are then compared with the median value in adults to compute the UF_H.[52,63–65]

These approaches may allow subpopulation (e.g., lifestage) distributions to be developed for populations of concern in an assessment. Depending on the needs of the assessment, it may also be possible to combine the results for multiple subgroups to develop a distribution for the overall population. In such a "whole population" approach, simulations are conducted for all the individuals and subgroups, as per the demographic data. In this regard, Valcke and Krishnan[66] constructed virtual populations consisting of specified proportions of adults, elderly, children, neonates, and pregnant women as per Canadian demographics (see Table 4-2). Using physiological parameters for these subgroups based on the literature and P3M software, they obtained the UF_H by dividing the simulated upper percentile value (99th, 95th) of the internal dose by the median value for the entire population. The percentage of each subpopulation protected by a "whole population" UF_H was determined by computing the number of individuals in each subpopulation exhibiting an internal dose metric that was lower than the entire population's upper percentile value underlying the UF_H calculation (i.e., 95th or 99th percentile value). Under the conditions of this study, the modeling results

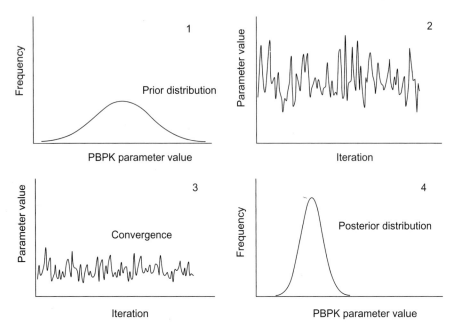

FIGURE 4-8 Illustration of the Markov Chain Monte Carlo approach. In this procedure, (1) the prior distributions of model parameters are defined, (2) parameter distributions are updated following Markov chain iterations, (3) a new distribution is produced, and (4) the posterior distribution is collected from the chain at convergence. *Reproduced with permission from Krishnan and Johanson.[69]*

indicated that, for example, in the case of benzene, only 57% of the neonates, 78% of the pregnant women, and 89% of the infants were covered by the 95th percentile-based UF_H.[66] Overall, these analyses indicated that the difference in the extent of coverage afforded by the "whole population" versus "distinct subpopulation" simulation approaches would depend on the proportion of the most sensitive individuals in the target population of a risk assessment. The applicability and accuracy of these simulation approaches can potentially be improved by refining the specification of population distributions of input parameters on the basis of the Bayesian approach.

Bayesian Approach in PBPK Modeling

Effective implementation of the Monte Carlo simulation approach with the PBPK models or steady-state algorithms requires the characterization of the distribution of input parameters.[55,63–66] While one is applying such an approach to characterize the population distribution of internal dose, it is important to specify the distributions of input parameters adequately to represent the target population. In this regard, Bayesian approaches are useful in facilitating the modeler to improve the knowledge, or specification of distribution characteristics, of an input parameter on the basis of information contained in new data. Thus, when limited, subject-specific kinetic data are available, that information can be used to improve on the input parameter distributions for PBPK modeling of specific subgroups or an entire

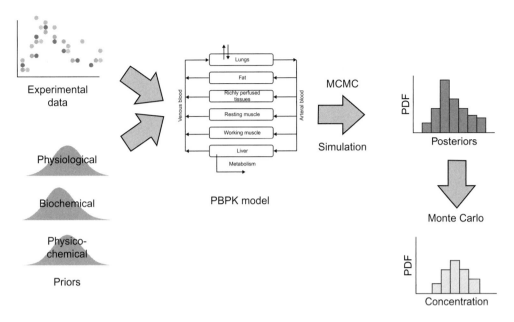

FIGURE 4-9 Illustration of the use of the Markov Chain Monte Carlo (MCMC) approach in refining input parameters and generating population distributions of dose metrics using a PBPK model for methyl *tert*-butyl ether. *From Nong et al.*[75]

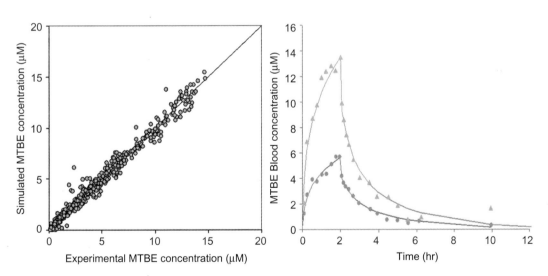

FIGURE 4-10 (Left panel) Comparison of model predictions with experimental data on individual measurements of methyl *tert*-butyl ether (MTBE) blood concentrations. (Right panel) Comparison between the PBPK model simulations using the posteriors and the experimental data for methyl *tert*-butyl ether (MTBE) obtained from the literature for exposure levels of 25 ppm and 50 ppm. *From Nong et al.*[75]

population. The Markov Chain Monte Carlo technique has been successfully used for performing Bayesian analysis of PBPK models.[67–70] This analysis randomly and iteratively samples the parameter values until they show convergence and are consistent with both new experimental data as well as prior distributions (see Figure 4-8). The resulting probability distributions (referred to as "posteriors") are then consistent not only with the new data analyzed but also with the existing prior distributions, due to the fact that the posteriors are a function of the likelihood of the new data and prior probability distributions of the input parameters. The posterior distributions of the input parameters of PBPK models can then be applied with Monte Carlo simulation for variability analysis. Several authors have assessed the population distribution of internal dose of chemicals using PBPK models and Markov Chain Monte Carlo (MCMC) simulation.[71,72]

Nong et al.[75] applied the MCMC analysis with a PBPK model for methyl *tert*-butyl ether (see Figure 4-9). The comparisons of model simulations of MTBE blood concentrations with experimental data from individual subjects are depicted in Figure 4-10. An interindividual variability factor for the general population was subsequently calculated as the ratio of the 95th to 50th percentile values for internal doses such as maximal blood concentration (C_{max}) or area under the blood concentration versus time curve (AUC). The U.S. EPA applied a similar approach in its recent Toxicological Reviews of trichloroethylene and dichloromethane.[76,77] In both cases, human PBPK models were calibrated to individual human experimental data using a Bayesian approach. For trichloroethylene, *in vitro* data on variability in metabolism were incorporated in prior distributions, which were then updated via MCMC with *in vivo* data.[74] For dichloromethane, information on GST polymorphisms was explicitly incorporated as part of the analysis.[73] These posterior distributions were then used by the EPA in a Monte Carlo analysis of human variability. Instead of calculating a UF_H for toxicokinetics, the EPA directly calculated the human equivalent dose or concentration (using a dose metric appropriate for the endpoint under consideration) for an individual more sensitive than 99% of the simulated population to serve as the POD.[76,77] For dichloromethane, the EPA incorporated additional information on how model parameters change with life stage and gender in its population analysis.[76]

ANIMAL PBPK MODELS FOR EVALUATING SENSITIVE SUBPOPULATIONS

Rodents are commonly used as surrogates for humans in assessing the health risks. In order to assess the rodent's suitability as a surrogate for pediatric studies, one should characterize developmental changes in both rodents and humans throughout each stage of life. Such information for specific chemicals may aid in evaluation of whether developmental effects may be more or less pronounced in humans when compared to rodents.

As the young organisms mature, "windows of vulnerability" may exist that could cause variations in the activity of a compound due to developmental changes in mechanistic determinants of kinetics and/or dynamics. PBPK models are capable of integrating developmental changes such as physiology, enzymatic activity, and tissue composition[78,79] to provide information on the pharmacokinetics and tissue dose of a chemical. PBPK models are becoming increasingly important in risk assessment not only for characterizing the

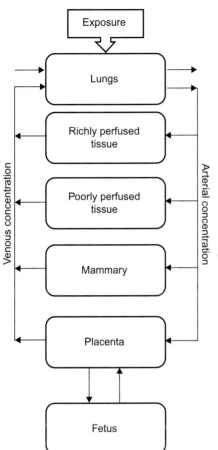

FIGURE 4-11 A generalized schematic of a pregnancy PBPK model in the rat.

tissue distribution of chemicals in humans based on data from animal studies but also for interpreting developmental toxicology studies on the basis of internal dose metric data.

PBPK models have been developed for a wide range of compounds and scenarios including pregnant dams (see Figure 4-11), lactating pups, and developing rodents (see Figure 4-12). Olanoff and Anderson[80] created the first pregnancy model for the compound tetracycline. Tetracycline is an antibiotic that affects the development of bones and teeth during which exposure occurs *in utero*. The initial pregnancy model created the template for future maternal models. In 1992, O'Flaherty et al.[81] constructed a PBPK model that depicted the entire gestational phase of rodent development and included organogenesis. The model accounted for maternal and fetal growth. Models that simulate the kinetics and disposition of compounds during pregnancy account for maternal and fetal exposures via placental transfer. During pregnancy, certain compounds (e.g., lipophilic organic compounds) pass through the placenta barrier and reach the fetus.

Fisher et al.[82] created algorithms for integration within PBPK models, to account for physiological changes occurring in maternal and neonatal rodents during lactation. The

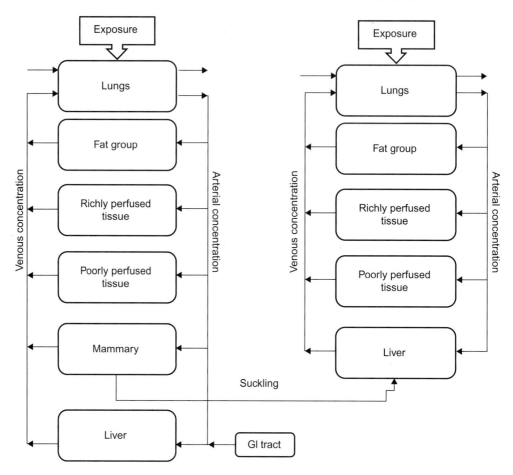

FIGURE 4-12 A generalized schematic of a maternal and lactational PBPK model in the rat.

resulting models predicted the exposure of trichloroethylene through the inhalation and ingestion of drinking water in lactating dams, and via inhalation and ingestion of maternal milk for nursing pups. The lactational transfer of compounds in rodents is a dominant route of exposure during early postnatal life. During the first 2 weeks of life, when the pup relies on maternal milk as the source of nutrients, its exposure to compounds is dependent on maternal and chemical characteristics. For example, the amount of compound the dam is exposed to (maternal characteristic) and the physiochemical properties, such as the molecular weight, pH, degree of ionization, capacity to bind to the plasma protein, and lipophilic properties (chemical characteristic), affect the transfer of compounds from the maternal milk to the nursing pup. Nonpolar lipophilic compounds, such as volatile organic compounds (VOCs), acquire easy passage into maternal milk.

In addition to pregnancy and lactation models, PBPK models have also been used to predict the kinetics of compounds during postnatal development, focusing on the weaning period (e.g., PND 17). Rodriguez et al.[83] developed PBPK models for PND 10, 2-month-old, and 22-month-old Sprague–Dawley rats. In this study, PBPK models were used to predict the effect of aging on the inhalation pharmacokinetics of perchloroethylene, trichloroethylene, benzene, chloroform, methylene chloride, and methyl ethyl ketone in SD rats. PBPK models for juvenile animals, in the past, were sometimes developed by scaling physiological data obtained in adult animals; however, age-specific tissue volumes and tissue blood flow rates are now available and compiled for consistent use in PBPK models. For example, Schoeffner et al.[84] determined the body and organ weight of commonly used tissues in developing Sprague–Dawley rats aged 4–40 weeks and in young Fischer 344 rats aged 9–10 weeks. Likewise, Gentry et al.[85] developed a data set of previously published information, which included the tissue weight, tissue blood flows, and intake for developing rats and mice from birth (PND 0) through PND 60. Despite the available information, few studies include a complete publication of various organ weights throughout development. Due to this lack of data, Mirfazaelian and Fisher[79] developed a generalized Michaelis–Menten model to determine the organ weight of developing rats using data obtained from Schoeffner et al.[84]

Additional parameters for PBPK modeling relate to those of metabolic and renal clearance. During development, the activity of hepatic enzymes involved in phase I and phase II metabolism has been shown to vary with age in rodents,[86,87] and the immaturity of the renal system and renal functions, compared to adult animals, may affect the excretion of compounds.[88-90] Similarly, the development of blood-brain barrier and the brain growth need to be accounted for in the PBPK models appropriately to simulate the kinetics of chemicals during various development stages in comparison with adult animals.[91,92] The use of simple allometric equations is unlikely to be adequate in capturing the qualitative and quantitative changes occurring in the various physiological parameters during development in rodents.[93]

CONCLUDING REMARKS

Sensitive subgroups require a particular attention in risk assessments, due to their peculiar toxicokinetic and toxicodynamic characteristics. The use of physiologically based pharmacokinetic (PBPK) models can account for changes in mechanistic determinants, thus facilitating the simulation of the distribution of internal dose in various subpopulations (e.g., elderly, children, pregnant women). Thus, they are useful for characterizing the magnitude of the interindividual variability factor in the health assessment of specific chemicals. The advantage of PBPK models is that they allow predictions of variables that cannot be easily accessed or measured in susceptible subpopulations with the available methodologies. The purpose-specific evaluation of these models should be conducted along a broad range of topics including (i) model purpose, (ii) model structure, (iii) mathematical representation, (iv) parameter estimation, (v) computer implementation, (vi) predictive capacity, and (vii) specialized analyses (i.e., sensitivity, variability, and uncertainty analyses).[59,94] Sensitivity analysis would be particularly useful in identifying those areas of uncertainty that should be the focus of further study or those key input parameters of the PBPK models that affect the risk numbers. When human data and PBPK models are lacking, emerging *in vitro* and other 21st century data streams (e.g., genetically defined human cell lines, genetically diverse rodent models, human omic profiling, and genome-wide association studies) may also offer opportunities to assess human variability, but challenges remain in incorporating such data in human health assessments of environmental chemicals.[95]

DISCLAIMER

The views in this chapter are those of the authors and do not necessarily reflect the views or policies of the U.S. Environmental Protection Agency.

References

[1] U.S. EPA. Report to Congress. EPA studies on sensitive subpopulations and drinking water contaminants. Office of Water, EPA 815-R-00-015, December 2000. Volume 63, Number 61. Tuesday, March 31, 1998. Proposed Rules. 15673-15692. Cited by Griffiths, J. K. (2001). Sensitive subpopulations. Rural water partnership fund. White paper. Duncan, OK: NRWA; 2000. p. 32.

[2] Hines RN, Sargent D, Autrup H, Birnbaum LS, Brent RL, Doerrer NG, et al. Approaches for assessing risks to sensitive populations: lessons learned from evaluating risks in the pediatric population. Toxicol Sci 2010;113:4–26.

[3] Dourson ML, Felter SP, Robinson D. Evolution of science-based uncertainty factors in noncancer risk assessment. Regul Toxicol Pharmacol 1996;24:108–20.

[4] Burin GJ, Saunder DR. Addressing human variability in risk assessment—the robustness of the intraspecies uncertainty factor. Regul Toxicol Pharmacol 1999;3:209–16.

[5] Price PS, Keenan RS, Schwab B. Defining the interindividual (intraspecies) uncertainty factor. Human Ecol Risk Assess 1999;5:1023–33.

[6] Vermeire T, Stevenson H, Peiters MN, Rennen M, Slob W, Hakkert BC. Assessment factors for human health risk assessment: a discussion paper. Crit Rev Toxicol 1999;5:439–90.

[7] Meek ME, Renwick A, Ohanian E, Dourson M, Lake B, Naumann BD, et al. International Programme on Chemical Safety: guidelines for application of chemical-specific adjustment factors in dose/concentration-response assessment. Toxicology 2002;181-182:115–20.

[8] IPCS (International Programme on Chemical Safety). Chemical-Specific Adjustment Factors (CSAFs) for interspecies differences and human variability: guidance document for the use of data in dose/concentration-response assessment. WHO/IPCS/01.4, Geneva, Switzerland; 2005. p. 1–96.

[9] Dorne JL, Walton K, Renwick AG. Human variability in glucuronidation in relation to uncertainty factors for risk assessment. Food Chem Toxicol 2001;39:1153–73.

[10] Dorne JL, Walton K, Renwick AG. Uncertainty factors for chemical risk assessment. Human variability in the pharmacokinetics of CYP1A2 probe substrates. Food Chem Toxicol 2001;39:681–96.

[11] Dorne JL, Walton K, Renwick AG. Human variability in CYP3A4 metabolism and CYP3A4-related uncertainty factors for risk assessment. Food Chem Toxicol 2003;41:201–24.

[12] Dorne JL, Walton K, Renwick AG. Polymorphic CYP2C19 and N-acetylation: human variability in kinetics and pathway-related uncertainty factors. Food Chem Toxicol 2003;41:225–45.

[13] Dorne JL, Walton K, Renwick AG. Human variability in the renal elimination of foreign compounds and renal excretion-related uncertainty factors for risk assessment. Food Chem Toxicol 2004;42:275–98.

[14] Brent RL. Utilization of juvenile animal studies to determine the human effects and risks of environmental toxicants during postnatal developmental stages. Birth Defects Res B Dev Reprod Toxicol 2004;71:303–20.

[15] Price K, Haddad S, Krishnan K. Physiological modeling of age-specific changes in the pharmacokinetics of organic chemicals in children. J Toxicol Environ Health 2003;66(**A**):417–33.

[16] Cayler GG, Rudolph AM, Nadas AS. Systemic blood flow in infants and children with and without heart disease. Paediatrics 1963;32:186–201.

[17] Sholler GF, Celermajer JM, Whight CM, Bauman AE. Echo doppler assessment of cardiac output and its relation to growth in normal infants. Am J Cardiol 1987;60:1112–6.
[18] Alcorn J, McNamara PJ. Pharmacokinetics in the newborn. Adv Drug Delivery Rev 2003;55:667–86.
[19] DeWoskin RS, Thompson CM. Renal clearance parameters for PBPK model analysis of early lifestage differences in the disposition of environmental toxicants. Regul Toxicol Pharmacol 2008;51:66–86.
[20] Clewell HJ, Teeguarden J, McDonald T, Sarangapani R, Lawrence G, Covington T, et al. Review and evaluation of the potential impact of age- and gender-specific pharmacokinetic differences on tissue dosimetry. Crit Rev Toxicol 2002;32:329–89.
[21] Hytten FE. Physiological changes in the mother related to drug handling. In: Krauer B, Krauer F, Hytten FE, del Pozo E, editors. Drugs and pregnancy: maternal drug handling: fetal drug exposure. New York: Academic Press; 1984. p. 7–17.
[22] Krauer B. Physiological changes and drug disposition during pregnancy. In: Nau H, Scott WJ, editors. Pharmacokinetics in teratogenesis, vol. 1. Boca Raton, FL: CRC Press; 1987. p. 3–12.
[23] Mattison DR, Blann E, Malek A. Physiological alterations during pregnancy: impact on toxicokinetics. Fundam Appl Toxicol 1991;16:215–8.
[24] Juchau MR. Enzymatic bioactivation and inactivation of chemical teratogens and transplacental carcinogens/mutagens. In: Juchau MR, editor. The biochemical basis of chemical teratogenesis. New York: Elsevier/North Holland; 1981. p. 63–94.
[25] Juchau MR, Faustman-Watts EM. Pharmacokinetic considerations in the maternal-placenta unit. Clin Obstet Gynecol 1983;26:379–90.
[26] Boyd E. Normal variability in weight of the adult human liver and spleen. Arch Pathol 1933;16:350–72.
[27] Swift CG, Homeida M, Halliwell M, Roberts CJ. Antipyrine disposition and liver size in the elderly. Eur J Clin Pharmacol 1978;14:149–52.
[28] Marchesini G, Bua V, Brunori A, Bianchi G, Pisi P, Fabbri A, et al. Galactose elimination capacity and liver volume in aging man. Hepatology 1988;8:1079–83.
[29] Bach B, Hansen JM, Kampmann JP, Rasmussen SN, Skovsted L. Disposition of antipyrine and phenytoin correlated with age and liver volume in man. Clin Pharmacokinet 1981;6:389–96.
[30] Hunt CM, Strater S, Stave GM. Effect of normal aging on the activity of human hepatic cytochrome P450IIE1. Biochem Pharmacol 1990;40:1666–9.
[31] Schmucker DL, Woodhouse KW, Wang RK, Wynne H, James OF, McManus M, et al. Effects of age and gender on *in vitro* properties of human liver microsomal monooxygenases. Clin Pharmacol Ther 1990;48:365–74.
[32] Shimada T, Yamazaki H, Mimura M, Inui Y, Guengerich FP. Interindividual variations in human liver cytochrome P-450 enzymes involved in the oxidation of drugs, carcinogens and toxic chemicals: studies with liver microsomes of 30 Japanese and 30 caucasians. J Pharmacol Exp Ther 1994;270:414–23.
[33] Woodhouse KW, Mutch E, Williams FM, Rawlins MD, James OF. The effect of age on pathways of drug metabolism in human liver. Age Ageing 1984;13:328–34.
[34] Wynne HA, Mutch E, James OF, Wright P, Rawlins MD, Woodhouse KW. The effect of age upon the affinity of microsomal mono-oxygenase enzymes for substrate in human liver. Age Ageing 1988;17:401–5.
[35] Zoli M, Iervese T, Abbati S, Bianchi GP, Marchesini G, Pisi E. Portal blood velocity and flow in aging man. Gerontology 1989;35:61–5.
[36] Zoli M, Magalotti D, Bianchi G, Gueli C, Orlandini C, Grimaldi M, et al. Total and functional hepatic blood flow decrease in parallel with ageing. Ageing 1999;28:29–33.
[37] Jassal SV, Oreopoulos D. The aging kidney. Geriatr Nephrol Urol 1998;8:141–7.
[38] Abraham K, Mielke H, Huisinga W, Gundert-Remy U. Elevated internal exposure of children in simulated acute inhalation of volatile organic compounds: effects of concentration and duration. Arch Toxicol 2005;79:63–73.
[39] Bogaards JJ, Hissink EM, Briggs M, Weaver R, Jochemsen R, Jackson P, et al. Prediction of interindividual variation in drug plasma levels *in vivo* from individual enzyme kinetic data and physiologically based pharmacokinetic modeling. Eur J Pharm Sci 2003;12:117–24.
[40] Byczkowski JZ, Fisher JW. A computer program linking physiologically-based model with cancer risk assessment for breast-fed infants. Comput Methods Programs Biomed 1995;46:155–63.
[41] Clewell HJ, Gentry PR, Covington TR, Sarangapani R, Teeguarden JG. Evaluation of the potential impact of age and gender-specific pharmacokinetic differences on tissue dosimetry. Toxicol Sci 2004;79:381–93.

[42] Corley RA, Mast TJ, Carney EW, Rogers JM, Daston GP. Evaluation of physiologically based models of pregnancy and lactation for their application in children's health risk assessments. Crit Rev Toxicol 2003;33:137–211.

[43] Fisher J, Mahle D, Bankston L, Greene R, Gearhart J. Lactational transfer of volatile chemicals in breast milk. Ind Hyg Assoc J 1997;58:425–31.

[44] U.S. EPA. Drinking Water Perchlorate Supplemental Request for Comments. Fed. Reg. 74:41883–93.

[45] Clewell RA, Merrill EA, Gearhart JM, Robinson PJ, Sterner TR, Mattie DR, et al. Perchlorate and radioiodide kinetics across life stages in the human: using PBPK models to predict dosimetry and thyroid inhibition and sensitive subpopulations based on developmental stage. J Toxicol Environ Health A 2007;70:408–28.

[46] Gentry PR, Hack CE, Haber L, Maier A, Clewell III HJ. An approach for the quantitative consideration of genetic polymorphism data in chemical risk assessment: examples with warfarin and parathion. Toxicol Sci 2002;70:120–39.

[47] Haddad S, Restieri C, Krishnan K. Characterization of age-related changes in body weight and organ weights from birth to adolescence in humans. J Toxicol Environ Health 2001;64(**A**):453–64.

[48] Lipscomb JC, Teuschler LK, Swartout J, Popken D, Cox T, Kedderis GL. The impact of cytochrome P450 2E1-dependent metabolic variance on a risk-relevant pharmacokinetic outcome in humans. Risk Anal 2003;23:1221–38.

[49] Nong A, McCarver DG, Hines RN, Krishnan K. Modeling interchild differences in pharmacokinetics on the basis of subject-specific data on physiology and hepatic CYP2E1 levels: a case study with toluene. Toxicol Appl Pharmacol 2006;214:78–87.

[50] Johnsrud EK, Koukouritaki SB, Divakaran K, Brunengraber LL, Hines RN, McCarver DG. Human hepatic CYP2E1 expression during development. J Pharmacol Exp Ther 2003;307:402–7.

[51] Price PS, Conolly RB, Chaisson CF, Gross EA, Young JS, Mathis ET, et al. Modeling interindividual variation in physiological factors used in PBPK models of humans. Crit Rev Toxicol 2003;33:469–503.

[52] Punt A, Jeurissen SM, Boersma MG, Delatour T, Scholz G, Schilter B, et al. Evaluation of human interindividual variation in bioactivation of estragole using physiologically based biokinetic modeling. Toxicol Sci 2010;113:337–48.

[53] Ferson S. What Monte Carlo methods cannot do. Hum Ecol Risk Assess 1996;2:990–1007.

[54] Ferson S, Root WT, Kuhn R. RAMAS risk calc: risk assessment with uncertain numbers. Setauket, New York: Applied Biomathematics; 1999.

[55] Nong A, Krishnan K. Estimation of interindividual pharmacokinetic variability factor for inhaled volatile organic chemicals using a probability-bounds approach. Regul Toxicol Pharmacol 2007;48:93–101.

[56] Andersen ME. Pharmacokinetics of inhaled gases and vapors. Neurobehav Toxicol Teratol 1981;3:383–9.

[57] Csanady GA, Filser JG, Kreuzer PE, Kessler W. Toxicokinetic models for volatile industrial chemicals and reactive metabolites. Toxicol Lett 1995;82/83:357–66.

[58] Pelekis M, Krewski D, Krishnan K. Physiologically based algebraic expressions for predicting steady-state toxicokinetics of inhaled vapors. Toxicol Meth 1997;7:205–25.

[59] Chiu WA, White P. Steady-state solutions to PBPK models and their applications to risk assessment I: route-to-route extrapolation of volatile chemicals. Risk Anal 2006;26:769–80.

[60] Chiu WA, Barton HA, DeWoskin RS, Schlosser P, Thompson CM, Sonawane B, et al. Evaluation of physiologically based pharmacokinetic models for use in risk assessment. J Appl Toxicol 2007;3:218–37.

[61] U.S. EPA. Guiding principles for Monte Carlo analysis. Washington, DC: U.S. Environmental Protection Agency, Risk Assessment Forum; 1997 [EPA/630/R-97/001]

[62] Thomas RS, Bigelow PL, Keefe TJ, Yang RS. Variability in biological exposure indices using physiologically based pharmacokinetic modeling and Monte Carlo simulation. Am Ind Hyg Assoc J 1996;57:23–32.

[63] Valcke M, Krishnan K. Evaluation of the impact of the exposure route on the human kinetic adjustment factor. Regul Toxicol Pharmacol 2011;59:258–69.

[64] Valcke M, Krishnan K. Evaluation of the impact of physico-chemical and biochemical characteristics on the human kinetic adjustment factor for systemic toxicants. Toxicology 2011;286:36–47.

[65] Valcke M, Krishnan K. Assessing the impact of the duration and intensity of inhalation exposure on the magnitude of the variability of internal dose metrics in children and adults. Inhal Toxicol 2011;23:863–77.

[66] Valcke M, Nong A, Krishnan K. Modeling the human kinetic adjustment factor for inhaled volatile organic chemicals: whole population approach versus distinct subpopulation approach. J Toxicol 2011;2012:404329. doi:10.1155/2012/404329.

[67] Bois FY, Jackson ET, Pekari K, Smith MT. Population toxicokinetics of benzene. Environ Health Perspect 1996;104(6):1405–11.

[68] Bernillon P, Bois FY. Statistical issues in toxicokinetic modeling: a Bayesian perspective. Environ Health Perspect Suppl 2000;108:883–93.

[69] Krishnan K, Johanson G. Physiologically-based pharmacokinetic and toxicokinetic models in cancer risk assessment. J Environ Sci Health 2005;23(C):31–53.

[70] Jonsson F, Johanson G. The Bayesian population approach to physiological toxicokinetic-toxicodynamic models—an example using the MCSim software. *Toxicol Lett* 2003;138:143–50.

[71] Jonsson F, Bois F, Johanson G. A Bayesian analysis of the influence of GSTT1 polymorphism on the cancer risk estimate for dichloromethane. Toxicol Appl Pharmacol 2001;174:99–112.

[72] El-Masri HA, Bell DA, Portier CJ. Effects of glutathione transferase theta polymorphism on the risk estimates of dichloromethane to humans. Toxicol Appl Pharmacol 1999;158:221–30.

[73] David RM, Clewell HJ, Gentry PR, Covington TR, Morgott DA, Marino DJ. Revised assessment of cancer risk to dichloromethane II. Application of probabilistic methods to cancer risk determinations. Regul Toxicol Pharmacol 2006;45:55–65.

[74] Chiu WA, Okino MS, Evans MV. Characterizing uncertainty and population variability in the toxicokinetics of trichloroethylene and metabolites in mice, rats, and humans using an updated database, physiologically based pharmacokinetic (PBPK) model, and Bayesian approach. Toxicol Appl Pharmacol 2009;241:36–60.

[75] Nong A, Krishnan K, Ernstgard L, Johanson G. Bayesian analysis of the inhalation pharmacokinetics of methyl tert-butyl ether (MTBE) and its metabolite tert-butanol in humans. Toxicol Sci 2006;90:487 (#2382).

[76] U.S. EPA. Toxicological review of dichloromethane (CAS No. 75-09-2) in support of summary information on the Integrated Risk Information System (IRIS). Washington, DC: National Center for Environmental Assessment; 2011 [EPA/635/R-10/003F].

[77] U.S. EPA. Toxicological review of trichloroethylene (CASRN 79-01-6) in support of summary information on the Integrated Risk Information System (IRIS). Washington, DC: National Center for Environmental Assessment; 2011 [EPA/635/R-09/011F].

[78] Delp MD, Evans MV, Duan C. Effects of aging on cardiac output, regional blood flow, and body composition in Fischer-344 rats. J Appl Physiol 1998;85:1813–22.

[79] Mirfazaelian A, Fisher JW. Organ growth functions in maturing male Sprague–Dawley rats based on a collective database. J Toxicol Environ Health A 2007;70:1052–63.

[80] Olanoff LS, Anderson JM. Controlled release of tetracycline—III: a physiological pharmacokinetic model of the pregnant rat. J Pharmacokinet Biopharm 1980;8(6):599–620.

[81] O'Flaherty EJ, Scott W, Schreiner C, Beliles RP. A physiologically based kinetic model of rat and mouse gestation: disposition of a weak acid. Toxicol Appl Pharmacol 1992;112(2):245–56.

[82] Fisher JW, Whittaker TA, Taylor DH, Clewell HJ, Andersen ME. Physiologically based pharmacokinetic modeling of the lactating rat and nursing pup: a multiroute exposure model for trichloroethylene and its metabolite, trichloroacetic acid. Toxicol Appl Pharmacol 1990;102(3):497–513.

[83] Rodriguez CE, Mahle DA, Gearhart JM, Mattie DR, Lipscomb JC, Cook RS, et al. Predicting age-appropriate pharmacokinetics of six volatile organic compounds in the rat utilizing physiologically based pharmacokinetic modeling. Toxicol Sci 2007;98:43–56.

[84] Schoeffner DJ. Organ weights and fat volume in rats as a function of strain and age. J Toxicol Environ Health A 1999;56:449–62.

[85] Gentry PR, Haber LT, McDonald TB, Zhao Q, Covington T, Nance P, et al. Data for physiologically based pharmacokinetic modeling in neonatal animals: physiological parameters in mice and Sprague–Dawley rats. J Child Health 2004;2:363–411.

[86] Johri A, Dhawan A, Lakhan Singh R, Parmar D. Effect of prenatal exposure of deltamethrin on the ontogeny of xenobiotic metabolizing cytochrome P450s in the brain and liver of offsprings. Toxicol Appl Pharmacol 2006;214:279–89.

[87] Cui JY, Choudhuri S, Knight TR, Klaassen CD. Genetic and epigenetic regulation and expression signatures of glutathione S-transferases in developing mouse liver. Toxicol Sci 2010;116:32–43.

[88] Aperia A, Herin P. Development of glomerular perfusion rate and nephron filtration rate in rats 17–60 days old. Am J Physiol—Legacy Content 1975;228:1319–25.

[89] Mantovani A, Calamandrei G. Delayed developmental effects following prenatal exposure to drugs. Curr Pharm Des 2001;7:859–80.
[90] Zoetis T, Hurtt ME. Species comparison of anatomical and functional renal development. Birth Defects Res B 2003;68:111–20.
[91] Saunders NR, Habgood MD, Dziegielewska KM. Barrier mechanisms in the brain, II. Immature brain. Clin Exp Pharmacol Physiol 1999;26:85–91.
[92] Vidair CA. Age dependence of organophosphate and carbamate neurotoxicity in the postnatal rat: extrapolation to the human. Toxicol Appl Pharmacol 2004;196:287–302.
[93] White L, Haines H, Adams T. Cardiac output related to body weight in small mammals. Comp Biochem Physiol 1968;27:559–65.
[94] IPCS (International Programme on Chemical Safety). Chemical-Specific Adjustment Factors (CSAFs) for interspecies differences and human variability: Guidance document for the use of data in dose/concentration-response assessment. WHO/IPCS/01.4. Geneva, Switzerland; 2005. p. 1–96.
[95] Zeise L, Bois FY, Chiu WA, Hattis D, Rusyn I, Guyton KZ. Addressing human variability in next-generation human health risk assessments of environmental chemicals. Environ Health Perspect 2013;121:23–31.

CHAPTER 5

Integrated Systems Biology Approaches to Predicting Drug-Induced Liver Toxicity
A Dynamic Systems Model of Rat Liver Homeostasis Combined with *In Vitro* Measurements to Predict *In Vivo* Toxicity

Kalyanasundaram Subramanian
Strand Life Sciences P. Ltd., Bangalore, India

INTRODUCTION

Poor pharmacokinetics and compound toxicity are frequent causes of late-stage failures in drug development.[1,2] Predicting these effects early by using *in silico* or *in vitro* methodologies would be highly desirable since they are safe, reduce animal usage, and potentially reduce risks to humans in the clinic. Failures in post-marketing due to toxicity are especially problematic since they result in large economic and health-outcome related costs. Toxicity is often associated with molecular biotransformation[3] carried out by the cytochrome P450 family of enzymes that exhibit great inter-species variations,[4] and has been associated with the development of idiosyncratic drug toxicity in humans.[5] Due to the abundance of these enzymes in the liver and due to its portal location in the body making it receive the largest concentration of any orally ingested drug from the gut, drug-induced liver injury (DILI) is a major cause of drug withdrawal.[6] Almost 50% of the liver failures seen in the clinic are drug-induced. This makes the ability to accurately predict DILI critical and rather challenging. However, interspecies metabolic differences make extrapolation from animal studies to the human difficult.[7] In preclinical development, hepatic clearance is measured using microsomes to assess metabolic stability.[8,9] But the data are of limited value, and the method is quite cumbersome to use. More recently, functional genomic studies are becoming increasingly common as a way of understanding the mechanistic basis for toxicity and for deriving toxicity signatures.[10] These suffer from the need to use very high doses to invoke detectable transcriptional responses. In addition, they are unable to predict toxicity that arises due to responses that are nontranscriptional in origin. All of the above-mentioned methods suffer from their inability to discern the potential for idiosyncratic toxicity.

The pressing need is for an approach that integrates many facets of toxic pathways to a chemical in a systematic manner to provide a detailed mechanistic rationale for toxicity. In this chapter, an integrated systems approach that models pathways in the rat liver to create a detailed predictive platform is described. The integrated systems approach is based on the principle that if one can model normal liver homeostasis, one can understand toxicity as perturbations of this normal system.[11] This approach allows us to model the biology independent of the action of any drug and allows us to design a predictive system that can generalize and is not limited by chemical space.

GENERAL PRINCIPLES

Let us start by defining the scope of the prediction platform with the following two questions:

1. Can liver functions be accurately represented by a collection of biochemical pathways, and can this collection be mathematically modeled using kinetic data to mimic normal homeostasis? This corresponds to model building.
2. Can toxicity be represented as a set of perturbations of this homeostasis to understand multiple mechanisms? This encompasses model validation and usage.

To understand the pathways that one may need to potentially model, we need to understand the "typical" toxicity outcomes that are observed in DILI. If one studies the major toxicity endpoints observed in the liver, one immediately sees that toxicity can be described as falling into three major categories: hepato-cellular death, cholestasis, and steatosis. There are other toxicity effects such as hepato-cellular-carcinoma and inflammatory and immune-mediated responses, but for the purposes of this exercise we will focus on these three major categories, as they constitute the majority of toxicity events seen. If we study the underlying biochemical pathways that must suffer derangement in order to precipitate such liver injury, we would immediately see that these pathways are part of three major biochemical areas: ROS and antioxidant metabolism, bile salt homeostasis, and ATP homeostasis, respectively. This tells us that we need to focus on processes in the liver that are involved in lipid, bile, energy, and antioxidant metabolism in detail to create a generally applicable predictive system.

MODEL BUILDING

Since there is an intrinsic dynamic nature to the phenomena of toxicity, we decided to use an approach that can represent both the changes in the biology over time as well as the adaptive responses in the liver to the perturbations brought about by drugs. To achieve both of these goals, we set up a system of nonlinear ordinary differential equations by modeling the kinetics of the key enzymes involved in the pathways described previously. These equations have a general form: the rate of change of metabolite concentration in the liver is the difference between its rates of appearance and disappearance. What this implies is that under homeostatic conditions, the rates of appearance and disappearance of metabolites are zero; i.e., they are produced at the rate they are consumed, and their concentration is unchanged.

In the following section we describe each of the subsystems that constitute the model in detail and the equations that arise as a consequence. In all of the equations, V stands for the rate of a process defined by its subscript. The subscript c is used to denote cytosolic processes and m for mitochondrial ones.

ENERGY HOMEOSTASIS

Energy homeostasis can be briefly described as the net balance in the processes that are involved in producing and consuming adenosine triphosphate (ATP) in the liver cell. These processes can occur both in the cytoplasm as well as the mitochondria and arise as a consequence of nutrient metabolism. ATP production in the cytosol is given by

$$d/dt[ATP]_c = V_{PK} + V_{PGK} - 2^*V_{PFK} + V_{ANT}*(1/R_{cm}) + V_{ADK} - V_{ute} \qquad (5\text{-}1)$$

where R_{cm} is the ratio of cell volume by mitochondrial volume. PGK is phosphoglycerate kinase, PFK is phosphofructokinase, and PK is pyruvate kinase (all glycolytic enzymes); ANT is adenine nucleotide translocase, ADK is adenylate kinase, and ute stands for the nonspecific cytosolic utilization of ATP. The corresponding equation for adenosine diphosphate (ADP) is

$$d/dt[ADP]_c = V_{ute} - V_{PK} - V_{PGK} + 2^*V_{PFK} - V_{ANT}*(1/R_{cm}) - 2^*V_{ADK} \qquad (5\text{-}2)$$

Note that fluxes involved in ADP production are exactly the same and opposite to the ATP fluxes except in the case of ADK, where 2 molecules of ADP combine to produce 1 molecule of ATP and 1 of adenosine monophosphate (AMP). We make use of the fact the total amounts of adenosine phosphates are constant in the cytosol by writing out an algebraic constraint:

$$ATP_c + ADP_c + AMP_c = constant \qquad (5\text{-}3)$$

To account for the rate of change of inorganic phosphate, we write

$$d/dt[Pi]_c = V_{ute} - V_P - V_{PK} - V_{PGK} \qquad (5\text{-}4)$$

where V_P is the phosphate transporter. We need to take into account the other metabolites that are involved in the fluxes represented in cytosolic ATP consumption and utilization. The equations that describe them are

$$d/dt[F1,6P2]_c = V_{PFK} - V_{ALD} \qquad (5\text{-}5)$$

$$d/dt[G3P]_c = 2 * V_{ALD} - V_{GPDH} \tag{5-6}$$

$$d/dt[BPG]_c = V_{GPDH} - V_{PGK} \tag{5-7}$$

$$d/dt[PEP]_c = V_{PGK} - V_{PK} \tag{5-8}$$

where *ALD* is aldolase and *GPDH* is glycerol-3-phosphate dehydrogenase. F1,6P2 is fructose-1,6-bisphosphate, *G3P* is glyceraldehyde 3-phosphate, *BPG* is bisphosphoglycerate, and *PEP* is phospho*enol* pyruvic acid.

Now let us take a look at the *ATP* production in the mitochondria. In the mitochondria, *ATP* is produced by the process of oxidative phosphorylation and transported to the cytosol by the transport *ANT*:

$$d/dt[ATP]_m = V_{f0f1} - V_{ANT} \tag{5-9}$$

where f0f1 stands for *ATP* synthase.

In the mitochondria, the total amount of adenine nucleosides is constant. This gives us

$$ATP_m + ADP_m = constant \tag{5-10}$$

We then write the other metabolites that influence and are involved in the kinetics of *f0f1* and *ANT*:

$$d/dt[NADH]_m = -V_{res} + V_{DH} \tag{5-11}$$

$$d/dt[\Delta\Psi] = (V_{res\_H} - V_{f0f1\_H} - V_{ANT} - V_{leak} - 3 * V_{uni})/C_{\_mito} \tag{5-12}$$

where $\Delta\Psi$ is the mitochondrial membrane potential, *NADH* is nicotinamide adenine dinucleotide, V_{leak} is rate of proton leak across the mitochondrial membrane, V_{res} is rate of respiration (Complex I), and V_{uni} is the transmembrane sodium-calcium transport.

There are some additional constraints on the system. In the mitochondria, we assume the *NADH* and *NAD* can only interconvert. This gives the following expression:

$$NADH_m + NAD_m = constant \tag{5-13}$$

We also impose the constraint that the total amount of adenosine nucleosides in the cell is constant, i.e., nothing synthesized, consumed, or transported across the cell wall:

$$3 * ATP_c + 2 * ADP_c + AMP_c + P_{ic} + (3 * ATP_m + 2 * ADP_m + P_{im})/R_{cm} = constant \tag{5-14}$$

Note that the algebraic constraints (x) and (xiv) obviate the need for writing differential equations for *ADP* and phosphate in the mitochondria. The kinetics for each flux in the preceding equations and the rate constants can be found in references.[12–26]

GLUTATHIONE HOMEOSTASIS

A key metabolite involved in the maintenance of cell health is glutathione, the major antioxidant in the cell. Glutathione (GSH) is responsible for dealing with any oxidative stress experienced in either the cytosol or mitochondria, and also in the detoxification of drug metabolites. Reduced levels of *GSH* in the cell imply high levels of stress and will lead eventually to cell death. In the cytosol, *GSH* is produced by the action of glutathione synthase on gamma glutamyl cysteine (γ-*GC*). In addition *GSH* can be transported to the mitochondria or oxidized to *GSSG*:

$$d/dt[GSH] = V_{GS} + V_{GR} - 2 \times V_{GPx} - V_{GSH\ efflux\ to\ sinusoid} - V_{GSH\ efflux\ to\ canaliculus} - V_{GST}$$
$$+ V_{GSH\ transport\ from\ mitochondria\ to\ cytosol}/R_{cm} - V_{GSH\ transport\ from\ cytosol\ to\ mitochondria} \quad (5\text{-}15)$$

where *GPx* is glutathione peroxidase, *GS* is glutathione synthase, *GR* is glutathione reductase, and *GST* is glutathione transferase. Other important metabolites that need to be considered are

$$d/dt[\gamma GC] = V_{GCS} - V_{GS} \quad (5\text{-}16)$$

where *GCS* is gamma glutamyl cysteine synthetase. The equation for the oxidized form of glutathione (*GSSG*) is

$$d/dt[GSSG] = V_{GPx} - V_{GR}/2 - V_{GSSG\ efflux\ to\ canaliculus} \quad (5\text{-}17)$$

$$d/dt[H_2O_2] = V_{H2O2prod} - V_{GPx} \quad (5\text{-}18)$$

where $V_{H2O2prod}$ is the rate of hydrogen peroxide production. In the mitochondria, *GSH* can be oxidized reversibly and affected by the amount of oxidative stress (H_2O_2):

$$d/dt[GSH] = V_{GR} - 2 \times V_{GPx} - V_{GST} - V_{GSH\ transport\ from\ mitochondria\ to\ cytosol} + R_{cm}$$
$$\times V_{GSH\ transport\ from\ cytosol\ to\ mitochondria} \quad (5\text{-}19)$$

$$d/dt[GSSG] = V_{GPx} - V_{GR}/2 \quad (5\text{-}20)$$

$$d/dt[H_2O_2] = V_{H2O2prod} - V_{GPx} \quad (5\text{-}21)$$

The kinetics for each flux in the preceding equations and the rate constants can be found in references.[27–41]

FATTY ACID METABOLISM

Modeling the fate of free fatty acids is important in the context of drug-induced liver injury. One of the major modes of injury is via high levels of triglyceride deposition in the liver, called steatosis or fatty liver. If severe, this condition can disrupt cell constituents, and may cause cell rupture or death. Free fatty acid modeled here as palmitate enters the hepatocyte via the plasma. In the cytosol, it is activated by chemical transformation and eventually moves to the mitochondria and microsomes to produce triglycerides and phospholipids via several cycles of activation followed by elongation. A part of the activated palmitate is also ß-oxidized to produce energy. The equation-set that governs these processes is presented here:

$$d/dt[Palmitate]_c = V_{influx} + V_{FAS} - V_{FACS} \quad (5\text{-}22)$$

$$d/dt[PalmitoylCoA]_c = V_{FACS} - V_{CPT1} - V_{GPATm} - V_{ACAT} - V_{LPAT} - V_{DGAT} \tag{5-23}$$

$$d/dt[Palmitoylcarnitine]_m = R_{cm} * V_{CPT1} - V_{CPT2} \tag{5-24}$$

$$d/dt[PalmitoylCoA]_m = V_{CPT2} - V_{AcylCoAdehydrogenase} \tag{5-25}$$

$$d/dt[EnoylCoA]_m = V_{AcylCoAdehydrogenase} - V_{EnoylCoAhydratase} \tag{5-26}$$

$$d/dt[HydroxyacylCoA]_m = V_{EnoylCoAhydratase} - V_{HydroxyacylCoAdehydrogenase} \tag{5-27}$$

$$d/dt[ketoacylCoA]_m = V_{HydroxyacylCoAdehydrogenase} - V_{AcetylCoAacetyltransferase} \tag{5-28}$$

$$d/dt[AcetylCoA]_m = 8 * (V_{AcetylCoAacetyltransferase}) - V_{CS} - 2 * V_{Acetoacetyltransferase} - V_{HMGSYN} + V_{HMGLYASE} \tag{5-29}$$

$$d/dt[AcetoacetylCoA]_m = V_{Acetoacetyltransferase} - V_{HMGSYN} \tag{5-30}$$

$$d/dt[HydroxymethylglutarateCoA]_m = V_{HMGSYN} - V_{HMGLYASE} \tag{5-31}$$

$$d/dt[Acetoacetate]_m = V_{HMGLYASE} - V_{Hydroxybutyratedehydrogenase} - V_{efflux\_acetoacetate} \tag{5-32}$$

$$d/dt[Hydroxybutyrate]_m = V_{Hydroxybutyratedehydrogenase} - V_{efflux\_\_hydroxybutyrate} \tag{5-33}$$

$$d/dt[sn1palmitoylglycerol3phosphate]_c = V_{GPAT} - V_{LPAT} \tag{5-34}$$

$$d/dt[sn12dipalmitoylglycero3phosphate]_c = V_{LPAT} - V_{PAP} \tag{5-35}$$

$$d/dt[Dipalmitoylglycerol]_c = V_{PAP} - V_{DGAT} - V_{DCPT} \tag{5-36}$$

$$d/dt[Triglyceride]_c = V_{DGAT} - V_{TGSEC} \tag{5-37}$$

$$d/dt[Phosphatidylcholine]_c = V_{DCPT} - V_{PLSEC} \tag{5-38}$$

$$d/dt[MalonylCoA]_c = V_{ACC} - 7 * V_{FAS} - V_{MYD} \tag{5-39}$$

Influx is transport of fatty acids from plasma to the hepatocyte, FAS is fatty acid synthase, ACS is fatty acyl CoA synthase, GPAT is mitochondrial glycerol-3-phosphate O-acyltransferase, ACAT is acyl-CoA cholesterol acyltransferase, LPAT is 1-acylglycerol-3-phosphate O-acyltransferase, DGAT is diacylglycerol O-acyltransferase, CPT1 is carnitine O-palmitoyltransferase1, CPT2 is carnitine O-palmitoyltransferase2, CS is citrate synthase, HMGSYN is hydroxymethylglutaryl CoA synthase, HMGLYASE is hydroxymethylglutaryl CoA lyase, Efflux_acetoacetate is transport of betahydroxybutyrate from the hepatocyte to the plasma, Efflux_ hydroxybutyrate is transport of betahydroxybutyrate from hepatocyte to the plasma, PAP is phosphatidate phosphatase, DCPT is diacylglycerol cholinephosphotransferase, TGSEC is transport of triglyceride from the hepatocyte to the plasma, and PCSEC is transport of phosphatidylcholine from the hepatocyte to the plasma.

These fluxes, their equations, kinetics, and rate constants are described in references.[42-66]

BILE SALT METABOLISM AND TRANSPORT

To complete the story, we need to model the fate of bile salts in the liver. Bile aids the process of lipid-digestion in the small intestine. Upon eating, bile acting as a surfactant helps emulsify the fats in the food. Drugs may induce blockage of bile flow, causing bile accumulation in the blood and the liver. If the condition is severe, liver failure may result. To model this phenomenon, we consider the normal metabolism of the major bile salt, cholic acid (CA). It is produced in the hepatocyte by the metabolism of cholesterol, converted to taurocholic acid (TCA), and secreted into the gut to aid digestion. Most of the synthesized TCA is recycled back to the liver via the blood. The part lost via excretion is replaced by synthesis. The equations describing these processes are given by

$$d/dt[CA]_h = V_{synthesis} + V_{uptake} - V_{conjugation} - V_{reflux} \tag{5-40}$$

$$d/dt[TCA]_h = V_{conjugation} + V_{uptake} - V_{reflux} - V_{secretion} - V_{sulphation} \tag{5-41}$$

In the blood, the bile salts are either reabsorbed from the gut or the kidney, as shown in the following equations:

$$d/dt[CA]_{blood} = V_{reabsorption\_int} + V_{reflux} - V_{uptake} - [GFR - V_{renal\ reabsorption}]/0.02 \tag{5-42}$$

$$d/dt[TCA]_{blood} = V_{reabsorption\_int} + V_{reflux} - V_{uptake} - [GFR - V_{renal\ reabsorption}]/0.02 \tag{5-43}$$

Bile salts secreted in the canaliculus eventually are used in the gut:

$$d/dt[TCA]_{bile} = V_{secretion} \tag{5-44}$$

When there is an incidence of cholestasis, the amount of bile salts in the blood increases and dark urine is seen, owing to the high amount of bile salts excreted. The rate of bile salt appearance in the urine is given by

$$d/dt[CA]_{urine} = GFR - V_{renal\ reabsorption} \tag{5-45}$$

$$d/dt[TCA]_{urine} = GFR - V_{renal\ reabsorption} \tag{5-46}$$

In the preceding equations, [metabolite]$_{blood}$ is the concentration of the metabolite in the blood, [metabolite]$_h$ is the concentration of the metabolite in the hepatocyte, [metabolite]$_{bile}$ is the concentration of the metabolite in bile, [metabolite]$_{urine}$ is the concentration of the metabolite in the urine, and GFR is the glomerular filtration rate. Reflux is the movement of bile salts from hepatocyte to blood, uptake is the movement of bile salts from blood to hepatocyte, synthesis is synthesis of cholic acid from cholesterol in the hepatocyte, conjugation is amidation of cholate to taurocholate, secretion is secretion of bile salts into the canaliculus to form bile, sulphation is the conversion of taurocholate to taurocholate sulphate, and renal reabsorption is the reabsorption of bile salts by the kidney. In the equation, 0.02 represents the ratio of the volumes of the hepatocyte to the plasma compartment and hence the dilution of the bile salts as they reach the plasma from the hepatocyte. Further details on each of these fluxes, their equations, kinetics, and rate constants can be found in references.[67–81]

SOLVING THE EQUATION-SET

Equation 5-1 to 5-46 form a coupled, nonlinear differential algebraic system of equations that cannot be solved analytically. Numerical solutions can be achieved by using any standard differential equation solution package. For the simulations shown in the next section, we used the software Avadis Systems Biology® version 1.2 © Strand Life Sciences, Bangalore, India.[82] In addition to the equation-set, one needs to assign initial-value guesses for each of the metabolites modeled. A good starting point is to choose values that are in the range of values reported in the literature. For metabolites whose values have not been reported, a good starting point is to assume that most metabolites operate in the range of their K_m.

After one chooses initial values or guesses, one can convert the ODE set into an algebraic set by setting the assuming homeostasis or steady state and setting the rate of change of every metabolite to zero. The resultant set of algebraic equations can be solved numerically to achieve steady state or homeostasis. Under this condition, the model should output concentrations of various metabolites and flux values similar to what is measured in the liver under normal conditions. This state is then the starting point for every other simulation. We can then perturb the model in various ways, either to represent the interactions of drugs with the underlying biochemistry of the liver or to represent other forms of damage to the liver, and compare our predictions against published observations of toxicity and liver damage.

MODEL VALIDATION AND PREDICTIONS

Since we built the model to represent normal homeostasis, we can first start by comparing model predictions to observed normal liver values. A set of comparisons is shown in Table 5-1. There is good concordance both at the metabolite levels as well as flux level.

TABLE 5-1 Comparison of the Simulations of Metabolite Concentrations and Fluxes with Their Experimental Values

Metabolite	Simulated Value	Experimental Value	Reference
GSH (cytosolic)	7.96 mM	5–10 mM	99
ATP (cytosolic)	2.95 mM	2.76 mM	100
Phosphate (cytosolic)	3.375 mM	3.340 mM	100
ATP from glycolysis	33%	38%	88
ATP from oxidative phosphorylation	66%	57%	88
Fraction of fatty acids influx in oxidation (Fed)	28%	34%	97
Acetyl CoA (Mitochondrial, Fed)	62 μM	33–67 μM	98
Fraction of fatty acids influx in oxidation (Fasted)	70%	70%	97
Acetyl CoA (Mitochondrial, Fasted)	116 μM	91–200 μM	98

The cytosolic and mitochondrial partitioning of the energy production is well modeled as well as the impact of nutrient input on fat metabolism. Under fed conditions, one tends to convert fatty acids to triglycerides for storage while using them for energy production under fasted conditions, a phenomenon well reproduced by the model. The major hepatocellular antioxidant, glutathione, is predicted to be in the right concentration range. Given that normal liver function is reasonably well represented, we can test model prediction for the effect of a spectrum of drugs whose mode of action is known.

Under conditions of mitochondrial impairment, oxidative phosphorylation is affected and can predispose the cell to ATP depletion and hence necrotic damage. This is commonly observed when patients are treated using nonsteroidal anti-inflammatory drugs (NSAIDs).[83] It is believed that drugs that are lipophilic in nature typically target the mitochondria,[84] changing the balance of energy production in the liver. Under normal conditions, glycolysis produces only a third of cellular ATP, the rest being produced by oxidative phosphorylation.[85] When the mitochondrial synthetic capacity is inhibited, simulations predict that ATP levels drop significantly, as observed in the literature[86,87] with a concomitant upregulation in glycolysis (Figure 5-1) that is compensatory in nature. Figure 5-2 shows that the ability to compensate is critical; if one runs a simulation in which the glycolysis rate is fixed and not allowed to rise, similar levels of mitochondrial impairment will cause cellular ATP to deplete completely. The limited ability to compensate may explain the observation that drugs which induce necrosis cause perivenous

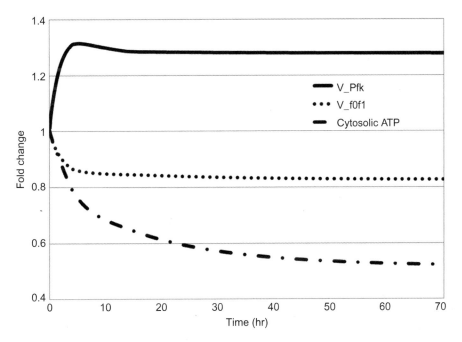

FIGURE 5-1 Mitochondrial impairment in f0f1 ATPase causes a net reduction in ATP levels in the hepatocytes. This results in a compensatory increase in the rate of glycolysis (PFK) to maintain ATP levels in the hepatocyte.

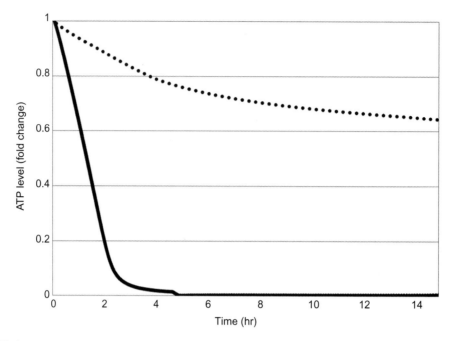

FIGURE 5-2 In the absence of glycolytic upregulation, the cellular ATP depletes fully if there is mitochondrial impairment (bold line). Glyoclytic upregulation allows the ATP levels to only partially deplete (dotted line).

damage,[3] an area of the liver lobule characterized by low oxygen concentrations along with high drug metabolism activity.

It is known that oxidative stress is a major cause of necrosis in the liver.[88] Excessive conjugation is one route by which glutathione (GSH), a key cellular antioxidant, is depleted, leading to the onset of necrosis. By setting high levels of $V_{GSHefflux}$, one can perform simulations that mimic the effect of drugs such as ethacrynic acid (EA) that cause excessive conjugation. Figure 5-3 shows that the prediction of EA exposure on the levels of cellular glutathione is in excellent agreement with literature values.[89]

Cholestasis is a complex process that eventually results in the accumulation of bile salts in the blood and their excretion in the urine. In pharmaceutical R&D, prediction of the cholestatic potential of compounds is based on their *in vitro* inhibition of the bile salt export pump (BSEP) in canalicular vesicles.[90] Although BSEP is a major contributor to the process of bile efflux and whose disruption can indeed cause cholestasis, other phenomena also need to be considered in order to make physiologically relevant predictions. Figure 5-4 is a simulation of the effect of cyclosporine A, which inhibits both the mitochondrial ATP production as well as the BSEP transporter function, and hence the bile efflux, to the canaliculus. The model predicts a profound rise in serum TCA and a fall in cellular ATP as observed in the literature.[91] A simulation in which only the corresponding BSEP inhibition is affected predicts much lower TCA accumulation, showing the importance of ATP-dependent transport as reported in the literature.[92] This implies that even a partial inhibition of BSEP along with mitochondrial impairment can lead to cholestasis.

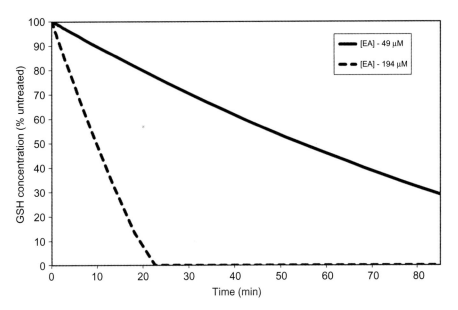

FIGURE 5-3 Simulation showing that increases in the dose of perfused ethacrynic acid causes GSH depletion due to high levels of conjugation matching experimental observations.

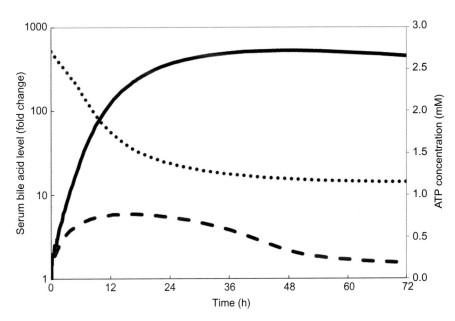

FIGURE 5-4 Cyclosporine A causes serum bile salt accumulation (bold line) by its synergistic effects on BSEP inhibition and mitochondrial ATP generation, leading to lower cellular ATP (dotted line). A comparable effect on the BSEP transporter alone would not lead to significant bile salt accumulation (dashed line).

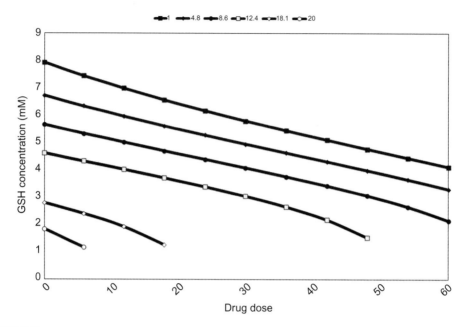

FIGURE 5-5 Simulation showing the effect of individual variations on drug-induced toxicity. Each line corresponds to a class of patients whose MnSOD enzyme is active to a certain degree. A fraction of individuals (shown by the empty diamonds and circles) who have polymorphisms in MnSOD causing it to work inefficiently will have severe GSH depletion at low doses of troglitazone manifesting as idiosyncratic toxicity.

Such complex phenomena cannot be understood by a simple *in vitro* test that may, in some cases, overstate the toxicity potential of a lead compound while in other cases dangerously understate it.

Very often post-marketing DILI occurs not in the general population, but rather in a small fraction of the patients who suffer deleterious effects of a drug. Usually, one's genetic makeup, disease/health status, and environmental conditions all have an effect on how one reacts to a drug, leading to individual variations in drug response. Figure 5-5 demonstrates how the impact of variations can be assessed by systems modeling. A drug such as troglitazone increases the levels of oxidative stress in the liver. Manganese superoxide dismutase (MnSOD) is an enzyme that catalyzes the dismutation of superoxide into oxygen and hydrogen peroxide and is hence important as an antioxidant defense mechanism. Polymorphisms in this enzyme can make individuals more susceptible to oxidative stress. If we combine various levels of MnSOD activity along with different doses of troglitazone, the model predicts that a fraction of individuals will experience high levels of glutathione depletion at relatively low doses of troglitazone. The results correlate well with the observation that variations in superoxide dismutase and glutathione reductase are correlated with the predisposition to liver disease.[93] Similarly, it is conceivable the other processes in the liver are likely to have variations making them susceptible to specific drugs in an idiosyncratic manner.

In all the simulations presented here, we fed in data on how drugs inhibit (or enhance) certain processes at an enzyme level leading ultimately to a toxic outcome. It is also

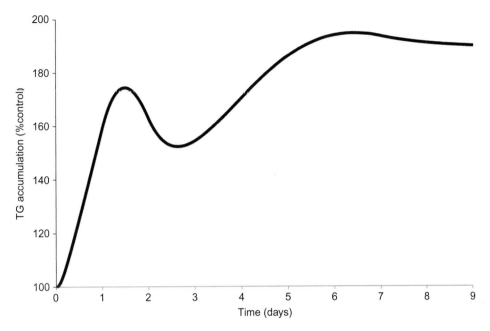

FIGURE 5-6 Simulation predicting the effect of valproic acid on intracellular triglyceride concentration using gene expression data as input. The predictions of intracellular TG rise are in excellent concordance with experimentally measured levels.

possible to integrate other data sources with this model to understand mechanisms. As an illustration, we can look at microarray expression data obtained from rats exposed to 200 mg/kg of valproate[94] (data accessible at NCBI GEO database,[95] accession GSE2303). A gene-set enrichment analysis of the data shows that a subset of processes affected by this treatment over the 9-day experimental period contains lipogenesis, β-oxidation, and ketone body synthesis in the lipid metabolism pathway. If we use mRNA levels as surrogates for protein activity, we can input fold changes in the enzyme list as inputs to our model. Figure 5-6 shows the results of this exercise. The simulations predict a rise in liver triglycerides comparable to experimental measurements,[96] indicating that our model can indeed utilize alternate data sources.

CONCLUSIONS

A systems approach allows us to simulate and predict the effect of drugs, develop hypotheses for idiosyncratic toxicity, and build virtual models of disease. What we have observed is that an approach based upon modeling basic processes within the liver such as antioxidant, bile, energy, and nutrient metabolism to achieve liver homeostasis and its perturbed state by drugs is reasonably versatile and accurate. Since our methodology does not use any drug structure information, it is likely to generalize and not be limited to certain classes of drugs.

We have described a relatively small model that consists of 46 states. One can increase the depth and complexity of the model to capture additional phenomena, signaling systems, etc. We have, however, shown that even a relatively modestly sized model can predict the effect of multiple drugs with varied targets, handle different data types, and be useful for hypothesis generation. Hence, models don't need to be very complex to be useful.

All biological systems upon perturbation respond actively and adapt to resist the change. An understanding of this is possible only if one builds models using methods that allow dynamic analysis such as ordinary differential equations. The methodology that has been employed to create this system is generic and can be applied to other organ and physiological systems as well.

At present, there is a lacuna in our ability to accurately predict toxicity, particularly idiosyncratic toxicity in humans. Since *in vitro* systems cannot reproduce variations in biochemistry due to genetic and environmental factors, they are not equipped to handle idiosyncratic toxicity. In our approach, the state of the system defines the starting point of the simulation. We can easily modify the model to represent disease conditions and immunological starting states to mimic disease-specific and patient-specific effects. If we believe that hepatotoxicity is a complex outcome of factors pertaining to drug, disease, and patient, idiosyncrasy is then not unpredictable, but rather a logical outcome of this complexity that can be hypothesized and modeled to discover new sources of patient variability.

Starting from the homeostasis described by the model, we can perform parameter sensitivity analyses to identify pathways and processes that perturb the system significantly. This computational "mining" will allow one to design *in vitro* assays that measure the impact of a drug on the pathways, enzymes, and processes that have the greatest potential impact on the liver. This is truly an integrative approach—one that integrates, on one hand, multiple measurements (assays) with, on the other hand, multiple methodologies (*in silico* and *in vitro*).

Our current state of the art where toxicology is an observational discipline of effects on animals is nonideal. We can combine *in vitro* and *in silico* approaches, understand the basic mechanisms that lead to evolution of toxicity, and subsequently define "confirmatory" experiments that may be more targeted and useful. The methodology presented in this section is a step in this direction and has the potential of being able to provide a quantitative and mechanistic assessment of toxic liabilities of chemical entities in the liver.

References

[1] Kennedy T. Managing the drug discovery/development interface. Drug Discov Today 1997;2:436—44.
[2] Wishart DS. Improving early drug discovery through ADME modelling: an overview. Drugs R D 2007;8:349—62.
[3] Zimmermann H. Hepatotoxicity—The adverse effects of drugs and other chemicals on the liver. Philadelphia: Lippincott and Williams; 1999.
[4] Wojnowski L, Kamdem LK. Clinical implications of CYP3A polymorphisms. Expert Opin Drug Metab Toxicol 2006;2:171—82.
[5] Tribut O, Lessard Y, Reymann JM, Allain H, Bentué-Ferrer D. Pharmacogenomics. Med Sci Monit 2002;8: RA152—63.

REFERENCES

[6] Lee WJ. Drug-induced hepatotoxicity. NEJM 2003;349:474–85.

[7] Pushparajah DS, Umachandran M, Plant KE, Plant N, Ioannides C. Differential response of human and rat epoxide hydrolase to polycyclic aromatic hydrocarbon exposure: studies using precision-cut tissue slices. Mutat Res 2008;640(1-2):153–61.

[8] Obach RS, Baxter JG, Liston TE, Silber BM, Jones BC, MacIntyre F, et al. The prediction of human pharmacokinetic parameters from preclinical and *in vitro* metabolism data. J Pharmacol Exp Ther 1997;283(1):46–58.

[9] Lu C, Li P, Gallegos R, Uttamsingh V, Xia CQ, Miwa GT, et al. Comparison of intrinsic clearance in liver microsomes and hepatocytes from rats and humans: evaluation of free fraction and uptake in hepatocytes. Drug Metab Dispos 2006;34(9):1600–5.

[10] Barros SA, Martin RB. Predictive toxicogenomics in preclinical discovery. Methods Mol Biol 2008;460:89–112.

[11] Subramanian K, Raghavan S, Rajan Bhat A, Das S, Bajpai Dikshit J, Kumar R, et al. A systems biology based integrative framework to enhance the predictivity of *in vitro* methods for drug-induced liver injury. Expert Opin Drug Saf 2008;7(6):647–62.

[12] Aon MA, Cortassa S. Coherent and robust modulation of a metabolic network by cytoskeletal organization and dynamics. Biophys Chem 2002;97:213–31.

[13] Lambeth MJ, Kushmerick MJ. A computational model for glycogenolysis in skeletal muscle. Ann Biomed Eng 2002;30(6):808–27.

[14] Cortassa S, Aon MA, Marbán E, Winslow RL, O'Rourke B. An integrated model of cardiac mitochondrial energy metabolism and calcium dynamics. Biophys J 2003;84(4):2734–55.

[15] Magnus G, Keizer J. Minimal model of beta cell mitochondrial Ca2+ handling. Am J Physiol 1997;273: C717–33.

[16] Korzeniewski B, Liguzinski P. Theoretical studies on the regulation of anaerobic glycolysis and its influence on oxidative phosphorylation in skeletal muscle. Biophys Chem 2004;110(1-2):147–69.

[17] Korzeniewski B. Theoretical studies on the regulation of oxidative phosphorylation in intact tissues. Biochim Biophys Acta 2001;1504(1):31–45.

[18] Korzeniewski B, Zoladz JA. A model of oxidative phosphorylation in mammalian skeletal muscle. Biophys Chem 2001;92(1-2):17–34.

[19] Korzeniewski B. Simulation of oxidative phosphorylation in hepatocytes. Biophys Chem 1996;58(3):215–24.

[20] Reinhart GD, Lardy HA. Rat liver phosphofructokinase: kinetic activity under near-physiological conditions. Biochemistry 1980;19(7):1477–84.

[21] Reinhart GD, Lardy HA. Rat liver phosphofructokinase: use of fluorescence polarization to study aggregation at low protein concentration. Biochemistry 1980;19(7):1484–90.

[22] Reinhart GD, Lardy HA. Rat liver phosphofructokinase: kinetic and physiological ramifications of the aggregation behavior. Biochemistry 1980;19(7):1491–5.

[23] Furuya E, Uyeda K. An activation factor of liver phosphofructokinase. Proc Natl Acad Sci USA 1980;77 (10):5861–4.

[24] Albe KR, Butler MH, Wright BE. Cellular concentrations of enzymes and their substrates. J Theor Biol 1990;143(2):163–95.

[25] Wan B, LaNoue KF, Cheung JY, Scaduto Jr RC. Regulation of citric acid cycle by calcium. J Biol Chem 1989;264(23):13430–9.

[26] Robb Gaspers LD, Burnett P, Rutter GA, Denton RM, Rizzuto R, Thomas AP. Integrating cytosolic calcium signals into mitochondrial metabolic responses. EMBO J 1998;17:4987–5000.

[27] Chance B, Sies H, Boveris A. Hydroperoxide metabolism in mammalian organs. Physiol Rev 1979;59:527–605.

[28] Yip B, Rudolph FB. The kinetic mechanisms of rat kidney gamma glutamyl cysteine synthetase. J Biol Chem 1976;251:3563–8.

[29] Segel IH. Enzyme kinetics. New York: John Wiley and Sons; 1975.

[30] Orlowski M, Meister. Isolation of highly purified gamma glutamyl cysteine synthetase from rat kidney. Biochemistry 1971;10(3):372–80.

[31] Luo J, Huang CS, Babaoglu K, Anderson ME. Novel kinetics of mammalian glutathione synthetase: characterization of gamma glutamyl substrate cooperative binding. Biochem Biochem Res Commun 2000;275:577–81.

[32] Griffith OW, Meister A. Origin and turnover of mitochondrial glutathione. Proc Natl Acad Sci 1985;82:4668–72.
[33] Carlberg I, Mannervik B. Purification and characterization of the flavoenzyme glutathione reductase from rat liver. J Biol Chem 1975;250(14):5475–80.
[34] Antunes F, Salvador A, Marinho HS, Alves R, Pinto RE. Lipid peroxidation in mitochondrial inner membranes. I. An integrative kinetic model. Free Radic Biol Med 1996;21(7):917–43.
[35] Wendel A. Glutathione peroxidase. Methods Enzymol 1981;77:325–33.
[36] Ookhtens M, Hobdy K, Corvasce MC, Aw TY, Kaplowitz N. Sinusoidal efflux of glutathione in the perfused rat liver. Evidence for a carrier mediated process. J Clin Invest 1985;75:258–65.
[37] Akerboom TPM, Sies H. Transport of glutathione, glutathione disulfide and glutathione conjugates across the plasma membrane. Methods Enzymol 1989;173:523–34.
[38] Akerboom TPM, Bilzer M, Sies H. The relationship of biliary glutathione disulfide efflux and intracellular glutathione disulfide content in the perfused rat liver. J Biol Chem 1982;257(8):4248–52.
[39] Rinaldi R, Eliasson E, Swedmark S, Morgenstern R. Reactive intermediates and the dynamics of glutathione-S-transferases. Drug Metab Dispos 2002;30(10):1053–8.
[40] Fernandez-Checa JC, Garcia Ruiz C, Ookhtens M, Kaplowitz N. Impaired uptake of glutathione by hepatic mitochondria from chronic ethanol fed rats. J Clin Invest 1991;87:397–405.
[41] Martensson J, Lai JCK, Meister A. High affinity transport of glutathione is a part of a multicomponent system essential for mitochondrial function. Proc Natl Acad Sci 1990;87:7185–9.
[42] Berk PD, Zhou S, Bradbury MW. Increased hepatocellular uptake of long chain fatty acids occurs by difference mechanisms in fatty livers due to obesity or excess ethanol use, contributing to the development of steatohepatitis in both settings. Trans Am Clin Climatol Assoc 2005;116:335–45.
[43] Philipp DP, Parsons P. Isolation and purification of long chain fatty acyl coenzyme a ligase from rat liver mitochondria. J Biol Chem 1979;254:10776–84.
[44] Bremer J, Norum KR. The mechanism of substrate inhibition of palmityl coenzyme A: carnitine palmityltransferase by palmityl coenzyme A. J Biol Chem 1967;242(8):1744–8.
[45] Ramsay RR, Derrick JP, Friend AS, Tubbs PK. Purification and properties of the soluble carnitine palmitoyl transferase from bovine liver mitochondria. Biochem J 1987;244:271–8.
[46] McKean MC, Herman FE, Mielke DM. General acyl-CoA dehydrogenase from pig liver kinetic and binding studies. J Biol Chem 1979;254:2730–5.
[47] Yang SY, Schulz H. Kinetics of coupled enzyme reactions. Biochemistry 1987;26(17):5579–84.
[48] Miyazawa S, Furuta S, Osumi T, Hashimoto T, Nobuo UP. Properties of peroxisomal 3-ketoacyl-CoA thiolase from rat liver. J Biochem 1981;90:511–9.
[49] Shepherd D, Garland PB. The kinetic properties of citrate synthase from rat liver mitochondria. Biochem J 1969;114:597–610.
[50] Middleton B. The kinetic mechanism and properties of the cytoplasmic acetoacetyl-coenzyme A thiolase from rat liver. Biochem J 1974;139:109–21.
[51] Lowe DM, Tubbs PK. 3-Hydroxy-3-methylglutaryl-coenzyme A synthase from ox liver purification, molecular and catalytic properties. Biochem J 1985;227:591–9.
[52] Stegink LD, Coon MJ. Stereospecificity and other properties, of highly purified beta-hydroxy-P-methylglutaryl coenzyme A cleavage enzyme from bovine liver. J Biol Chem 1968;243:5272–9.
[53] Tucker GA, Dawson AP. The kinetics of rat liver and heart mitochondrial P-hydroxybutyrate dehydrogenase. Biochem J 1979;179:579–81.
[54] Cobelli C, Nosadini R, Toffolo G, McCulloch A, Avogaro A, Tiengo A, et al. Model of the kinetics of ketone bodies in humans. Am J Physiol 1982;243(1):R7–17.
[55] Bates MW. Kinetics of ketone body metabolism in fasted and diabetic rats. Am J Physiol 1971;221(4):984–91.
[56] Hashimoto T, Numa S. Kinetic studies on the reaction mechanism and the citrate activation of liver acetyl coenzyme A carboxylase. Eur J Biochem 1971;18:319–31.
[57] Wang X, Stanley WC, Brunengraber H, Kasumov T. Assay of the activity of malonyl-CoA decarboxylase by gas chromatography-mass spectrometry. Anal Biochem 2001;298(1):69–75.
[58] Cox BG, Hammes GG. Steady-state kinetic study of fatty acid synthase from chicken liver (enzyme kinetics/multienzyme complex/enzyme mechanism). Proc Natl Acad Sci USA 1983;80:4233–7.

[59] Vancura A, Haldar D. Purification and characterization of glycerophosphate. Acyltransferase from rat liver mitochondria. J Biol Chem 1994;269:27209–15.
[60] Kvilekval K, Lin J, Cheng W, Abumrad N. Fatty acids as determinants of triglyceride and cholesteryl ester synthesis by isolated hepatocytes: kinetics as a function of various fatty acids. J Lipid Res 1994;35:1786–94.
[61] Yamashita S, Hosaka K, Numa S. Acyl-donor specificities of partially purified 1-acylglycerophosphate acyltransferase, 2-acylglycerophosphate acyltransferase and 1-acylglycerophosphorylcholine acyltransferase from rat-liver microsomes. Eur J Biochem 1973;38(1):25–31.
[62] Berglund L, Björkhem I, Angelin B, Einarsson K. Activation of rat liver cytosolic phosphatidic acid phosphatase by nucleoside diphosphates. Biochim Biophys Acta 1989;1002(3):382–7.
[63] Pontoni G, Manna C, Salluzzo A, del Piano L, Galletti P, De Rosa M, et al. Studies on enzyme-substrate interactions of cholinephosphotransferase from rat liver. Biochim Biophys Acta 1985;836(2):222–32.
[64] Andersson M, Wettesten M, Borén J, Magnusson A, Sjöberg A, Rustaeus S, et al. Purification of diacylglycerol acyl transferase from rat liver to near homogeneity. J Lipid Res 1994;35:535–45.
[65] Zammit VA, Lankester DJ, Brown AM, Park BS. Insulin stimulates triacylglycerol secretion by perfused livers from fed rats but inhibits it in livers from fasted or insulin-deficient rats. Implications for the relationship between hyperinsulinaemia and hypertriglyceridaemia. Eur J Biochem 1999;263:859–64.
[66] Yao Z, Vance DE. The active synthesis of phosphatidylcholine is required for very low density lipoprotein secretion from rat hepatocytes. J Biol Chem 1988;263(6):2998–3004.
[67] Kinugasa T, Uchida K, Kadowaki M, Takase H, Nomura Y, Saito Y. Effect of bile duct ligation on bile acid metabolism in rats. J Lipid Res 1981;22(2):201–7.
[68] Thompson MB, Davis DG, Morris RW. Taurine conjugate of 3 alpha, 6 beta, 7 beta trihydroxy-5 beta, 22-cholen-24-oic acid (tauro delta 22-beta-muricholate): the major bile acid in the serum of female rats treated with alpha-naphthylisothiocyanate and its secretion by liver slices. J Lipid Res 1993;34:553–61.
[69] Hata S, Wang P, Eftychiou N, Ananthanarayanan M, Batta A, Salen G, et al. Substrate specificities of rat oatp1 and ntcp: implications for hepatic organic anion uptake. Am J Physiol Gastrointest Liver Physiol 2003;285(5):G829–39.
[70] Lee HB, Blaufox MD. Blood volume in the rat. J Nucl Med 1985;25:72–6.
[71] Hirohashi T, Suzuki H, Takikawa H, Sugiyama Y. ATP-dependent transport of bile salts by rat multidrug resistance-associated protein 3 (Mrp3). J Biol Chem 2000;275(4):2905–10.
[72] Mosbach EH. Hepatic synthesis of bile acids. Biochemical steps and mechanisms of rate control. Arch Intern Med 1972;130(4):478–87.
[73] Uchida K, Okuno I, Takase H, Nomura Y, Kadowaki M, Takeuchi N. Distribution of bile acids in rats. Lipids 1978;13(1):42–8.
[74] Fiorucci S, Clerici C, Antonelli E, Orlandi S, Goodwin B, Sadeghpour BM, et al. Protective effects of 6-ethyl chenodeoxycholic acid, a farnesoid X receptor ligand, in estrogen-induced cholestasis. J Pharmacol Exp Ther 2005;313(2):604–12.
[75] Crocenzi FA, Pellegrino JM, Catania VA, Luquita MG, Roma MG, Mottino AD, et al. Galactosamine prevents ethinylestradiol-induced cholestasis. Drug Metab Dispos 2006;34(6):993–7.
[76] Uchida K, Nomura Y, Kadowaki M, Takase H, Takano K, Takeuchi N. Age-related changes in cholesterol and bile acid metabolism in rats. J Lipid Res 1978;19:544–52.
[77] Colman R. Biochemistry of bile secretion. Biochem J 1987;244:249–61.
[78] Kuipers F, Enserink M, Havinga R, van der Steen AB, Hardonk MJ, Fevery J, et al. Separate transport systems for biliary secretion of sulfated and unsulfated bile acids in the rat. J Clin Invest 1988;81(5):1593–9.
[79] Barnes S, Gollan JL, Billing BH. The role of tubular reabsorption in the renal excretion of bile acids. Biochem J 1977;166(1):65–73.
[80] Schlattjan JH, Winter C, Greven J. Regulation of renal tubular bile acid transport in the early phase of an obstructive cholestasis in the rat. Nephron Physiol 2003;95(3):49–56.
[81] Jobin J, Bonjour JP. Measurement of glomerular filtration rate in conscious unrestrained rats with inulin infused by implanted osmotic pumps. Am J Physiol Renal Physiol 1985;248:F734–8.
[82] Avadis® Systems Biology, Version 1.2, Strand Life Sciences Pvt. Ltd., Bangalore, India.
[83] Masubuchi Y, Nakayama S, Horie T. Role of mitochondrial permeability transition in diclofenac-induced hepatocyte injury in rats. Hepatology 2002;35:544–51.

[84] Wallace KB, Starkov AA. Mitochondrial targets of drug toxicity. Annu Rev Pharmacol Toxicol 2000;40:353–88.
[85] Ainscow EK, Brand MD. Top-down control analysis of ATP turnover, glycolysis and oxidative phosphorylation in rat hepatocytes. Eur J Biochem 1999;263:671–85.
[86] Petrescu I, Tarba C. Uncoupling effects of diclofenac and aspirin in the perfused liver and isolated hepatic mitochondria of rat. BBA 1997;1318:385–94.
[87] Moreno-Sánchez R, Bravo C, Vásquez C, Ayala G, Silveira LH, Martínez-Lavín M. Inhibition and uncoupling of oxidative phosphorylation by non-steroidal anti-inflammatory drugs. Biochem Pharmacol 1999;57:743–52.
[88] Reed DJ. Glutathione: toxicological implications. Annu Rev Pharmacol Toxicol 1990;30:603–31.
[89] Tirona RG, Pang KS. Bimolecular glutathione conjugation kinetics of ethacrynic acid in the rat liver: *in vitro* and perfusion studies. J Pharmacol Exper Ther 1999;290:1230–40.
[90] Horikawa M, Kato Y, Tyson CA, Sugiyama Y. Potential cholestatic activity of various therapeutic agents assessed by bile canalicular membrane vesicles isolated from rats and humans. Drug Metab Pharmacokinet 2003;18:16–22.
[91] Stone BG, Udani M, Sanghvi A, Warty V, Plocki K, Bedetti CD, et al. Cyclosporin A-induced cholestasis. The mechanism in a rat model. Gastroenterology 1987;93:344–51.
[92] Böhme M, Müller M, Leier I, Jedlitschky G, Keppler D. Cholestasis caused by inhibition of the adenosine triphosphate-dependent bile salt transport in rat liver. Gastroenterology 1994;107:255–65.
[93] Huang YS, Su WJ, Huang YH, Chen CY, Chang FY, Lin HC, et al. Genetic polymorphisms of manganese superoxide dismutase, NAD(P)H: quinone oxidoreductase, glutathione S-transferase M1 and T1, and the susceptibility to drug-induced liver injury. J Hepatol 2007;47(1):128–34.
[94] Jolly RA, Goldstein KM, Wei T, Gao H, Chen P, Huang S, et al. Pooling samples within microarray studies: a comparative analysis of rat liver transcription response to prototypical toxicants. Physiol Genomics 2005;22(3):346–55.
[95] Edgar R, Domrachev M, Lash AE. Gene expression omnibus: NCBI gene expression and hybridization array data repository. Nucleic Acids Res 2002;30(1):207–10.
[96] Tong V, Teng XW, Chang TK, Abbott FS. Valproic acid I: time course of lipid peroxidation biomarkers, liver toxicity, and valproic acid metabolite levels in rats. Toxicol Sci 2005;86(2):427–35.
[97] Ontko JA. Metabolism of free fatty acids in isolated liver cells. J Biol Chem 1972;247:1788–800.
[98] Herrera E, Freinkel N. Internal standards in the estimation of acetyl CoA in liver extracts. J Lipid Res 1967;8:515–8.
[99] Wahlländer A, Soboll S, Sies H, Linke I, Müller M. Hepatic mitochondrial and cytosolic glutathione content and the subcellular distribution of GSH-S- transferases. FEBS Lett 1979;97:138–40.
[100] Akerboom TP, Bookelman H, Zuurendonk PF, van der Meer R, Tager JM. Intramitochondrial and extramitochondrial concentrations of adenine nucleotides and inorganic phosphate in isolated hepatocytes from fasted rats. Eur J Biochem 1978;84:413–20.

CHAPTER 6

Computational Translation and Integration of Test Data to Meet Risk Assessment Goals

Luis G. Valerio Jr., Ph.D.

Science and Research Staff, Office of Pharmaceutical Science, Center for Drug Evaluation and Research, U.S. Food and Drug Administration, Silver Spring, MD, USA

INTRODUCTION

Scientific discoveries and new technologies are recognized to lead to improved medical products and better understanding of complex biological processes that underlie human diseases. Therefore, it is desirable to expedite the progression and application of discoveries and technology to the clinical level. By doing so, there is great potential to improve public health. However, in order for these innovations to be brought to the clinical setting, their scientific translation is needed in order to enable a pragmatic application. The investigative and intricate process of harnessing and converting innovative laboratory "bench" discoveries, associated data, and technologies into meaningful progress at the clinical level (i.e., patient "bedside") is known as translational research (a.k.a. translational science).

Translational research is an enabler for the applied sciences and is a truly integrated discipline. It requires specific subspecialities to provide input into all phases, including design, development, and implementation. Simply put, translational research is interdisciplinary and aims to drive the advancement of applied science. The most common public health topic addressed by translational research projects is support for expediting the development of new safe and effective human therapeutics. This is the case especially for human therapeutics that are needed in a time-sensitive way (e.g., oncology drug products). Translational research also plays an important role in the field of medical diagnostics. Of interest to this chapter is how translational research focused on computational toxicology tools can support risk assessment and how the integration of test data may help meet today's needs in risk assessment.

COMPUTATIONAL ANALYSIS AND TRANSLATIONAL RESEARCH

Computational analysis can be used to accomplish many tasks including but not limited to the calculation of physico-chemical properties of investigative molecules, simulation of tissue organ responses after exposure to substances, biological pathway analysis, elucidation of gene expression patterns to gain mechanistic inferences about compound toxicity, prediction of adverse effects of compounds on tissue organs, and in applications to organize and weigh the evidence from toxicological study data. These approaches can be implemented as translational research tools and are being evaluated by government agencies for supporting specific regulatory needs and use-case scenarios for risk assessments.[1-3] Given the prominent and important role of premarket risk assessments in protecting public health, there is great interest especially at public health agencies and consumer protection programs in technologies that provide reliable predictive data.[2,4,5] In this context, computerized models can offer support to risk assessment decision making through the prediction of chemical-induced toxicities,[6-9] and for providing descriptions for cell signaling pathways,[10,11] intracellular biological effects in perturbation of pathways,[12] and for gaining insight into on- and off-target effects of drugs and chemicals.[13] Computer models have several advantages over conventional risk assessment techniques. These advantages as well as disadvantages are covered throughout this chapter. The approach to integrate pre-existing chemical hazard information and associated exposure data with computational toxicology predictions from computer models should lead to a better understanding of toxicity pathways and prediction of risk.

An integrated approach to toxicological testing and assessment is at the cornerstone of new advancements in the field of toxicology. The paradigm that is emerging in recent years is integration of *in vivo* (animal) testing data with alternative methods such as *in vitro* target toxicity tests and *in silico* computer modeling and simulation. The data output would provide chemical-specific information needed for human health effects evaluations and environmental risk assessments. Currently, there are extensive research programs directed toward methods development in these areas, but equally important to the paradigm will be learning how an integrated approach to toxicological testing would be incorporated into present-day risk characterizations to support regulatory decisions on the human safety of substances. Although there are no clear-cut answers to this proposition presently, there are several approaches being investigated using established physiological modeling and emerging technologies.[14-16] Such research will provide insight into how far we have progressed and where scientific research gaps exist. In this chapter, the following areas are addressed in order to gain an understanding of how computational tools may translate information through the integration of test data to support risk assessment goals in the toxicological sciences.

In doing so, the following major points are discussed in the context of integration of test data:

- Toxicology-based structure-activity relationship (SAR) and quantitative structure-activity relationship (QSAR) computer models referred to as (Q)SARs
- Data mining for computational translation and integration of test data
- High-throughput screening for signal detection in risk assessment
- Integrating computational tools with test data for risk assessment

TOXICOLOGY-BASED (Q)SARS

Although (Q)SAR methodology has been in use for many decades, since 2007, a role for (Q)SAR approaches in review of regulated products has gradually increased. The increase in (Q)SARs relates to current demands for viable alternative approaches in prediction science. Research activities primarily center on developing new predictive models, understanding performance characteristics, ascertaining regulatory value, constructing models that have mechanistic insight, and optimizing (Q)SAR predictivity. From a drug regulatory standpoint, toxicology-based (Q)SARs have progressed into the realm of helping to qualify drug impurities for safety in order to protect patients from exposure to potential mutagenic and genotoxic chemical carcinogens found in clinical batches of drug products proposed for human trials. The U.S. Food and Drug Administration (FDA) issued a draft guidance in 2008 regarding its recommendations for addressing the potential genotoxicity of drug impurities in drug products and in this document is described the use of structure-based computational toxicology methods. These methods are namely (Q)SAR models to identify alerting structures for mutagenicity. However, this draft guidance does not provide extensive information regarding the nature and quality of a computational toxicology assessment. In order to begin to address this, in the spring of 2012, the Drug Information Agency held a public workshop, which was co-chaired by the FDA and involved panels of experts from the pharmaceutical industry.[17] Results included learning which computational software is frequently used by the pharmaceutical industry, including commercially available systems.[18] Some discussion on private computational software took place. However, the focus was on commonly used commercial systems. A major conclusion was that predictions from (Q)SAR systems should not be taken on face-value alone, but should be combined with expert knowledge in order to justify a prediction and improve accuracy.

The justification of an *in silico* prediction is important in order to appropriately use predictive data for safety assessment purposes. In the context of expert knowledge combined with (Q)SAR prediction, this may be relative and potentially subjective. Despite being context dependent, the integration of expert knowledge with (Q)SAR predictions is generally considered to be essential when being used in an applied way for protecting public health. Expert knowledge entails a high level of experience and supportive information needed to interpret the significance of a (Q)SAR-based prediction.[7,19–21] A (Q)SAR evaluation integrated with expert knowledge is indeed relevant, since appropriate interpretation is critical *in silico* structure-based predictions. Expert knowledge extends to a number of factors to consider.

For example, detailed chemical explanation on potential mechanism of action as well as literature data supporting or refuting the predictive data, assessment of the applicability of the model both from a chemical standpoint and a practical one, the model validation statistics and methodological aspects employed during construction of the training set, expert judgment, and understanding of the weaknesses of the (Q)SAR approach are all important aspects to consider. Chemical explanation refers to the known or computationally calculated physicochemical properties and chemical attributes or features (both mitigating and positively contributing) of the molecule. In addition to these factors, the public workshop discussed the contribution and utility of in-house computational systems in the pharmaceutical industry.

The use of in-house computational systems can be advantageous since models derived from them can be customized to the chemistry under development in the drug program. In this way, in-house data can also be more easily integrated into an in-house computational system. Another advantage of in-house computational systems is that the predictive performance characteristics can be optimized to meet product development needs and improve upon commercially available models if those are found to be unsatisfactory. For example, there may be situations when higher specificity is warranted with a predictive model during drug development so that one is confident in a "positive" prediction because of the low false positive rate that is carried with a high specificity performing model. Such an approach makes sense when considering drug developers would not want to discard a lead on the basis of an *in silico* prediction for a therapeutic indication that is difficult to find innovative therapeutics knowing that eventually high-throughput *in vitro* screening will produce empirical data for the endpoint. From a regulatory perspective, predictive models that are being used to evaluate safety endpoints such as the potential of a compound to be a mutagen or carcinogen make sense in the interest of protecting public health. Therefore, the utilization of *in silico* models that have predictive performance characteristics with high sensitivity is desired for regulatory application.[7,18] When high sensitivity is targeted, this results in models focused on high negative predictivity. High negative predictivity equates to the probability of predicting a negative compound accurately. This means that a prediction that a compound would not be toxic (i.e., predicted negative) is held with greater confidence because it was produced from a model that bears a low false negative rate (i.e., high sensitivity model). The net effect is a safety benefit to patients who would have been exposed to a mutagenic compound during a clinical trial if it were predicted negative under the context of a model with a high false negative rate (i.e., high specificity model). Thus, high confidence in a negative prediction is needed for regulatory reasons of protecting public health. Generally, this is a conservative approach necessary for regulatory review and is logical, since the goal is to minimize patient exposure to genotoxic carcinogens. However, in order to construct a model with performance characteristics of high negative predictivity (sensitivity), it is usually at the expense of positive predictivity (specificity). Therefore, the risk evaluator has to be highly selective and should be knowledgeable of the model(s) used to generate predictions. Such selectivity should include evaluation of the performance and validation results in order to determine if they are appropriate for the regulatory use-case scenario at hand. It is worth noting that it is possible to observe balanced performance characteristics of an *in silico* model, but this usually is not the norm. The important point is that a high degree of scrutiny and caution should be used before considering any *in silico* approach.[7,20]

The International Conference on Harmonisation (ICH) is developing an M7 document that is anticipated to indicate that *in silico* structure-based computational toxicology assessments involving either QSAR and/or SAR knowledge-based approaches can be performed to assess the mutagenic potential of drug impurities for the purpose of qualifying these substances found present in drug substance and product batches proposed to be used in clinical trials to ensure patient safety. If the ICH M7 recommends the use of *in silico* structure-based computational toxicology assessments under certain circumstances, it would be the first internationally harmonized guidance for pharmaceuticals recognizing and recommending *in silico* (Q)SAR techniques to substitute for a regulatory empirical safety test assay. For ICH M7, the *in silico* structure-based computational toxicology assessment is geared toward predicting the Ames test. Despite the regulatory interest in use of *in silico* structure-based approaches for predicting genotoxicity of molecules occurring in drug substances and products, there remain several challenges.

Frequently, there are misunderstandings regarding the difference between QSAR and SAR approaches. SAR approaches represent human knowledge-based systems and frequently are referred to as "expert systems" or systems and *in silico* methods using "expert knowledge" as the basis for a prediction. *In silico* SARs are also known as structural alert identification systems, which evolved as an extension of the classic Ashby and Tennant structural alerts (Figure 6-1).[22,23] The Ashby–Tennant structural alerts have been well known in a multitude of chemical risk assessment areas for decades, as they have been well studied to be linked to specific chemical genotoxic mechanisms involving direct interaction with DNA leading to mutagenesis.

The Ames bacterial reverse mutation assay as a measure of DNA reactivity as a result of exposure to a chemical substance is frequently associated with carcinogenic potential in rodents.[24] Since Ashby and Tennant and others have found that the presence of mainly electrophilic substructures in a chemical substance provides a mechanistic interpretation as far as chemical reactivity, and these substructures correlated well with positive outcomes from the bacterial reverse mutation assay and carcinogenic potential in rodents,[21,25] the alerts have become a common practical tool for toxicologists in raising "red flags" in applied safety science settings such as premarket risk assessments of chemical substances where there is anticipated human exposure.[18,26–28] Thus, the Ashby–Tennant structural alerts have formed the basis as chemical information that has served a purpose in safety evaluation and hazard identification that a substance bearing the alert may have a chemical genotoxic mechanism that directly interacts with DNA, and thus portends carcinogenic activity. Furthermore, such an interaction with DNA has a mechanistic basis and would be assumed to occur with or without metabolic activation. Of course, there are other structural alert classification schemes that expand upon the Ashby–Tennant alerts, and some of these have been tested using *in silico* techniques for their sensitivity and accuracy in detecting known mutagenic and nonmutagenic chemicals in the *Salmonella t.* assay.[29] Such studies suggest structural alerts for mutagenicity may be oversensitive and not as accurate as statistical QSARs for predicting bacterial mutagenicity outcomes. However, depending upon one's use-case scenario for risk assessment, what chemical data set is tested, including the study calls and quality of the Ames test data being considered (e.g., purity of the test article, cytotoxicity observed, GLP or non-GLP study), the expert knowledge-based system one uses, and interpretation of the prediction data, there are

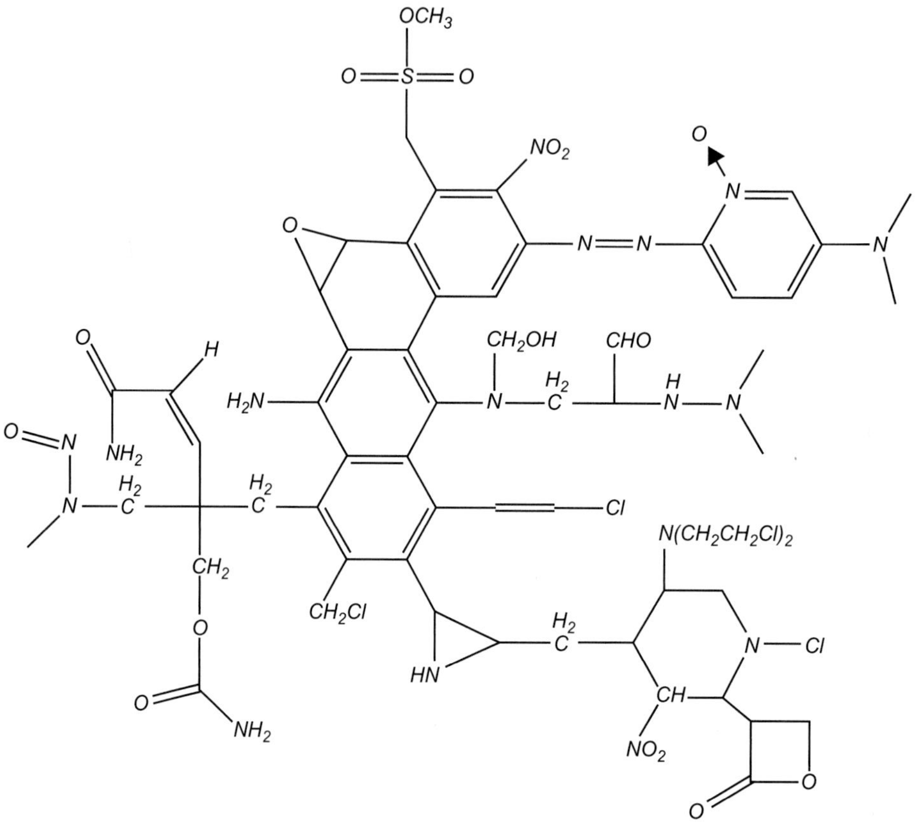

FIGURE 6-1 Chemical structures of toxicophores identified to correlate with mutagenicity to *Salmonella* and rodent carcinogenicity outcomes. The toxicophores represent classic structural alerts, which served as one basis for development of *in silico* (computational) toxicology knowledge-based systems. *Adapted from reference 23.*

without question different desirable traits and a wide range of variability in the predictive performance results for the *in silico* (Q)SAR approach for this endpoint. For example, some studies have found *in silico* SAR approaches for detecting alerts on chemicals as a predictive technique are dependent upon the nature of the data set being tested. Hillebrecht et al.[30] found that the knowledge-based Derek system performed with high sensitivity of 80.9% in predicting mutagenic substances available from a large public database of 2,630 chemicals,[31] yet had only 43.4% sensitivity in predicting known mutagens from a private data set of 2,335 chemicals from Hoffman–La Roche Ltd. Likewise, in the same study, the European Commission's publically available open source knowledge-based system Toxtree, which also identifies structural alerts for genotoxicity and carcinogenicity but uses a different rule-based system for deriving alerts in a decision-tree approach, performed very well with the public data set in predicting mutagens. Toxtree had a sensitivity value of 85.2% but only 42.9% sensitivity with the private Roche data. In an earlier study, and as a consequence of using an earlier version of the software, the Derek system performed with suboptimal sensitivity of 52% with public data.[32] However, a limited number of compounds were evaluated in that study. In a more recent study comparing Ashby–Tennant and related human expert structural alerts, and Derek and Toxtree knowledge-based systems to statistical QSAR approaches, it was shown that statistical QSARs employing k-nearest neighbors and naïve Bayesian classification models with fingerprints as descriptors had higher sensitivity (87%), specificity (91%), and concordance (90%) compared to Derek (67% sensitivity, 79% specificity, 74% concordance), Toxtree (76% sensitivity, 70% specificity, 73% concordance), and human expert alerts (71% sensitivity, 70% specificity, 70% concordance) using a large public data set of 4,971 chemicals as the validation test set.[33] This study, like other investigations,[30,34–36] found that by integrating the various predictive methods (i.e., human expert structural alerts, knowledge-based systems, and statistical QSAR) in a consensus modeling approach, an improvement in accuracy and confidence in predictions is produced and is superior to use of any of the individual methodologies alone. The combined results of five methods achieved sensitivity of 94% and specificity of 96%. The integration of different in silico methods appears to be favorable to enhancing predictivity at least for the mutagenicity and carcinogenicity endpoint despite some studies showing improvement is not substantial.[18] Thus, the question is when do returns start to diminish through use of a multimodeling consensus methodology? To check if the methods were simply data dependent by the test set, the authors performed a second external validation using an independent data set of 881 compounds, and this resulted in sensitivity of 87%, 91% specificity, and 90% overall accuracy. This result discounts the possibility of bias within the context of the data set tested by the models. Overall, several in silico prediction systems for genotoxicity are approaching the accuracy of inter- and intra-laboratory reproducibility of the Ames assay estimated at 87%.[37] It is important to underscore that these examples are not comprehensive of all validation studies in the public literature for these software programs and endpoint for that matter. There are many other software programs available both commercially and cost-free. The point of these examples, however, is to demonstrate there can be significant variability of several-fold in the predictive performance results for in silico toxicology-based (Q)SARs. Generally, validation results are context dependent upon the data and model used to generate predictions. It is even context dependent upon the method of validation (e.g., percent leave-group-out vs. external validation). Because of this variability in predictive performance characteristics, toxicologists have suggested using a standardized or benchmark data set to validate a (Q)SAR model.[31] This notion holds merit from a view of concept; however, given that structures in a public validation set would likely be contained in another commercially available (Q)SAR model being compared, it is inevitable that duplicate structures would occur, and thus have to be removed from a "benchmark" validation test set to avoid a test using compounds used to train a model. Upon removal of the duplicate compounds, this leads to changes in the number and nature of chemicals in the validation test set. As a consequence, the validation test set that would be submitted to each (Q)SAR system to learn of predictive performance will be different, and so there will not be an exactly equitable comparison. This makes it questionable how practical and accurate it will be to compare predictive performances across different software programs using a proposed standardized/benchmark data set for validation purposes.

Statistical machine learning QSAR approaches have exploded in recent years, employing well-recognized data mining algorithms as outlined by Valerio and Choudhuri.[3] The accessibility of statistical data mining algorithms has grown to a level that finding a suitable method for a particular type of data is not an issue. Thus, a modeler should first determine if the algorithm being considered is suitable for the data set in search of associations between biological and toxicological effects and structural attributes based on computer-calculated descriptors, fingerprints, fragments, and features. In recent years, one of the most popular algorithms happens to be one of the oldest known. This is the algorithm based on the Bayesian theorem. The Bayes theorem is named after Thomas Bayes, an 18th century mathematician who first introduced the concept to evaluate the probability of a hypothesis. His work was updated by Richard Price and posthumously read at the Royal Society of London.[38] The modern interpretation is about Bayesian probability, which evaluates the probability of a hypothesis. The simplest form is expressed in Eq. 8-1, which illustrates the basic definition through the evaluation of the probability (P) of two events (A and B) occurring:

$$P(A|B) = \frac{P(B|A)P(A)}{P(B)} \tag{6-1}$$

There are several deviations of this probabilistic analysis and many excellent reviews of Bayesian statistics and applied uses.[39–41] Nowadays with modern computing, implementation of the Bayes probability theorem is very straightforward; it has been applied in many fields, including engineering, economics, and medicine. Bayesian statistics can be found implemented in many commercial and noncommercial software programs. Despite the availability of these reliable algorithms, there are still questions in the scientific community regarding which computational toxicology software is more accurate, statistical QSARs or human expert knowledge-based SAR approaches.

Case example: To illustrate the capability of *in silico* QSAR and human expert knowledge-based systems, the hypothetical toxicophore structure in Figure 6-1 was screened in various systems for mutagenicity. Figure 6-1 illustrates the Ashby–Tennant alerts clustered together on one molecular backbone, and this toxicophoric molecule was screened *in silico* to provide illustrative examples using a *Salmonella* mutagenicity QSAR model[29] and human knowledge-based *in silico* systems.[2,42] We find the statistical QSAR prediction is 100% probability that the compound is a mutagen (Figure 6-2). Since the toxicophore molecule is not in its entirety one of the structures in the training set of the model, we can conclude that the QSAR model possesses features predictive of the classic Ashby–Tennant alerts.

However, going beyond a computer output prediction is especially important these days in order to provide scientifically defensible evidence on chemical hazard. Thus, being able to explain a prediction is of heightened importance for any *in silico* toxicology model, and fortunately many although not all computational toxicology software programs provide a transparent method for interpretation. The Derek Nexus system will provide the references to literature and examples of the structural alert which back up the knowledge base, including positive predictive performance for the alert. Hence, the ability to make mechanistic inferences is possible with the Derek Nexus prediction so that human expert

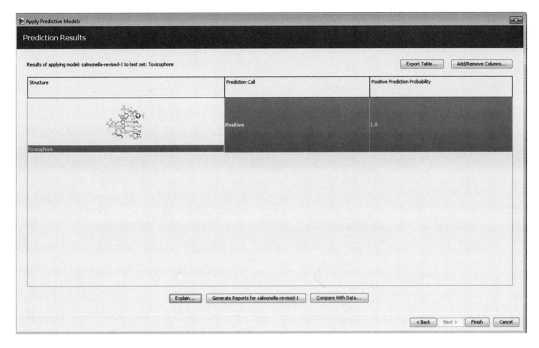

FIGURE 6-2 *In silico* prediction of the assembled Ashby–Tennant toxicophore molecule (Figure 6-1) using a validated *Salmonella* mutagenicity QSAR model.

interpretation is important to an overall decision on the hazard. However, in evaluating the human knowledge-based predictive data, it is still a generalized association between the presence of a matching substructural feature in a queried chemical and that in the knowledge base regarding a toxicological endpoint. Although the alert goes through a reasoning engine to produce a qualitative plausibility to the prediction, expert interpretation is still needed. For that reason, it is recommended that expert analysis of a positive prediction be conducted and not simply accepting the *in silico* prediction on face value. The expert interpretation might involve assessing the relevance of the substructural alert to "local" substructures on the molecule being screened *in silico*. In other words, an evaluation of potentially mitigating (de-risking) factors should be conducted alongside an *in silico* prediction, since such de-risking factors can allay concern that the entire substance in question is toxic for the modeled endpoint. Another important factor to consider is an assessment of physico-chemical properties, and the plausibility of the chemical reactivity or molecular attributes on the substance, and potentially findings of empirical data obtained through literature review that may serve as evidence to override the *in silico* prediction. The software Leadscope® has an enabling function for explaining the prediction in terms of the features and properties that contributed to the prediction. In the case of the toxicophore we screened in the *Salmonella* mutagenicity QSAR model in Figure 6-1, we find that the top features used in the model that match structures on the toxicophore were the alkyl-sulfonate and aziridine substructural features (Figure 6-3).

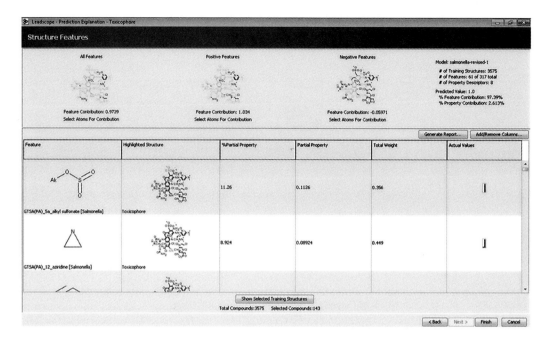

FIGURE 6-3 Illustrating alkyl-sulfonate and aziridine as top positive structural features in the *in silico* *Salmonella* mutagenicity QSAR model that match features on the toxicophore.

However, a total of 61 features used to construct the model were identified that match features present on the toxicophore. If we query the software to show us the chemicals in the training set of the *Salmonella* mutagenicity QSAR model that have a match with the aziridine substructure (i.e., a substructural search), it produces 15 chemicals (Figure 6-4).

When we screen the toxicophore molecule with Derek Nexus (version 2.0.3), we obtain plausibility that the structure would be mutagenic, and numerous specific alerts are identified with supporting references (Figure 6-5).

The evaluation of predictivity of *in silico* toxicology platforms has been a scientific issue that has not undergone an extensive objective investigation. For the prediction of genotoxicity, there is an array of studies that evaluate predictive performance either on a single software and model basis[29,42–47] or comparatively using various types of data sets.[2,19,21,30,32,33,35] However, validation studies performed alone by the software developer should be considered questionable given obvious potential for conflict of interest.[43] Nevertheless, gaining scientific acceptance of a toxicology-based QSAR model or expert system is a challenge with variables spanning from the user familiarity or lack thereof for a system, interpretability or understandability of a prediction, ability to make mechanistic inferences, confidence in the modeling approach, confidence in the prediction itself, transparency of an algorithm, reliability, and overall predictivity. Predictivity is thought to be context dependent.[37] This means the chemotype under *in silico* evaluation should be in the same context or domain space as the chemicals used to train the model that is generating the prediction. Other factors contributing to predictivity involve biological and physico-chemical properties of the model versus the test

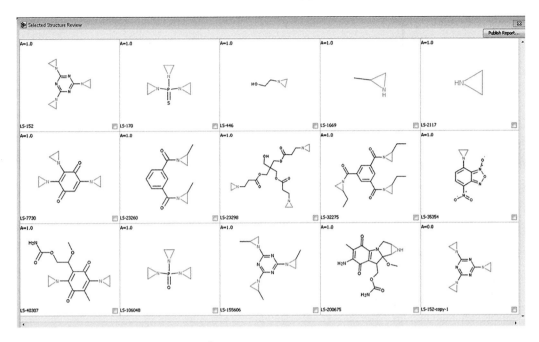

FIGURE 6-4 Aziridine was identified as a top positive structural feature in the *in silico Salmonella* mutagenicity QSAR model that matches a feature on the toxicophore. Using aziridine as the substructural feature to search the QSAR model, we are able to identify all the compounds in the training set of the model that have the matching aziridine feature. Fifteen chemicals in total are found through this search.

set. Plausibility of a prediction given all the available evidence to consider and how to deal with conflicting results have also risen as important factors in explaining a prediction. Consensus prediction approaches have been proposed in some cases for dealing with conflicting or complementary predictions or to improve overall accuracy.[33,36,48–52] Although these approaches have great merit, they are first investigations that have not yet undergone further scientific evaluation and testing, so there is no unified scientifically accepted method to date. In fact, the current FDA regulatory guidance for assessment of genotoxic impurities does not assess consensus or conflicting predictions, and it is hoped that the new ICH M7 will address this issue.

Another matter that remains unresolved is the methodologies that could best assess similarity between two compounds. The concept of chemical similarity has implications for how close in structure two compounds are not only to infer toxicity, but also to determine whether a compound under evaluation for toxicity risk and safety is structurally similar enough to the compounds in the training set of a prediction model. The latter aspect is often referred to as an applicability domain assessment. The reliability and confidence in a prediction is often based on how representative the test substance is to compounds in the model prediction space (chemical diversity or chemical space) of the model.[53,54] There are a multitude of methods to assess the applicability domain of a model, and that may be one reason why it is difficult to find scientific consensus as to what defines a model's

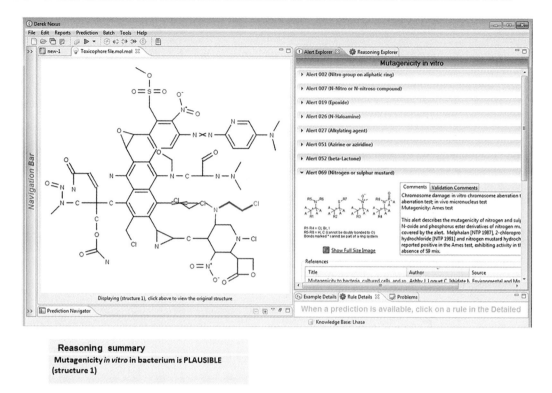

FIGURE 6-5 The toxicophore molecule in Figure 6-1 was screened using Derek Nexus 2.0.3 for the mutagenicity and carcinogenicity endpoints, and the result is the finding of numerous structural alerts detailed with rational, plausibility, and specific examples from the literature to support the alert.

prediction space. However, Naven et al.[37] provided a general definition for the applicability domain using two nonexclusive points. These are that the applicability domain is the region of chemical or response space related to the model training set, and is the region of chemical or response space where a model makes an acceptable prediction error. The assumptions are that the predictions from a model that are based on interpolations are generally reliable, but predictions based on extrapolations are less reliable. Moreover, the authors argue that a subset of compounds in a training set will be misclassified, so similarity to such compounds is no guarantee of a reliable prediction. Given the wide range of methods for defining the applicability domain (e.g., fragment-based, fingerprint-based, multiple property-based, descriptor-based, and more),[55] the diversity of training sets for a particular toxicity endpoint, the potential for misclassification of some of the compounds in the training set (i.e., source of prediction error), and the large range of methods for measuring the model's prediction space, we can see that similarity between a test molecule and the training set can truly be model dependent. We conclude that presently there is no single or standardized approach for determining confidence in an *in silico* prediction.

As Naven et al.[37] and others suggest,[56] it should be fitted to the model. From a product quality perspective, it is clear that safety-based assessments would benefit from establishing a standard for reliable predictions. By building quality standards as a critical component of accepting *in silico* data as evidence for safety evaluation, we will better qualify and set science-based standards for confidence in both positive and negative toxicity predictions. This area remains in need of further research and testing.

READ-ACROSS

Read-across is an integrated process of computer analysis and expert decision making whereby chemical hazard data from one chemical are used to predict the chemical hazard of another chemical, and the two chemicals are established to have structural similarity based on matching substructural features deemed relevant to the risk assessment. Thus, read-across is an integrative and nontesting procedure. The basis for performing read-across is usually that experimental data are not available for a given endpoint on a chemical being evaluated in a risk assessment and/or there is a lack of confidence behind a (Q)SAR prediction. Unfortunately, there are no universally accepted methods for performing read-across. However, the Organisation for Economic Co-operation and Development (OECD) has issued guidance on the grouping of chemicals based on structure and read-across.[57] OECD has an application toolbox for grouping chemicals and recognizes there are qualitative read-across processes and quantitative ones. In qualitative read-across, the identification of a chemical substructure or mode of mechanism of action is made between two substances and assumes the presence of a chemical property/activity can be inferred from the presence (or absence) of the same property/activity to the analogous substance. In quantitative read-across, the identification of a chemical substructure or mechanism of action that is common to two analogues is considered under the assumption that the known value of a property for one substance can be used to estimate the unknown value of the same property for another substance. OECD also recognizes that expert judgment is needed when inter- and extrapolating read-across data.

In regulatory practices and in the literature, there are various approaches described using read-across in chemical risk assessment. In FDA Center for Food Safety and Applied Nutrition (CFSAN) regulatory safety assessments of food contact materials (a.k.a., indirect food additives), read-across analysis is usually performed on a case-by-case basis and in concert with (Q)SAR prediction data for the genotoxicity and carcinogenicity endpoints.[4,8,58] However, in contrast to FDA CFSAN, read-across has not been performed at the FDA Center for Drug Evaluation and Research (CDER) for the purpose of supporting regulatory assessments of the mutagenic potential of drug impurities.[59] According to these authors, it is thought to be impractical and labor intensive, requiring that experimental data be available for structurally similar compounds for this use.[59] It appears that both human knowledge-based SAR systems and statistical QSAR modeling are advocated for assessments of the genotoxicity of drug-related impurities.[29,60] In the literature, there are many approaches and research investigations on read-across. In Europe, the **R**egistration, **E**valuation, **A**uthorisation and Restriction of **Ch**emicals (REACH) legislation on industrial chemicals requires that existing toxicological information is collected and considered in chemical risk assessments.[61] When it is found that there are limited data for a toxicity endpoint, an Integrated Testing Strategy (ITS) can be put forth in which all evidence, including physico-chemical property data, is weighed. When data gaps still exist, there arise the opportunities for toxicology-based (Q)SARs and read-across analysis. Recently, an integrative assessment scheme (IAS) was developed under this context and proposed a read-across analysis of ecotoxicological information.[62] The IAS is aligned to meeting REACH's General Decision Making Framework (GDMF), whereby risk assessors need to document the reliability of the read-across toxicity data, validity of the method employed, and regulatory needs of the endpoint. Embedded within these three principles are a series of questions that need to be addressed as part of the IAS. These questions are analogous to the OECD principles for regulatory acceptability of (Q)SARs. For example, the first OECD principle for QSARs is there must be a defined endpoint for a model. Thus, in the IAS framework, a question under the reliability of data principle is whether the defined endpoint is the same as the one used in the method. Several steps in the proposed IAS framework lead to determination of the weight and decision on the adequacy, partial adequacy, or inadequacy of the ecotoxicology data. Lastly, a determination has to be made on the information gaps under the IAS key principles for utility in the ITS, whether it is possible to make an informed decision for the REACH endpoint. In an innovative approach to read-across, investigators have proposed "omics" technologies such as metabolomics that might help to optimize traditional chemical structure-based SAR grouping by supplying biological-based criteria to the grouping process and toxicological equivalence.[63] The method was coined by the authors as quantitative biological activity relationship (QBAR) analysis. The biological data proposed from QBARs may derive from *in vitro* or animal studies correlating metabolite patterns and mode of action profiles. Thus, the "omics" technologies may serve as a translational tool to improve the quality of chemical read-across and predictive toxicology-based QSARs.

DATA MINING FOR COMPUTATIONAL TRANSLATION AND INTEGRATION OF TEST DATA

One of the most important aspects to computational translation of high-quality data and its integration into meaningful information for risk assessment is selection of the appropriate data mining algorithm. From the standpoint of building models and performing data mining through new software applications for specific risk assessment and hazard identification needs, the available approaches can be divided into two general categories: classical techniques and "next generation" techniques.

Classical data mining techniques involve more mature methods developed decades ago, and rely heavily on statistics. These methods have been thoroughly tested and applied across different business and research areas. The more classical techniques include classification and regression trees (CART), linear regression analysis, chi square automatic interaction detection (CHAID), nearest neighbor, and clustering (hierarchical and nonhierarchical).

Next generation techniques are more recently developed and mainly refer to approaches including but not limited to decision tree technologies, artificial neural networks, Kohonen feature maps, and rule-based analysis with reasoning. The complexity can be envisioned by analysis of artificial neural network (ANN) methods. ANN attempts to mimic the human brain by creating neural networks trying to detect patterns and not only make predictions from them but learn to make the best prediction. Essentially, this approach embellishes the premise that machines can be led to "think" as long as we input the highest quality information. Still, ANNs have disadvantages that are a consequence of their complexity. A disadvantage is that their ease of use and implementation can be difficult. However, there are some commercial platforms that have crossed the bridge between difficult to use and implement to basically user-friendly and straightforward to implement. One example is the computational software ADMET Predictor™. This software uses ANN and has a battery of sampling techniques for external validation, including Kohonen maps. ANNs use input, hidden, and output nodes in processing data. Nodes represent "neurons" and are connected by links (axons, synapses) such as in the simple illustration in Figure 6-6.

The ANN works by accepting incoming predictor values and performs calculations on those values to generate new quantifiable values for each node. The links store values used to generate weights applied to predictors in the nearest neighbor method and move across to the right, where they are added and go into the output node. A threshold function is applied to the output node, and the result is the prediction from the network model. Hidden nodes have a rather obscure interpretation but are very important in carrying out the learning function of ANN models. The values in hidden nodes are not visible to the user, which becomes a drawback for interpretation and transparency of the entire model system.

Most all of the aforementioned techniques can be applied toward resolving problems in data preprocessing, prediction, data exploration, and classification discovery, to name a few. The fundamental questions are when to use a data mining technique and which one to use. The answers depend upon the nature of the data, the type of question being asked to find a solution, what information is needed in the decision-analysis, the availability of expertise and technology capable of calculating the technique, and interpretability of the model and the output from the data mining toward the scientific issue. In general, though, the newer techniques are more complex and difficult to interpret or explain, and typically require high-performance scientific computing. There are also available hybrid techniques that use a combination of two different algorithms designed to improve performance. One example is the software Symmetry® (Prous Institute for Biomedical Research, Spain), which uses a freely available FDA National Center for Toxicology Research molecular descriptor package called Mold2.[64] Symmetry® has the capability to build predictive

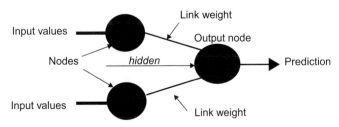

FIGURE 6-6 Diagram showing the fundamental architecture of the artificial neural network method, a newer statistical algorithm used in computational toxicology tools. Nodes represent "neurons" connected by weighted links (axons) to arrive at an output node (model).

TABLE 6-1 Representative List of Data Mining Techniques and Algorithms That Have Been Investigated or Applied as Integrated Approaches to Risk Assessments in Chemical Toxicology

Technique	Reference
Bayesian networks and learning	65; 66; 67
Stochastic gradient boosting machine learning	68
Artificial neural networks	69; 70
Discriminant analysis	44; 71; 72
Logistic regression and similarity analysis combined	73
Partial least squares (PLS) and chemistry building blocks	29
Categorical Bayesian Integration and Partial Least Squares Discriminant Analysis	33
Random forest	31; 74; 75; 76
Support vector machines	31; 77; 78
Decision tree	79; 80; 81
k Nearest neighbor	31; 77

QSAR models using a combined algorithm technique. The combined algorithm involves creating two models from the training set and annealing them into one model. One model is a logistic regression, and the second one is a similarity analysis model. These models are based on the Mold2 descriptors selected using a genetic wrapper optimization algorithm in searching for the best descriptors. It is likely that in the near future computational structure-based modeling techniques such as QSAR will involve the use of hybrid approaches in order to increase model predictive performance characteristics and enhance the efficiency of data mining. There are other approaches such as combining Bayesian categorical integration techniques with Partial Least Squares Discriminant Analysis to achieving predictivity superior to a single method.[33] A reference list of representative types of data mining techniques and algorithms that have been investigated or actually applied successfully in scientific risk assessments is provided in Table 6-1. Although this list is not exhaustive, it does show the diversity of methods and provides insight into all the choices modelers and toxicology risk assessors have to make in considering a computational toxicology method.

HIGH-THROUGHPUT SCREENING FOR SIGNAL DETECTION IN RISK ASSESSMENT

There are a number of initiatives under way that are performing high-throughput screening using computational analysis to detect safety signals for risk assessment purposes. A four-party memorandum of understanding (MOU) is in place to strengthen collaborations between the U.S. National Institutes of Health (NIH) National Institute of Environmental Health Sciences/National Toxicology Program (NIEHS/NTP), the NIH National Human Genome Research Institute (NHGRI) National Center for Chemical Genomics Center (NCGC), the U.S. Environmental Protection Agency (EPA) Office of Research and Development, and the U.S. Food and Drug Administration (FDA). Central to the MOU is the exploration of high-throughput screening (HTS) assays and tests using lower animal species and high-throughput genomic methods to assess mechanisms of toxicity of nominated substances.[82] Over 10,000 substances will be screened in cell-and biochemical-based assays. The data that result from the screening program are to be provided to risk assessors for use in protection of public health. The U.S. EPA's Office of Research (ORD) and Development Computational Toxicology Research Program will coordinate the research on chemical screening and prioritization, informatics, and systems modeling. Interestingly, the U.S. EPA's ORD recently estimated the economic cost savings of predictive models developed from HTS data from 36 *in vitro* assays mapped to 8 genes for reproductive toxicity under different screening scenarios.[83] A cost efficiency of $1.87 million was yielded in one scenario. Other goals of this initiative, better known as Tox21, are to prioritize chemicals for more extensive toxicological studies such as those needed to be performed in animals, and develop predictive methods involving the integration of the *in vitro* assay data from the screening program into computer-based models. A desirable outcome from the entire initiative is to reduce or even potentially replace the need for animal studies in evaluating the large numbers of substances that the program will screen. Despite this transformative effort, challenges remain with respect to how the biological responses identified in the *in vitro* assays can be translated to human phenotypic outcomes, especially long-term ones, and genomic outcomes.[84] Other challenges will be incurred in translation of high-throughput *in vitro* data to support required regulatory data packages for tested substances. Since HTS data are not normally submitted as part of regulatory filings by sponsors, it is likely that sponsors will need to interpret HTS data and translate the safety signals on a single substance for regulators. This notwithstanding, it is reassuring that federal government public health agencies are exploring ways to harness technology and examine its massive data repositories on the toxicological testing of substances with widespread human exposure for the greater good of public health.

HTS is clearly a data-driven approach that is normally used to collect vast amounts of information for screening and prioritization of a given goal, such as development of probes, leads, optimization of leads, and testing of drug candidates in established assays useful for supporting safety assessments. From an investigative standpoint, HTS can also be used to explore how modifications to compound structural feature space may affect biological responses at the cellular level. For example, cancer stem cells thought to drive metastasis and tumor formation were tested in a miniature (1,536-well) HTS assay using a collection of oncology drugs and clinical candidates to find cytotoxic agents against these aggressive cells.[85] This new mini-HTS assay may help elucidate which pathways need to be targeted to kill the cancer stem cells. The obvious advantage of HTS is the automated capacity to analyze a large volume of compounds in a time-efficient manner. Moreover, HTS provides a strategy for discovery of toxicity pathway analysis and mechanism-based phenotypic responses of compound series for supporting determination of the toxicity of structures. The disadvantage is clearly the uncertainty of the relevance of local cellular responses to *in vivo* human effects, also referred to as *in vitro–in vivo* extrapolation (IVIVE). One method reported to overcome the challenge of IVIVE is to integrate non-biased *in vitro* kinetic data relevant to target tissue with *in vitro* system modeling for the kinetic inputs generating mechanistically informed PBPK-IVIVE models with enhanced predictive power.[86]

INTEGRATING COMPUTATIONAL TOOLS WITH TEST DATA FOR RISK ASSESSMENT

A broad assessment of advances using *in silico* toxicology was performed and found that most approaches involved prediction based on scientific support of a theoretical possibility.[7] It has become apparent that computer-aided methods to predict potential hazards of chemicals are not as well accepted by the scientific community if based on computer output alone. This may be due to the computer-generated information which is considered under the context that the toxicity analysis is only a theoretical possibility.[18,37] Calls for mechanistic or mode-of-action–based models to enhance the ability to make mechanistic inferences from prediction data have been made.[2,4,87] Such investigations are becoming more and more visible in the literature and in demand for investigation. With the potential exception of the example of predicting genotoxic drug impurities, obtaining a simple prediction alone from a computational tool does not seem to be enough evidence in its entirety to justify a risk management decision of not having to perform a toxicological study without integrating other experimental data or exposure information into the equation. Put another way, although desirable for many reasons, computational toxicology tools clearly have not matured as of yet so as to be able to carry out a full-blown toxicological risk assessment based only on the information these tools can provide. Moreover, there are questions centered around their reliability.[88] A virtual toxicology data package has not yet arrived with reliable information that would permit risk evaluators to make regulatory or product safety decisions solely based on computational analysis (e.g., data mining and prediction), although that may be the vision of some initiatives which do seem to point to such a futuristic setting.[84] Currently, there are computational toxicology software programs that do offer an entire suite of prediction models for nonclinical assays and organ toxicities, and human clinical adverse effects,[89,90] and certainly there are enough approaches out in the public domain to cover the large number of toxicity endpoints that are normally considered in a risk and safety assessment.[12,91] The question is how realistic is this proposition. Certainly, it is worth investigating and testing. New methods and assessments directed toward identifying problem areas and correcting them to better enable the computational sciences to be used more pragmatically would be worthwhile. The existence of large amounts of data and chemicals on toxicity endpoints is not enough to justify the creation of reliable computational toxicology models. In recognizing the limitations of theoretical predictions and diversity of evidence (lack of causality based only on association), attempts to improve the understanding and applicability of toxicological prediction data have been made. Many recent efforts have gone the way of integrating actual experimental test data with computational tools. This strategic direction of improving our understanding of toxicity of substances through assessments of mechanisms and pathways, and also using integrative approaches with empirical evidence, seems to be a science-based path that would better enable risk management decisions.

The pharmaceutical industry employs a practice of testing compounds in defined safety panels to identify undesirable activity.[92] Undesirable activity may be interaction of a compound series with receptors (e.g., 5-hydroxytryptamine 2B) that mediate physiologic functions leading to adverse effects. Of course, this is considered under estimated human exposure (C_{max} protein-bound drug) for proposed use. This information is used to optimize compounds to minimize safety liabilities. The point here is that these data provide a base of information for subsequent data mining and integration for risk assessment. Studies have shown that both structural features and properties can be mined from these types of data to distinguish between compounds of low and high promiscuity.[93,94] One observation by Nigsch et al. was that promiscuity of a compound can be due to either the inherent physico-chemical properties rather than biological action which can interfere with screening assay technology readouts.[95] They give the example of quercetin as nonspecific in protein binding and in a nonstoichiometric way, and because of this inherent property, it can lead to false positives, which would warrant caution in drawing conclusions from a single method. Alongside *in vitro* assay test data, animal toxicology study data represent a large data source to collect, mine, and integrate into computational systems. Animal toxicology study data, however, have various limitations. One limitation is the sourcing. If the data are not, for example, from regulatory submissions to agencies, it is likely they were not standardized and not conducted under the rigor of good laboratory practices (GLPs). Non-GLP studies are more challenging to reproduce, and thus, the quality of those study data varies. Nonclinical pharmacology and toxicology study data for pharmaceutical regulatory review are of higher quality and are in abundance and mostly privately held. However, some databases are publically available that hold such studies with confidential information redacted but require a license for access. One pharmaceutical company estimated in 2011 that it archives nonclinical animal study data on more than 14,000 compounds.[95] If we were to assume the average animal study report holds 250 pages of text available for data mining, and that is a conservative estimate since chronic studies like carcinogenicity bioassays hold much more, we can estimate 3.5 million pages that would need to be screened. Multiply this by the hundreds if not thousands of pharmaceutical companies, and one can easily envision a deep sea of data available for mining that could be put to scientific use. Thus, at a regulatory agency such as the U.S. Food and Drug Administration (FDA), there is a gold mine of nonclinical and premarket clinical data that could be harnessed for use in computational analysis. In fact, this area is one of the FDA's Critical Path opportunities and also consistent with its initiatives in advancing regulatory science.[96,97] Recent reviews and opinion have covered how the FDA is approaching the important point of computational sciences.[1,98] Another limitation of animal toxicology studies as a data source of information for computational modeling and prediction is that nonclinical models are in fact models in themselves. Thus, if one creates a prediction model for a toxicological endpoint that can actually be measured in humans either via an experimental study or diagnostically (e.g., blood in urine), then essentially one has created a model based upon another model or a useless model in which a simple noninvasive diagnostic test can detect. When one creates a model of a model, the net effect is that uncertainty is added into the prediction, which reduces accuracy of potentially useful knowledge to support risk assessments on human adverse effects. For those endpoints that cannot be tested in humans because of ethical reasons (teratogenicity, carcinogenicity, mutagenicity, reproductive toxicity), computational modeling based on animal data is well justified.

Based on the observation that development of a computational (*in silico*) model from other models of nonclinical toxicology/pharmacology study data has its limitations with uncertainty, an FDA Critical Path Initiative project was recently undertaken to develop *in silico* predictive models based on high-quality clinical safety data.[99] In this innovative approach, *in silico* models were constructed using different computer analytical technologies to predict QTc prolongation and torsade de pointes without any regard to nonclinical study outcomes. Only human data were used to train the predictive *in silico* models for these cardiac safety endpoints. The basis was that extrapolation of risk from *in vitro* assay and animal data has inherent challenges. Since a predictive model of these predictive model data could lead to added uncertainty, the choice made was to select the very best human clinical data from which to model. The models that were constructed were risk-based, and the human study data were held to regulatory standards and ICH E14 clinical guidance on QT/QTc interval prolongation and risk for proarrhythmia. The computerized models did not have to outpredict *in vitro*/animal models but should predict as well as or in some cases identify proarrhythmic risk when conventional *in vitro* and animal study data could not. Although *in vitro* assays such as hERG (human ether-a-go-go K + channel) and other safety pharmacology studies are indispensable and recommended under the International Conference on Harmonisation (ICH) S7B guidance as part of integrated risk assessments to identify potential of a drug to be proarrhythmic, it is still challenging for these studies to succeed all of the time. Good science tells us that no method or model is perfect. In fact, many models can be and will be wrong. The science of their use today still centers upon knowing when they are wrong. In the FDA Critical Path Initiative project, the *in silico* models were coined hypothesis-based. The reason is that the training data set for the models constitutes human expert determinations on the medical toxicity outcomes from carefully controlled and randomized clinical trials and post-market surveillance data (e.g., drugs withdrawn from the market). Clinical experts at the FDA with regulatory knowledge of risk on QT prolongation assessments were responsible for selecting drugs for the training sets of the models, and computational toxicologists and chemists constructed the models using structure-based predictive technologies available to the FDA. As a consequence of this integrated approach with very high-quality data, a hypothesis on proarrhythmic risk of each drug was formulated based on the clinical evidence. These hypotheses were used to classify drugs for entrance into the training sets of the models, and the models are being tested under clinical regulatory review activities on new drugs at the FDA. An illustration of hypothesis-based data integration to generate prediction models is shown in Figure 6-7.

Expert decision making is critical in the initial data standards setting and appraisal of study data, and in the final steps of deciding if the models are of scientific value. Given that a thorough QT/QTc study may cost upward of $5 million, and the trainings sets of the models contained over 200 clinically tested drugs, the study data used to construct a single model in the FDA Critical Path Initiative are estimated to have cost approximately

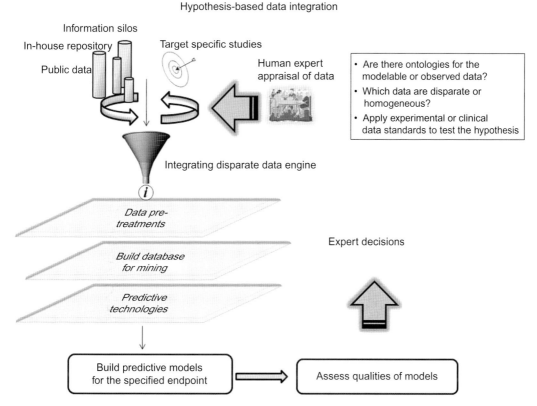

FIGURE 6-7 This diagram illustrates the major components of hypothesis-based data integration to generate prediction models. As shown in the panel, several questions need to be addressed initially. Critical steps are the human expert decision making on the quality of the data and standards of evidence used toward the data. In addition, after a prediction model is built, expert decisions need to be made, including the need to assess overall quality attributes of the models (e.g., predictive performance characteristics), interpretability, and utility of the translational tools for risk assessment.

$1 billion. Also, in considering the low cost of computational software, perhaps running less than $50,000, the cost savings and efficiency in running a prediction on a single drug are enormous, even as in this case the prediction is for decision support and not substitutive of a clinical study. In essence, the entire construction of the models reflects the aims and goals of the FDA Critical Path Initiative: detect toxicity earlier in drug development and regulatory review prior to significant human exposure and harness the power of computer analytics and chemical informatics. Also demonstrated was how to use the right expertise in order to integrate experimental data appropriately with computational tools for risk assessment purposes.

To date, there are several notable initiatives to integrate and mine animal and *in vitro* toxicology study data in order to make advances in predictive toxicology, and maintain the development of alternative methods that reduce, refine, or replace the use of animals in

TABLE 6-2 Representative List of Data Mining and Standards Initiatives to Integrate Information into Computational Toxicology Systems and Approaches

Initiative	Reference
Tox21	100
eTOX—Expert Systems in Toxicology	101
COSMOS—European Union project. Integrated *in silico* models for the prediction of human repeated dose toxicity of cosmetics to optimize safety	102
U.S. FDA 2011–2015 Strategic Priorities—Modernization of Toxicology	103
U.S. FDA Critical Path Initiative	96
U.S. FDA Standard for Exchange of Nonclinical Data	104
U.S. National Institutes of Health, National Center for Advancing Translational Sciences	105
Wyss Institute, Harvard University, Tissue-Organs on a Micro-Chip	106
U.S. EPA DSSTox	5
U.S. EPA Virtual Liver	107
OpenTox	108
ToxML	4

hazard and risk assessments. Moreover, these initiatives envision increasing efficiency in risk assessments that could lead to more expeditious risk management decisions, and such strategies are intended to enable human safety assessments with higher throughput analysis on hazard and risk using data. The data are desired to be from shared or standardized controlled vocabularies and toxicology ontologies. A representative nonexhaustive list of initiatives and data standardization efforts embellishing these goals is presented in Table 6-2.

One of these initiatives called OpenTox is a framework that supports development and application of interoperable predictive toxicology software and uses ontologies as a solution for organizing data, study terms, and protocols.[108] By application of ontologies, it supports the integration of toxicology data in a way that allows construction of higher-level queries regarding the data. Through use of ontologies, the aggregation of data on the toxicity of substances including mechanistic knowledge can be stored in hidden and nonhidden data relationships for the purpose of finding previous experimental test data or facilitating discovery. The OpenTox ontology provides a shared information model that contains approximately five study types: carcinogenicity, *in vitro* bacterial mutagenesis, *in vivo* micronucleus, repeat-dose toxicity studies (chronic, subchronic, and subacute), and aquatic toxicity. These study types correspond to endpoint classifications as detailed by the European REACH guidance so that database fields will be amenable to endpoints reviewed by the European Chemicals Agency (ECHA). OpenTox seems to be intended to support a dynamic linking of models, databases, and toxicology study data through an interface that retrieves relevant information on a queried chemical, searching all available databases and computational models. There is a ToxPredict (http://toxpredict.org) application to send queries on compounds to its ontology service. Currently, there are 12 organ systems in the OpenTox Organ ontology service. The computational toxicity models run by ToxPredict may be hosted remotely by different organizations participating in the OpenTox collaboration. Running multiple services from remote sites integrates models and descriptors that are developed by different scientific organizations into an extensive meta-prediction and data mining results useful for risk assessments as described by REACH.

A classic example of integrating diverse data into computational models and tools for extrapolation to support chemical risk assessments is the integration of key input parameters in physiologically based pharmacokinetic (PBPK) modeling. Most PBPK models represent a systematic set of ordinary differential equations requiring input variables, of which the computation can be carried out by various simulation software programs. PBPK models consider uptake routes, compound storage compartments (e.g., fat, muscle, liver) and binding, metabolism, excretion, and target tissue. PBPK models can predict plasma and tissue concentration-time profiles as well as gain insight into mechanistic considerations for a compound.[109] In order to make predictions, PBPK models require values of compound clearance, tissue-to-plasma partition coefficients, and the rate and extent of absorption. Thus, species-specific data are integrated with a simulation model and are essential for obtaining compound-specific predictions on the plasma concentration-time profile. *In vitro* hepatocyte models have been used as data parameters for PBPK models to predict clearance.[110] Scaling procedures are required for human models.[111] Some of the computational software that can carry out PBPK simulations tailored for pharmaceuticals includes but is not limited to SimCYP, GastroPlus™, Chloe®PK, and PKSIM®. The SimCYP approach is able to perform quantitative integration of *in vitro* metabolism and inhibition data for prediction of drug–drug interactions.[112] It should be noted that a crucial approach for developing a PBPK model relies upon integration of test data after kinetic data are compared to experimental values through the design and conduct of validation experiments, and subsequently, those data are used to refine the model.

In summary, by fostering the development of innovative methods and technologies to integrate data across areas via computational systems approaches, we can use a more evidentiary base of information to accelerate the necessary processes for chemical safety and risk assessments for the benefit of patients and the general public. It can be concluded that expert judgment and interpretation of computational toxicology methods will remain a critical component in assuring accuracy, reliability, confidence, and success of these tools. The future of not only computational toxicology but also the entire field of toxicology will rely heavily upon the use of innovative strategies integrating test data with computational scientific technologies to achieve the best translation of valuable information to meet risk assessment goals.

DISCLAIMER

The content of this publication does not necessarily reflect the views or policies of the Department of Health and Human Services, nor does mention of trade names, commercial products, or organizations imply endorsement by the U.S. Government. This paper reflects the current thinking and experience of the author.

References

[1] Valerio Jr. LG. Application of advanced *in silico* methods for predictive modeling and information integration. Expert Opin Drug Metab Toxicol 2012;8:395–8.

[2] Mostrag-Szlichtyng A, Zaldívar Comenges J-M, Worth AP. Computational toxicology at the European Commission's Joint Research Centre. Expert Opin Drug Metab Toxicol 2010;6:785–92.

[3] Valerio Jr. LG, Choudhuri S. Chemoinformatics and chemical genomics: potential utility of in silico methods. J Appl Toxicol 2012;32:880–9.

[4] Yang C, Valerio LG, Arvidson KB. Computational toxicology approaches at the US Food and Drug Administration. Altern Lab Anim 2009;37:523–31.

[5] Richard AM, Yang C, Judson RS. Toxicity data informatics: supporting a new paradigm for toxicity prediction. Toxicol Mech Methods 2008;18:103–18.

[6] Valerio LG. Computational science in drug metabolism and toxicology. Expert Opin Drug Metab Toxicol 2010;6:781–4.

[7] Valerio LG. In silico toxicology for the pharmaceutical sciences. Toxicol Appl Pharmacol 2009;241:356–70.

[8] Arvidson KB. FDA toxicity databases and real-time data entry. Toxicol Appl Pharmacol 2008;233:17–9.

[9] Boyer S. The use of computer models in pharmaceutical safety evaluation. ATLA 2009;37:467–75.

[10] Brennan RJ, Nikolskya T, Bureeva S. Network and pathway analysis of compound-protein interactions. In: Edgar J, editor. Chemogenomics methods and applications, vol. 575. New York, NY, USA: Humana Press; 2009. p. 225–47. 934 vols.

[11] Wu Y, Lu Y, Chen W, Fu J, Fan R. *In silico* experimentation of glioma microenvironment development and anti-tumor therapy. PLoS Comput Biol 2012;8:e1002355.

[12] Geenen S, Taylor P, Snoep J, Wilson I, Kenna J, Westerhoff H. Systems biology tools for toxicology. Arch Toxicol 2012;86:1251–71.

[13] Scheiber J, Chen B, Milik M, Sukuru SCK, Bender A, Mikhailov D, et al. Gaining insight into off-target mediated effects of drug candidates with a comprehensive systems chemical biology analysis. J Chem Inf Model 2009;49:308–17.

[14] Woodhead JL, Howell BA, Yang Y, Harrill AH, Clewell HJ, Andersen ME, et al. An analysis of n-acetylcysteine treatment for acetaminophen overdose using a systems model of drug-induced liver injury. J Pharmacol Exp Ther 2012;342:529–40.

[15] Thomas RS, Clewell III HJ, Allen BC, Yang L, Healy E, Andersen ME. Integrating pathway-based transcriptomic data into quantitative chemical risk assessment: a five chemical case study. Mutat Res/Genet Toxicol Environ Mutagen 2012;746:135–43.

[16] Howell B, Yang Y, Kumar R, Woodhead J, Harrill A, Clewell H, et al. *In vitro* to *in vivo* extrapolation and species response comparisons for drug-induced liver injury (DILI) using DILIsym™: a mechanistic, mathematical model of DILI. J Pharmacokinet Pharmacodyn, 2012;39(5):1–15.

[17] Drug Information Association. DIA/FDA quantitative structure-activity relationship (Q)SAR approaches for assessing genotoxic impurities in pharmaceuticals. April 7, 2011 Bethesda, MD.

[18] Dobo KL, Greene N, Fred C, Glowienke S, Harvey JS, Hasselgren C, et al. *In silico* methods combined with expert knowledge rule out mutagenic potential of pharmaceutical impurities: an industry survey. Regul Toxicol Pharmacol 2012;62:449–55.

[19] Naven RT, Louise-May S, Greene N. The computational prediction of genotoxicity. Expert Opin Drug Metab Toxicol 2010;6:797–807.

[20] Tropsha A. Best practices for QSAR model development, validation, and exploitation. Mol Info 2010;29:476–88.

[21] Benigni R, Bossa C. Predictivity and reliability of QSAR models: the case of mutagens and carcinogens. Toxicol Mech Methods 2008;18:137–47.

[22] Ashby J, Tennant RW, Zeiger E, Stasiewicz S. Classification according to chemical structure, mutagenicity to *Salmonella* and level of carcinogenicity of a further 42 chemicals tested for carcinogenicity by the U.S. Toxicology Program. Mutat Res 1989;223:73–103.

[23] Ashby J, Tennant RW. Definitive relationships among chemical structure, carcinogenicity and mutagenicity for 301 chemicals tested by the U.S. NTP. Mutat Res 1991;257:229–306.

[24] Zeiger E. Identification of rodent carcinogens and noncarcinogens using genetic toxicity tests: premises, promises, and performance. Regul Toxicol Pharmacol 1998;28:85–95.

REFERENCES

[25] Zeiger E. Carcinogenicity of mutagens: predictive capability of the *Salmonella* mutagenesis assay for rodent carcinogenicity. Cancer Res 1987;47:1287—96.
[26] Mayer J, Cheeseman MA, Twaroski ML. Structure—activity relationship analysis tools: validation and applicability in predicting carcinogens. Regul Toxicol Pharmacol 2008;50:50—8.
[27] Devillers J, Mombelli E, Samserà R. Structural alerts for estimating the carcinogenicity of pesticides and biocides. SAR QSAR Environ Res 2011;22:89—106.
[28] Benigni R. Alternatives to the carcinogenicity bioassay for toxicity prediction: are we there yet? Expert Opin Drug Metab Toxicol 2012;8:407—17.
[29] Valerio Jr. LG, Cross KP. Characterization and validation of an *in silico* toxicology model to predict the mutagenic potential of drug impurities. Toxicol Appl Pharmacol 2012;260:209—21.
[30] Hillebrecht A, Muster W, Brigo A, Kansy M, Weiser T, Singer T. Comparative evaluation of *in silico* systems for Ames test mutagenicity prediction: scope and limitations. Chem Res Toxicol 2011;24:843—54.
[31] Hansen K, Mika S, Schroeter T, Sutter A, ter Laak A, Steger-Hartmann T, et al. Benchmark data set for in silico prediction of Ames mutagenicity. J Chem Inf Model 2009;49:2077—81.
[32] Snyder RD, Smith MD. Computational prediction of genotoxicity: room for improvement. Drug Discov Today 2005;10:1119—24.
[33] Modi S, Li J, Malcomber S, Moore C, Scott A, White A, et al. Integrated in silico approaches for the prediction of Ames test mutagenicity. J Comput Aided Mol Des 2012;26(9):1017—33.
[34] Mazzatorta P, Tran L-A, Schilter B, Grigorov M. Integration of structure — activity relationship and artificial intelligence systems to improve *in silico* prediction of Ames test mutagenicity. J Chem Inf Model 2006;47:34—8.
[35] Snyder RD. An update on the genotoxicity and carcinogenicity of marketed pharmaceuticals with reference to *in silico* predictivity. Environ Mol Mutagen 2009;50:435—50.
[36] Matthews EJ, Kruhlak NL, Benz RD, Contrera JF, Marchant CA, Yang C. Combined use of MC4PC, MDL-QSAR, BioEpisteme, Leadscope PDM, and Derek for Windows software to achieve high-performance, high-confidence, mode of action—based predictions of chemical carcinogenesis in rodents. Toxicol Mech Methods 2008;18:189—206.
[37] Naven RT, Greene N, Williams RV. Latest advances in computational genotoxicity prediction. Expert Opin Drug Metab Toxicol 1—9.
[38] Bayes T, Price R. An essay towards solving a problem in the doctrine of chance. By the late Rev. Mr. Bayes, communicated by Mr. Price, in a letter to John Canton, M.A. and F.R.S. Philos Trans R Soc Lond B Biol Sci 1763;53:370—418.
[39] Beerenwinkel N, Siebourg J. In: Anisimova M, editor. Probability, statistics, and computational science evolutionary genomics, vol. 855. New York, NY: Humana Press; 2012. p. 77—110.
[40] Berry DA. Bayesian clinical trials. Nat Rev Drug Discov 2006;5:27—36.
[41] Hoff PD. In: Hoff PD, editor. A first course in bayesian statistical methods. 2nd ed. London: Springer; 2009.
[42] Marchant CA, Briggs KA, Long A. In silico tools for sharing data and knowledge on toxicity and metabolism: Derek for Windows, meteor, and vitic. Toxicol Mech Methods 2008;18:177—87.
[43] Saiakhov RD, Klopman G. Benchmark Performance of MultiCASE Inc. Software in Ames Mutagenicity Set. J Chem Inf Model 2010;50:1521.
[44] Contrera JF, Matthews EJ, Kruhlak NL, Benz RD. *In silico* screening of chemicals for genetic toxicity using MDL-QSAR, nonparametric discriminant analysis, e-state, connectivity, and molecular property descriptors. Toxicol Mech Methods 2008;18:207—16.
[45] Helma C. Lazy structure-activity relationships (lazar) for the prediction of rodent carcinogenicity and *Salmonella* mutagenicity. Mol Divers 2006;10:147—58.
[46] Fellows MD, Boyer S, O'Donovan MR. The incidence of positive results in the mouse lymphoma TK assay (MLA) in pharmaceutical screening and their prediction by MultiCase MC4PC. Mutagenesis 2011;26:529—32.
[47] Devillers J, Mombelli E. Evaluation of the OECD QSAR application toolbox and toxtree for estimating the mutagenicity of chemicals. Part 1. Aromatic amines. SAR QSAR Environ Res 2010;21:753—69.
[48] Arvidson KB, Valerio LG, Diaz M, Chanderbhan RF. *In silico* toxicological screening of natural products. Toxicol Mech Methods 2008;18:229—42.
[49] Contrera JF, Kruhlak NL, Matthews EJ, Benz RD. Comparison of MC4PC and MDL-QSAR rodent carcinogenicity predictions and the enhancement of predictive performance by combining QSAR models. Regul Toxicol Pharmacol 2007;49:172—82.

[50] Kar S, Roy K. First report on development of quantitative interspecies structure−carcinogenicity relationship models and exploring discriminatory features for rodent carcinogenicity of diverse organic chemicals using OECD guidelines. Chemosphere 2012;87:339−55.

[51] Hewitt M, Cronin MTD, Madden JC, Rowe PH, Johnson C, Obi A, et al. Consensus QSAR models: do the benefits outweigh the complexity? J Chem Inf Model 2007;47:1460−8.

[52] Sushko I, Novotarskyi S, Körner R, Pandey AK, Cherkasov A, Li J, et al. Applicability domains for classification problems: benchmarking of distance to models for Ames mutagenicity set. J Chem Inf Model 2010;50:2094−111.

[53] Tetko IV, Sushko I, Pandey AK, Zhu H, Tropsha A, Papa E, et al. Critical assessment of QSAR models of environmental toxicity against *Tetrahymena Pyriformis*: focusing on applicability domain and overfitting by variable selection. J Chem Inf Model 2008;48:1733−46.

[54] Fjodorova N, Novič M, Roncaglioni A, Benfenati E. Evaluating the applicability domain in the case of classification predictive models for carcinogenicity based on the counter propagation artificial neural network. J Comput Aided Mol Des 2011;25:1147−58.

[55] Ellison CM, Sherhod R, Cronin MT, Enoch SJ, Madden JC, Judson PN. Assessment of methods to define the applicability domain of structural alert models. J Chem Inf Model 2011; [Epub ahead of print].

[56] Dragos H, Gilles M, Alexandre V. Predicting the predictability: a unified approach to the applicability domain problem of QSAR models. J Chem Inf Model 2009;49:1762−76.

[57] OECD. Guidance on Grouping of Chemicals. ENV/JM/MONO(2007)28. Environment directorate. Joint meeting of the Chemicals Committee and The Working Party on Chemicals, Pesticides and Biotechnology. Series on Testing and Assessment, 2007;80:1097.

[58] Arvidson KB, Chanderbhan R, Muldoon-Jacobs K, Mayer J, Ogungbesan A. Regulatory use of computational toxicology tools and databases at the United States Food and Drug Administration's Office of Food Additive Safety. Expert Opin Drug Metab Toxicol 2010;6:793−6.

[59] Kruhlak NL, Benz RD, Zhou H, Colatsky TJ. (Q)SAR modeling and safety assessment in regulatory review. Clin Pharmacol Ther 2012;91:529−34.

[60] Kruhlak NL, Contrera JF, Benz RD, Matthews EJ. Progress in QSAR toxicity screening of pharmaceutical impurities and other FDA regulated products. Adv Drug Deliv Rev 2007;59:43−55.

[61] The European Parliament and the Council of the European Union. Regulation (EC) No 1907/2006 of the European Parliament and of the Council of 18 December 2006, concerning the Registration, Evaluation, Authorisation and Restriction of Chemicals (REACH), establishing a European Agency, amending Directive 1999/45/EC and Repealing Council Regulation (EEC) No 793/93 and Commission Regulation (EC) No 1488/94 as well as Council Directive 76/769/EEC and Commision Directives 91/155/EEC, 93.67/EEC, 93/105/EC and 2000/21/EC. Off J Eur Union 2007;50:3−280.

[62] Hulzebos E, Gunnarsdottir S, Rila J-P, Dang Z, Rorije E. An Integrated Assessment Scheme for assessing the adequacy of (eco)toxicological data under REACH. Toxicol Lett 2010;198:255−62.

[63] van Ravenzwaay B, Herold M, Kamp H, Kapp MD, Fabian E, Looser R, et al. Metabolomics: a tool for early detection of toxicological effects and an opportunity for biology based grouping of chemicals—from QSAR to QBAR. Mutat Res/Genet Toxicol Environ Mutagen 2012;746:144−50.

[64] Hong H, Xie Q, Ge W, Qian F, Fang H, Shi L, et al. Mold2, Molecular Descriptors from 2D Structures for Chemoinformatics and Toxicoinformatics. J Chem Inf Model 2008;48:1337−44.

[65] Jaworska J, Gabbert S, Aldenberg T. Towards optimization of chemical testing under REACH: a Bayesian network approach to integrated testing strategies. Regul Toxicol Pharmacol 2010;57:157−67.

[66] Langdon S, Mulgrew J, Paolini G, van Hoorn W. Predicting cytotoxicity from heterogeneous data sources with Bayesian learning. J Cheminformatics 2010;2:11.

[67] Zhang J, Bailer AJ, Oris JT. Bayesian approach to estimating reproductive inhibition potency in aquatic toxicity testing. Environ Toxicol Chem 2012;31:916−27.

[68] Martin R, Rose D, Yu K, Barros S. Toxicogenomics strategies for predicting drug toxicity. Pharmacogenomics 2006;7:1003−16.

[69] LeDonne N, Rissolo K, Bulgarelli J, Tini L. Use of structure-activity landscape index curves and curve integrals to evaluate the performance of multiple machine learning prediction models. J Cheminformatics 2011;3:7.

[70] Marjan V. Kohonen artificial neural network and counter propagation neural network in molecular structure-toxicity studies. Curr Comput Aided Drug Des 2005;1:73−8.

REFERENCES

[71] Contrera JF, Matthews EJ, Kruhlak NL, Benz RD. In silico screening of chemicals for bacterial mutagenicity using electrotopological E-state indices and MDL QSAR software. Regul Toxicol Pharmacol 2005;43:313–23.

[72] Martin MT, Knudsen TB, Reif DM, Houck KA, Judson RS, Kavlock RJ, et al. Predictive model of rat reproductive toxicity from ToxCast high throughput screening. Biol Reprod 2011;85:327–39.

[73] Valerio Jr LG, Dixit A, Moghaddam S, Mora O, Prous J, Valencia A. QSAR modeling for the mutagenic potential of drug impurities with Symmetry®. In: Toxicologist T, editor. Society of toxicology, vol. Supp. San Francisco, CA: Oxford Journals; 2012, p. 2906.

[74] Guha R, Schürer S. Utilizing high throughput screening data for predictive toxicology models: protocols and application to MLSCN assays. J Comput Aided Mol Des 2008;22:367–84.

[75] Cassano A, Manganaro A, Martin T, Young D, Piclin N, Pintore M, et al. CAESAR models for developmental toxicity. Chem Cent J 2010;4:S4.

[76] Coull BA, Wellenius GA, Gonzalez-Flecha B, Diaz E, Koutrakis P, Godleski JJ. The toxicological evaluation of realistic emissions of source aerosols study: statistical methods. Inhal Toxicol 2011;23:31–41.

[77] Low Y, Uehara T, Minowa Y, Yamada H, Ohno Y, Urushidani T, et al. Predicting drug-induced hepatotoxicity using QSAR and toxicogenomics approaches. Chem Res Toxicol 2011;24:1251–62.

[78] Spjuth O, Eklund M, Ahlberg Helgee E, Boyer S, Carlsson L. Integrated decision support for assessing chemical liabilities. J Chem Inf Model 2011;51:1840–7.

[79] Pavan M, Worth AP. Publicly-accessible QSAR software tools developed by the Joint Research Centre. SAR QSAR Environ Res 2008;19:785–99.

[80] Blackburn K, Bjerke D, Daston G, Felter S, Mahony C, Naciff J, et al. Case studies to test: a framework for using structural, reactivity, metabolic and physicochemical similarity to evaluate the suitability of analogs for SAR-based toxicological assessments. Regul Toxicol Pharmacol 2011;60:120–35.

[81] Hammann F, Gutmann H, Jecklin U, Maunz A, Helma C, Drewe J. Development of decision tree models for substrates, inhibitors, and inducers of p-glycoprotein. Curr Drug Metab 2009;10:339–46.

[82] Shukla SJ, Huang R, Austin CP, Xia M. The future of toxicity testing: a focus on in vitro methods using a quantitative high-throughput screening platform. Drug Discov Today 2010;15:997–1007.

[83] Martin MT, Knudsen TB, Judson RS, Kavlock RJ, Dix DJ. Economic benefits of using adaptive predictive models of reproductive toxicity in the context of a tiered testing program. Syst Biol Reprod Med 2012;58:3–9.

[84] Bucher JR. Regulatory forum opinion piece: Tox21 and toxicologic pathology. Toxicol Pathol 2013;41(1):125–7.

[85] Mathews LA, Keller JM, Goodwin BL, Guha R, Shinn P, Mull R, et al. A 1536-well quantitative high-throughput screen to identify compounds targeting cancer stem cells. J Biomol Screen 2012;17:1231–42.

[86] Harwood MD, Neuhoff S, Carlson GL, Warhurst G, Rostami-Hodjegan A. Absolute abundance and function of intestinal drug transporters: a prerequisite for fully mechanistic in vitro–in vivo extrapolation of oral drug absorption. Biopharm Drug Dispos 2012; [Epub ahead of print]

[87] Hartung T, van Vliet E, Jaworska J, Bonilla L, Skinner N, Thomas R. Food for thought ... systems toxicology. Altex 2012;29:119–28.

[88] Combes RD. Challenges for computational structure–activity modelling for predicting chemical toxicity: future improvements? Expert Opin Drug Metab Toxicol 2011;7:1129–40.

[89] Valerio LG, Yang C, Arvidson KB, Kruhlak NL. A structural feature-based computational approach for toxicology predictions. Expert Opin Drug Metab Toxicol 2010;6:505–18.

[90] Myshkin E, Brennan R, Khasanova T, Sitnik T, Serebriyskaya T, Litvinova E, et al. Prediction of organ toxicity endpoints by QSAR modeling based on precise chemical-histopathology annotations. Chem Biol Drug Des 2012;80:406–16.

[91] Kavlock R, Chandler K, Houck K, Hunter S, Judson R, Kleinstreuer N, et al. Update on EPA's ToxCast program: providing high throughput decision support tools for chemical risk management. Chem Res Toxicol 2012;25:1287–302.

[92] Whitebread S, Hamon J, Bojanic D, Urban L. Keynote review: in vitro safety pharmacology profiling: an essential tool for successful drug development. Drug Discov Today 2005;10:1421–33.

[93] Azzaoui K, Hamon J, Faller B, Whitebread S, Jacoby E, Bender A, et al. Modeling promiscuity based on in vitro safety pharmacology profiling data. ChemMedChem 2007;2:874–80.

[94] Faller B, Wang J, Zimmerlin A, Bell L, Hamon J, Whitebread S, et al. High-throughput in vitro profiling assays: lessons learnt from experiences at Novartis. Expert Opin Drug Metab Toxicol 2006;2:823–33.

[95] Nigsch F, Lounkine E, McCarren P, Cornett B, Glick M, Azzaoui K, et al. Computational methods for early predictive safety assessment from biological and chemical data. Expert Opin Drug Metab Toxicol 2011;7:1497–511.
[96] FDA. FDA Critical Path Initiative. <http://www.fda.gov/ScienceResearch/SpecialTopics/CriticalPathInitiative/default.htm>. In Critical Path Website, vol. 2012. Silver Spring, MD: US Food and Drug Administration; 2012.
[97] FDA. Advancing Regulatory Science at FDA (Health and Human Services). Rockville, MD: FDA; 2011. p. 7–10.
[98] Valerio Jr. LG. In silico toxicology models and databases as FDA critical path initiative toolkits. Hum Genomics 2011;5:200–7.
[99] Valerio Jr. LG, Balakrishnan S, Fiszman ML, Kozeli D, Li M, Moghaddam S, et al. Development of cardiac safety translational tools for QT prolongation and torsade de pointes. Expert Opin Drug Metab Toxicol 2013.
[100] Sun H, Xia M, Austin C, Huang R. Paradigm shift in toxicity testing and modeling. AAPS J 2012;14:473–80.
[101] Briggs K, Cases M, Heard DJ, Pastor M, Pognan F, Sanz F, et al. Inroads to predict *in vivo* toxicology—an introduction to the eTOX project. Int J Mol Sci 2012;13:3820–46.
[102] Richarz A, Cronin MT, Neagu D, Yang C, Zaldivar-Comenges J, Berthold M. A new European project to develop computational models for the repeat dose toxicity of cosmetic ingredients. Altex 2011;28:39–40.
[103] FDA. Strategic Priorities. 2011–20155. Responding to the Public Health Challenges of the 21st Century, vol. 2011, pp. 2.0 Cross-Cutting Strategic Priorities. Silver Spring, MD: Department of Health and Human Services; 2011.
[104] FDA. Standard for Exchange of Nonclinical Data (SEND) <http://www.fda.gov/ForIndustry/DataStandards/StudyDataStandards/ucm155320.htm>. In Data Standards for Industry, vol. 2012. Silver Spring, MD, USA: U.S. Food and Drug Administration; 2012.
[105] Collins FS. Mining for therapeutic gold. Nat Rev Drug Discov 2011;10:397.
[106] Huh D, Matthews BD, Mammoto A, Montoya-Zavala M, Hsin HY, Ingber DE. Reconstituting organ-level lung functions on a chip. Science 2010;328:1662–8.
[107] Wambaugh J, Shah I. Simulating microdosimetry in a virtual hepatic lobule. PLoS Comput Biol 2010;6: e1000756.
[108] Tcheremenskaia O, Benigni R, Nikolova I, Jeliazkova N, Escher SE, Batke M, et al. OpenTox predictive toxicology framework: toxicological ontology and semantic media wiki-based OpenToxipedia. J Biomed Semantics 2012;3(Suppl. 1):S7.
[109] Jones HM, Dickins M, Youdim K, Gosset JR, Attkins NJ, Hay TL, et al. Application of PBPK modelling in drug discovery and development at Pfizer. Xenobiotica 2012;42:94–106.
[110] Houston JB. Utility of *in vitro* drug metabolism data in predicting *in vivo* metabolic clearance. Biochem Pharmacol 1994;47:1469–79.
[111] Jones HM, Gardner IB, Watson KJ. Modelling and PBPK simulation in drug discovery. AAPS J 2009;11:155–66.
[112] Jamei M, Marciniak S, Feng K, Barnett A, Tucker G, Rostami-Hodjegan A. The SimCYP population-based ADME simulator. Expert Opin Drug Metab Toxicol 2009;5:211–23.

CHAPTER 7

Computational Translation of Nonmammalian Species Data to Mammalian Species to Meet REACH and Next Generation Risk Assessment Needs

Edward J. Perkins[1] and Natàlia Garcia-Reyero[2]

[1]U.S. Army Engineer Research and Development Center, Vicksburg, MS, USA,
[2]Mississippi State University, Starkville, MS, USA

A CHANGING REGULATORY ENVIRONMENT

Regulatory programs, most notably the European Union Registration, Evaluation, and Authorization of Chemicals (REACH), legislation enacted in 2006, and chemical screening programs such as implemented by the U.S. Environmental Protection Agency (U.S. EPA) have set challenging goals to determine the potential hazards of thousands of chemicals over just a few years. REACH legislation has profoundly changed directions and requirements for chemical assessment in both the European Union (EU) and for chemicals produced in the United States for EU markets. This legislation has set a long list of data requirements for manufacturers and importers on fate and behavior, ecotoxicity, and toxicity for chemicals in different classes of production volumes. As of 2010, an estimated 145,000 substances were preregistered, indicating the very large scale of information that will be needed for approval of chemical usage. While the required data exist for a relatively small number of chemicals, REACH has targeted 2018 as the date for which chemicals to be used in the European Union will be required to have data on the potential for the chemicals to cause both mammalian (human) and ecological toxicity. Depending on the volume of production, information from a wide range of mammalian and ecotoxicological tests may be required under REACH. Data will be needed on the chemicals' potential to cause mammalian chronic toxicity, carcinogenicity, and reproductive toxicity, to name a few. Ecotoxicological tests may be required including short-term, long-term, and reproductive toxicity testing with a variety of species (plants, fish, algae and daphnia, terrestrial organisms, and birds).

While the United States does not have a similar regulatory program requiring data for all chemicals in use, the U.S. EPA, in collaboration with industry, has developed the High Production Volume (HPV) Challenge program. In the HPV program, chemical manufacturers, importers, and government entities voluntarily developed human health and environmental effects data on approximately 2,200 HPV chemicals (http://www.epa.gov/chemrtk/index.htm). Similar in scope to REACH is the U.S. EPA ToxCast high-throughput screening (HTS) effort to prioritize tens of thousands of chemicals by their potential to effect human health hazards.[1] Complementary to ToxCast are efforts to develop new risk assessment approaches, or "NexGen risk assessment," that incorporate systems biology information, data from *in silico, in vitro* screening; "omic" technologies; short-term *in vivo* studies; and traditional testing to accelerate the risk assessment process without losing quality.[2]

If traditional *in vivo* vertebrate testing methods were to be used to acquire data for REACH and U.S. EPA health hazard assessment efforts, even for moderate numbers of chemicals, the number of animals used for testing would vastly increase and the tests would be prohibitively costly and time consuming.[3] This, paired with increasing regulatory cost and social pressure, has prompted demands and regulations to refine and reduce the number of animals used in toxicological testing and ultimately replace vertebrate animal tests altogether with *in vitro* or *in silico* testing.[4]

In vitro and *in vivo* testing alternatives to standard mammalian testing species are needed to enable current hazard and risk assessment goals both in the European Union and the United States. Nonmammalian and nonvertebrate species have the potential to provide valuable and relevant information more rapidly, at lower cost, and with more statistical power than traditional species used for human chemical risk assessment. However, the most benefit will be realized from nonmammal, nonvertebrate, or *in vivo* tests using embryos, by computationally translating results to relevant human and mammalian health impacts. Through use of this approach, testing could be protective of both human and ecological health while simultaneously reducing numbers of test animals, both mammalian and nonmammalian, and cost.

This chapter describes computational approaches to extrapolate chemical effects on nonmammalian species and nonanimal models (nonvertebrates, early embryo vertebrates, or cell-based) to mammalian test animals in order to reduce, refine, and replace usage animals, determining chemicals' hazards to human health. We first review what alternative animals and nonanimal models are being used. We then present the species-independent context of pathway-based toxicology and risk assessment. The chapter reviews different approaches for translating impacts on different levels of adverse outcome pathways from an alternative species to human, characterizing dose-time-response relationships and extrapolation of dose response to inform human dose response. Finally, suggestions are made on how these data might be used in establishing risk.

NONMAMMALIAN SPECIES CAN HELP TO REDUCE, REFINE, AND REPLACE MAMMALIAN ANIMAL TESTING

The functional definition of "nonanimal testing models" was essentially set by the EU directive on the protection of animals used for scientific purposes which directed the elimination of testing using vertebrate animals, independently feeding larval forms, fetal forms of mammals in the last third of their normal development, and live cephalopods, unless there was a direct human health need.[5] Hence, research in Europe has focused mostly on replacement of vertebrate *in vivo* testing through development of nonanimal models for both mammals and nonmammals (*in vitro* and embryo testing) to replace vertebrate testing, whereas U.S. research has focused more on refinement and reduction of animal use, development of alternative vertebrate species, and human *in vitro* models.

In vivo testing for chemical effects is still necessary for many of the toxicity and ecotoxicity assays required by REACH or the U.S., as no nonanimal alternatives are currently available. The use of mammals can be reduced and refined by the use of integrated testing procedures and alternative species including aquatic vertebrates. Nonmammalian species can help to reduce, refine, and replace mammalian animal testing by providing *in vivo* models that integrate dose-response effects, pathways, and apical effects useful for assessing chemical risks to both human health and to other species. Nonanimal testing using nonmammalian species in embryonic assays, such as zebrafish embryo testing, provides an alternative to vertebrate testing while still incorporating some of the biological complexity present in live animals.[6-8]

PATHWAY-BASED HAZARD AND RISK ASSESSMENT

Efforts to reduce animal testing have focused on moderate to high-throughput *in vivo* and *in vitro* screening assays to generate data on toxicological pathways and dose-response effects that may be used in hazard prioritization and potentially rapid/preliminary risk assessments.[1,9-12] These efforts are ultimately based on a toxicological pathway or mode of action approach.[1,13] Using a pathway-based hazard assessment enables the focus of screening and assessment to center on pathways important in mammalian toxicity, such as reproduction, mutagenicity, and carcinogenicity. If enough information exists to causally link a pathway to an adverse effect such as reproductive toxicity, then perturbations in that pathway could be potentially indicative of a chemical that can cause reproductive impacts.

The rapid technological advances in sequencing and other "omics" technologies (high-content assays such as transcriptomics, proteomics, or metabolomics) have led to a greater knowledge of how toxicological and other pathways and key initiating events are conserved among species, thereby providing new choices in toxicological models useful for chemical risk assessment (Figure 7-1). For example, critical systems involving signaling and regulation are important elements in developmental programming and highly conserved across metazoans.[14] Disruption of signaling pathways such as Wnt, Notch, TGFb, Hedgehog, and Hippo can lead to common developmental abnormalities in mammal and

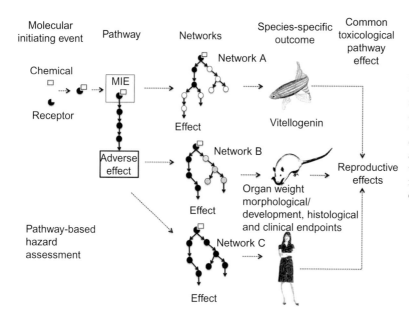

FIGURE 7-1 Toxicological pathways leading to toxic effects or adverse outcomes are started with a molecular initiation event such as receptor binding. Different species can have portions or all of a conserved pathway that leads to effects. These effects may be different across species, but seemingly unrelated effects can still reflect a common adverse effect.

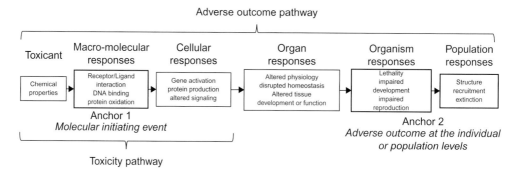

FIGURE 7-2 Conceptual diagram of key features of an adverse outcome pathway (AOP). Each AOP begins with a molecular initiating event in which a chemical interacts with a biological target (anchor 1) leading to a sequential series of higher-order effects to produce an adverse outcome with direct relevance to a given risk assessment context (e.g., survival, development, reproduction; anchor 2). The first three boxes are the parameters that define a toxicity pathway. *Adapted from reference 18.*

invertebrate models such as *Drosophila melanogaster*.[15,16] A high degree of conservation is also seen in genes and pathways expressed in oocytes from species as different as cow (*Bos taurus*), the African clawed frog (*Xenopus laevis*), and mouse (*Mus musculus*) where a substantial proportion of the genes expressed in mouse oocytes were conserved between the three species (74%, 7,275 genes) including core processes such as RNA metabolism and cell cycle.[17]

In a pathway-based approach, such as adverse outcome pathway (Figure 7-2), toxicological impacts of chemicals are defined by a molecular initiating event where a chemical interacts with a biological receptor.[2,18] If this event is sufficient to overcome compensatory mechanisms of an animal, it then causes a cascade of responses at the cellular, organ, individual, and population levels that are linked by signaling, regulatory, and metabolic pathways. The adverse outcome pathway (AOP) concept provides a useful framework in which critical events leading to toxicological effects are captured and can be used to extrapolate between species. This framework captures elements that should be included, or accounted for, in computational models for chemical hazard assessment. The AOP begins with a molecular initiating event (MIE) that details how a chemical interacts with an animal. The AOP is anchored to this event (anchor 1) and anchored again to an adverse outcome at the organism or population level (anchor 2). An AOP pathway causally or statistically links responses at multiple levels to the two anchors. Therefore, a model that computationally links an MIE linked to a critical event or an adverse outcome could be used to predict if a chemical could cause an effect such as reproductive toxicity. An adverse outcome pathway–based analysis *fosters* nontraditional approaches for hazard screening by opening numerous opportunities in applying nontraditional approaches and model species to assess the potential risks of chemical exposures.

Even if the apical effect measured due to chemical exposure is different among species, the conservation of MIEs and underlying toxicological pathways in an AOP can be exploited to use a species with an easily assessed effect to predict toxicological effects in other species that are difficult to assess. For example, a high degree of conservation has been established across vertebrate estrogen and androgen receptors (see alignment of estrogen receptor alpha, Figure 7-3), indicating that a screening assay for estrogen receptor activation utilizing fish (or fish receptors) might be as effective in predicting potential estrogenicity in humans as assays based on rodents.[19,20] Indeed, since pathways for signaling, control, and production of estrogen are highly conserved in the hypothalamus-pituitary-gonadal (HPG) axis in vertebrates, endocrine-disrupting effects in mammals are generally replicated in nonmammalian vertebrates. As a result, one could use nonmammalian species such as the fathead minnow (*Pimephales promelas*) to screen chemicals for endocrine-disrupting effects for both nonmammals and mammals, even if the apical effects measured are different (Figure 7-1). Comparison of screening results for endocrine-disrupting effects of 10 chemicals using traditional rat Hershberger and Uterotrophic assays to results obtained using the alternative species fathead minnow reproductive assay shows that the fish assay identified all chemicals found by the rat assays to have endocrine-disrupting effects.[21] Here, fish could reasonably serve as an effective alternative (to rats) for human health assessments.

As noted previously, one of the great desires in testing is to move away from *in vivo* testing to nonanimal models. These models may not faithfully reflect the hazard a chemical poses to adult animal models since the nonanimal model may not have a complete organ system or metabolic capacity. However, recent work has demonstrated that nonanimal models such as fish embryos can be predictive of *in vivo* chemical effects in adult animal models, both fish and rat. Chemical toxicity to zebrafish embryos has been shown to be well correlated to acute toxicity in adult fish with the exception of some chemicals that had differential toxicity possibly due to metabolic activation or other causes.[8] Enough conservation of toxicological pathways is present in 24-hour post-fertilized zebrafish embryos such that toxicity (LC50) caused by 60 different water-soluble chemicals in

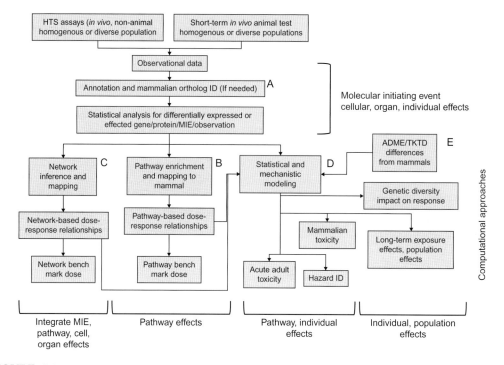

FIGURE 7-3 Flow chart for extrapolation of nonmammalian to mammalian species for risk assessment. Elements corresponding to the adverse outcome pathway (AOP) framework are highlighted with brackets. A. Computational translation of molecular initiating events linked to observed apical effects (e.g., egg protein production) between species using gene/protein homology. B. Pathway-level extrapolation between species. C. Network or integrative data-level extrapolation between species. D. Systems-level extrapolation between species.

96-hour exposures in the embryo was well correlated to toxicity (LD50) in rats although this correlation was dependent on the chemical class examined.[7] Zebrafish embryos also have complete pathways for thyroid hormone synthesis,[22,23] heart development, and many others.[24] Therefore, analysis and extrapolation of effects on embryos may be protective of both adults and mammals using an AOP framework.

TRANSLATING EFFECTS ON NONMAMMALIAN SPECIES TO MAMMALIAN SPECIES

The AOP concept and toxicological pathways that are causally linked to an adverse effect provide a clear framework in which to translate effects between species. Effects on nonmammalian species can be computationally translated to mammalian species at several different levels with increasing complexity and relevance to the adverse outcome being assessed. A general strategy is presented in Figure 7-3. The simplest and most obvious extrapolation between species is the comparison of sequences or structures of individual proteins involved in molecular initiating events (Figure 7-3A). More complex is

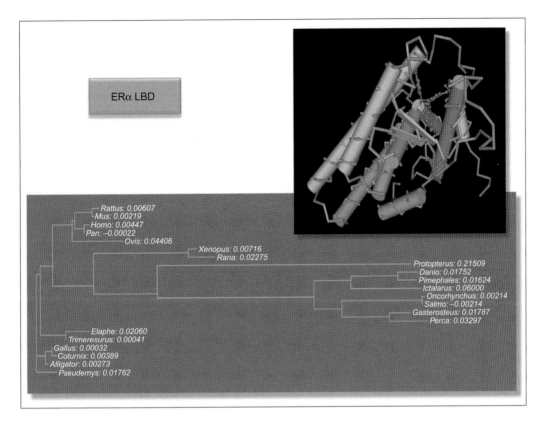

FIGURE 7-4 Alignment of the estrogen receptor alpha (ERα) ligand-binding domain sequences from 21 different species using Clustaw2: *Homo sapiens* (NM_000125.3), *Danio rerio* (NM_152959.1), *Rattus norvegicus* (NM_012689.1), *Mus musculus* (NM_007956.4), *Oncorhynchus mykiss* (NM_001124349.1), *Gallus gallus* (NM_205183.1), *Xenopus laevis* (NM_001089617.1), *Ictalurus punctatus* (NM_001200074.1), *Pan troglodytes* (XM_003311548.1), *Salmo salar* (NM_001123592.1), *Elaphe quadrivirgata* (AB548295.1), *Trimeserus flavoviridis* (AB548294.1), *Rana rugosa* (AB491673.1), *Ovis aries* (AY033393.1), *Pseudemys nelsoni* (AB301060.1), *Gasterosteus aculeatus* (AB330740.1), *Perca flavescens* (DQ984124.1), *Alligator mississippiensis* (AB115909.1), *Pimephales promelas* (AY727528.1), *Coturnix japonica* (AF442965.1), *Propopterus annectens* (AB435636.1). The numbers correspond to genetic distance from a common ancestor. Genetic distances were initially calculated as percent identity scores and were converted to distances by dividing by 100 and subtracting from 1.0 to give the number of differences per site.

comparison of a toxicological pathway between species that include multiple components (Figure 7-3B). The use of pathways linked to adverse effects is the most well-developed area and has already been applied to establish toxicological dose-response relationships. Since toxicity pathways are composed of many different interactions, they can be explored either as discrete pathways or as networks using cross-species comparative genomics (for a brief review, see reference 25; Figure 7-3C). Network-based dose-response relationships are likely to be more difficult to establish, as the notion of what is a normal network is difficult to determine at this point in time; therefore, this chapter does not delve too deeply into network analysis. Ultimately, the most realistic approach is to translate effects through computational modeling at the systems level where dynamic events are incorporated, such as homeostasis, effects over time, and species-specific parameters (e.g., absorption, distribution, metabolism, and excretion; Figure 7-3D).

TRANSLATING MOLECULAR INITIATING EVENTS: GENE/PROTEIN ANNOTATION AND MAMMALIAN ORTHOLOG IDENTIFICATION

Cross-species comparisons focusing on the conservation of molecular initiating events permit anchoring the AOP of a nonmammalian species to AOPs from mouse, rat, or human species through identification of common biological machinery. Chemical interactions with proteins can be extrapolated among species under the assumption that evolutionarily conserved protein sequences lead to conserved functions. Proteins that are derived from an ancestral gene and retain the same function are called orthologs. Due to the recent explosion of sequencing data, there is an ever-increasing amount of information on genomes of many different species in addition to much more detailed information on humans and the mammalian model species mouse and rat. Using these data, researchers have created databases and tools to identify orthologs across species that can be useful in characterizing whether a gene, protein, or molecular-initiating event in one species is similar to mammalian gene or proteins and is a potential AOP molecular-initiating event (Table 7-1; see reference 25). Note that updated resources for biological databases and tools for genes, pathways, and genome analysis can be found at http://metadatabase.org, a wiki-database of biological databases.[26]

Although still challenging, cross-species analysis of genomic/transcriptomic data can lead to insights that cannot be obtained from the analysis of a single species (reviewed in reference 27). When an ortholog to an MIE is found, for example, a receptor or enzyme, the extent of homology in overall sequence or the target-binding site of the protein could be used to predict potential relative sensitivities between a test species and desired (mammalian) species. Once an AOP construct has been created linking anchor 1, the MIE, to anchor 2, the adverse effect, the orthologous protein sequence for a known molecular target of a drug in an alternative species can be used to infer possible effects of a chemical on a mammalian species (Figure 7-2). Homology searches have been used to investigate the likelihood that pharmaceuticals and pesticides would also affect ecological species. These chemicals and target proteins readily fit into an AOP framework for cross-species comparison because the mechanisms of interaction and targeted proteins, or MIEs, are very well known. Comparison of target protein orthologs indicate that many receptors and enzymes such as hormone receptors (e.g., estrogen, androgen, progesterone), neurological receptors (e.g., gamma amino butyric acid, acetylcholine, opioid), and biosynthetic enzymes (e.g., cytochrome P450, aromatase, metabolic enzymes) are present in several vertebrate and nonvertebrate species and that the conservation of the ortholog to the human target decreased with the evolutionary distance, as seen in comparison of estrogen receptor alpha sequences across multiple species (Figure 7-4).[28,29]

TABLE 7-1 Tools for Extrapolation of Nonmammalian Species to Mammalian Species

Tool	Description
COMPARE[99]	COMPARE provides access to a wide array of information including genomic structure, expression data, annotations, pathways, and literature links for human and three widely studied animal models (zebrafish, *Drosophila,* and mouse). A consensus ortholog-finding pipeline combining several ortholog prediction methods allows accurate comparisons of data across species and has been utilized to transfer information from well-studied organisms to more poorly annotated ones. (http://compare.ibdml.univ-mrs.fr)
DIOPT and DIOPT-DIST[100]	Mapping of orthologous genes across species serves an important role in functional genomics by allowing researchers to develop hypotheses about gene functions in one species based on what is known about the functions of orthologs in other species. The Integrative Ortholog Prediction Tool facilitates rapid identification of orthologs among human, mouse, zebrafish, *Caenorhabditis elegans*, *Drosophila melanogaster*, and *Saccharomyces cerevisiae*, combining data from several ortholog and network interaction prediction approaches (Compara, Homologene, Inparanoid, Isobase, OMA, orthoMCL, Phylome, RoundUp, and TreeFam). (http://www.flyrnai.org/DRSC-OPT.html)
eggNOG[101]	eggNOG (evolutionary genealogy of genes: Nonsupervised Orthologous Groups) is a database of orthologous groups of genes. The orthologous groups are annotated with functional description lines (derived by identifying a common denominator for the genes based on their various annotations), with functional categories (i.e., derived from the original COG/KOG categories). eggNOG's database currently counts 721,801 orthologous groups in 1,133 species.
InParanoid7[102]	InParanoid7 provides a database of eukaryotic orthologs and in-paralogs from 100 species and tools to explore them. (http://inparanoid.sbc.su.se/cgi-bin/index.cgi)
OrthoInspector[103]	OrthoInspector features algorithms for detection of orthologs and paralogs. The data set includes 940,855 proteins from 59 eukaryotic species. (http://lbgi.igbmc.fr/orthoinspector/)
QuartetS-DB[104]	QuartetS-DB provides a database of orthologs and in-paralogs from 1,365 bacterial, 92 archaeal, and 164 eukaryotic species genomes. (https://applications.bioanalysis.org/quartetsdb)
Reactome[105]	Reactome is a collaborative, open source, bioinformatics database of human pathways and reactions with the capability to perform orthology-based inferences of pathways in 20 nonhuman species. (http://www.reactome.org)
Signalog[106]	This site combines orthology data from InParanoid and signaling pathway information from the SignaLink database to signaling components. The site contains an online tool and an interactive orthology network viewer for prediction and visualization of components of orthologous pathways focused on the nematode *Caenorhabditis elegans*, the fruit fly *Drosophila melanogaster*, and humans. (http://netbioldev.geobio.elte.hu/signalink_ortholog/)
WikiPathways[107]	WikiPathways is a public wiki for pathway curation.[101] New features include a zoomable pathway viewer and support for pathway ontology annotations. WikiPathways contains over 1,600 pathways and supports 21 species, including vertebrates, several plant species, bacteria, and model organisms such as *Caenorhabditis elegans*, yeast, and fruit fly. (http://www.wikipathways.org)

Conservation of genes is seen throughout biology as evidenced by the identification of dynamic expression patterns that are highly conserved across species ranging from bacteria to human.[30] Chen and Burgoyne[31] used an integrated analysis to identify shared regulators involved in neurodegeneration in various disease models in yeast, fly and *C. elegans*. Their analysis identified 950 genetic regulators, of which 624 were separate genes with orthologs in *C. elegans*. Furthermore, 34 of these genes had human orthologs and were found to overlap across all species. Of the common genetic regulators, some were involved in expected pathways, whereas others suggested novel targets for neurodegenerative disease therapy. Some of these invertebrate species, such as *C. elegans*, have been crucial for elucidating conserved pathways and processes relevant to human physiology and disease. Generally, evolutionary relationships inferred from sequence analysis are used to predict orthologs.[32] These genes, which share a conserved function between species, usually evolve from a common ancestor and have high sequence similarity. Indeed, many of the conserved functions for *C. elegans*–human orthologs have been verified experimentally. The identification of orthologs is crucial when we are trying to compare species. For instance, Shaye et al.[32] used several orthology prediction methods to infer a high confidence list of *C. elegans*–human orthologs, revealing more *C. elegans* orthologs, such as transcription factors, than previously described. Identification of conserved genes through ortholog analysis is a fundamental tool for annotating genes. It also permits identification of appropriate conserved elements from experimentally tractable animals or nonanimal systems like *C. elegans* that are necessary for future translation to potential effects on mammals.

ANNOTATION OF LARGE GENE SETS

Approaches such as transcriptomics, proteomics, and metabolomics can provide a genome-wide snapshot of changes in gene expression, protein abundance, and metabolic states in an animal. Transcriptomics, analysis of all the RNA expressed in a cell or animal, in particular has been useful in generating data for understanding how chemicals affect all species (vertebrate, nonvertebrate, eukaryote, or noneukaryote). In part, this has been due to the fact that DNA sequencing has matured from an expensive endeavor capable of generating a few thousand cDNA sequences to routine sequencing of RNA (via cDNA) libraries generating millions of sequences. As a result, it has become relatively routine to obtain good coverage of a transcriptome from which a custom microarray can be designed, and many microarrays are available for many alternative vertebrate and nonvertebrate species used (e.g., medaka, fathead minnow, zebrafish, earthworm, *Daphnia pulex*, *Daphnia magna*, *C. elegans*). Newer developments in whole transcriptome shotgun sequencing, or RNAseq, promise to replace microarrays in the future as per sample costs drop.

Different platforms—microarray,[33] bead-based,[34] or RNAseq[35]—can be used to detect what transcripts are being expressed in a cell or animal. However, regardless of platform, each gene probe on an array or found by RNAseq must be annotated in order to attach some functional significance to each gene. This step is also critical for identifying orthologs of genes in an alternative animal that are present in mammals and subsequent extrapolation of effects to mammals. For example, we have developed or co-developed microarrays for bobwhite quail,[36] zebra finch, Japanese quail, Western fence lizard (Gust et al., in preparation), earthworms,[37] fathead minnow,[38] largemouth bass,[39] and *Daphnia magna*[40] and used them to analyze chemical effects based on the flow chart in Figure 7-4.

Differential gene/protein expression studies produce lists of probes/genes/proteins as output. To know which pathways or biological functions are affected, one needs to get appropriate annotations for the probes on an array or gene set. This can be done using public databases or through mining literature. A simple approach to annotate sequences is to use the Basic Local Alignment Search Tool (BLAST)[41] to find sequence homologies to sequences of interest in another database. Additional information such as Entrez, GeneId, RefSeq, Gene Ontology (GO), and Gene Symbols is collected that can include the mammalian (rat, mouse, human) gene symbols and is attached to the gene information with BLAST. Specialized biological databases and tools such as OrthoDB and others in Table 7-1 can also be used to find and annotate genes of interest with ortholog genes in human or other species. This information can then be used to infer the biological functions of a gene.

GO is a structured vocabulary for describing biological processes, molecular functions, and cellular components of gene products.[42,43] Standardized terms in the Gene Ontology describe roles of genes and gene products in any organism. Gene products can have one or more molecular functions, can be associated with one or more cellular components, and can be used in one or more biological processes.[42] Gene Ontology itself does not contain gene products of any organism; rather, biologists annotate the roles of gene products using GO. Exploring Gene Ontology annotations is a common and widespread practice to get first insights into the potential biological meaning of the experiment in structured and controlled classifications. GO classification can be used to gain biological insights from a set of identified genes of interest to determine which GO terms or annotations are overrepresented among the genes in the set. Databases and tools such as DAVID (the database for annotation, visualization, and integrated discovery; http://david.abcc.ncifcrf.gov/),[44,45] Pantherdb (http://www.pantherdb.org/), or BiNGO (http://www.psb.ugent.be/cbd/papers/BiNGO/)[46] can be used to retrieve the relevant GO annotations. These tools test the significance of all GO labels present in the test set out of the reference set. Gene Ontology contains more than 7,000 terms for molecular functions and almost 5,000 biological processes arranged in 12-level-deep hierarchies. Although this level of detail provides a very rich vocabulary for functional annotations of gene products, a simpler ontology would benefit the scientific applications. Tools such as Panther, DAVID, and BinGO test for the significance of all GO labels in the data set to control the type I (false positive) rate. For each molecular function, biological process, cellular component, protein classes, or pathways in these databases, the genes associated with those terms are evaluated according to the likelihood that their numerical values were drawn randomly from the overall distribution of values. The data sets can then be mapped and viewed in many ways such as a pie or bar chart of functions, processes, or pathways.

PATHWAY-LEVEL COMPARISON/TRANSLATION

Analysis of pathways impacted in different species has generally been done through statistical analysis of pathway enrichment using gene annotation such as GO terms or ortholog gene names and symbols in mammalian species (see recent reviews in references 47–51).

When nonmammalian genes are annotated with their orthologous gene names in rat, mouse, or human, these genes can then be examined in the context of freely available mammalian pathway databases that include search tools such as Kyoto Encyclopedia of Genes and Genomes (KEGG; www.genome.jp/kegg/), Pantherdb, or WikiPathways (www.wikipathways.org), to name a few. Several software packages are available for mapping genes onto pathways and assessing enrichment, including ArrayTrack,[52] DAVID, GenMAPP,[53] Ingenuity Pathway Analysis (Ingenuity Systems, Redwood City, CA), and MetaCore (GeneGo, Carlsbad, CA). Updated and relatively comprehensive lists of databases and tools for pathway analysis can be found at http://www.pathguide.org and at http://www.pathwaycommons.org.[54]

PATHWAY-BASED EXTRAPOLATION TO MAMMALS IN DETERMINING CHEMICAL MODE OF ACTION

Perhaps the most straightforward application of computational extrapolation is in determining potential modes of action underlying toxicity. Identification of impacted pathways in a nonmammalian species can be used to hypothesize modes of action that can then be confirmed through experimentation and conservation of responses across species. An example of how cross-species extrapolation using pathways can help identify how a chemical might cause toxicity in addition to identifying potential adverse effects is provided by recent investigations into toxic effects of nitrotoluenes, chemicals used in energetics, propellants, dyes, and plasticizers. Many of these chemicals are produced in mixed formulations; e.g., technical grade Dinitrotoluene can contain 78% 2,4-dinitrotoluene (24DNT), 19% 2,6-dinitrotoluene (26DNT), and minor amounts of other dinitrotoluene isomers.[55]

Wintz et al.[56] used a 5000 cDNA shotgun microarray to explore the toxicity of 24DNT, a chemical used in dyes, propellants, and plasticizing agents, to fathead minnow. Mapping of fathead differentially expressed genes to 24DNT impacted mitochondrial respiration pathways involving hemoglobin and mitochondrial cytochrome oxidase consistent with methemoglobinemia and its associated effects (anemia, reticulocytosis, and increased number of Heinz bodies) that have been observed as a primary effect of nitrobenzene compounds and 24DNT in mammals.[57] The 24DNT was also observed to cause fatty livers, changes in the type and amounts of hepatic lipids, and downregulation of fatty acid metabolism pathways and peroxisome proliferator-activated receptor α (PPARα) expression, suggesting that 24DNT could also impact energy metabolism in exposed animals.

A second series of investigations in Northern bobwhite quail (*Colinus virginianus*) found that 26DNT caused toxicological effects similar to 24DNT impacting liver and kidney, leading to gastrointestinal distress, dehydration, and a reduction in body mass and feed consumption.[58,59] To understand potential mechanisms underlying these responses, Rawat et al.[60] developed a 15,000-probe oligonucleotide microarray based on massively parallel sequencing of quail cDNAs. To do this, they assembled 467,708 sequence reads of normalized cDNA libraries from multiple quail tissues using CAP3, resulting in 71,384 sequences (35,904 contigs and 35,480 singletons) that were considered putative unique genes.[61] Homologous genes were then identified in the unique gene set using Refseq ID for protein (http://www.ncbi.nlm.nih.gov). Northern bobwhite genes annotated with Refseq IDs were then functionally annotated with GO terms against the closely related *Gallus gallus* annotation using Web Gene Ontology Annotation Plot (WEGO).[62] The annotation derived was used to define second-level GO function categories within each primary GO level (molecular function, cellular component, and biological process). Quail raw sequences with significant BLAST matches, where $E > 10^{-5}$, were further annotated with KEGG pathways using the KEGG Orthology-Based Annotation System web server.[63] Sequences with significant KEGG orthology were compared against *G. gallus* as a reference organism. The complete annotated gene set was then used to construct a microarray containing a total of 8,454 nonredundant sequences with known in-frame orientations and a total of 3,272 unique Refseq IDs where the translation frame was not known. Both sense and antisense probes were developed for genes with an unknown translation frame yielding the 6,546 probe sequences. The custom oligonucleotide array was then used to measure the impact of 26DNT on quail livers. Differentially expressed quail genes were matched to chicken and mammalian pathways where significant impacts were seen on genes and pathways involved in prostaglandin synthesis and regulation, porphyrin and chlorophyll metabolism, generation of precursor metabolites, and energy that were consistent with observations of edema in the gastrointestinal tract, diarrhea and weight loss, and reduction in plasma glucose levels, respectively. The 26DNT downregulated expression of PPARγ, a key regulator of fatty acid storage and glucose metabolism in mammals,[64] and PPAR gamma coactivator 1 (PPARGC1), a key regulator of energy metabolism in mammals,[25,65] and the pathways that they regulate, including fatty acid biosynthesis and gluconeogenesis. Downregulation of PPARγ and PPARGC1 pathways may have led to reduced energy metabolism and loss of weight. Data from both fathead minnow and bobwhite quail support a mode of action where dinitrotoluenes act as antagonists to PPARs, thereby affecting lipid and energy metabolism. Since these pathways are highly conserved, dinitrotoluenes are likely to impact mammals in a similar manner. In fact, Deng et al.[66] showed that PPARα and PPAR/RXR signaling were affected in rat that had been exposed to 24DNT or 26DNT, corroborating the fact that the pathways are highly conserved and dinitrotoluenes can have similar adverse effects on phylogenetically distant species.

PATHWAY-BASED DOSE-RESPONSE RELATIONSHIPS

Pathway-based dose-response effects can be identified once differentially expressed genes are mapped onto canonical pathways in either the nonmammalian test species or a target species (e.g., human). These effects can then lead to generating hypotheses identifying a chemical mode of action, identifying different sensitivities between species, or, in an assessment context, a pathway-based point of departure and a possible benchmark-dose. Pathways whose perturbation is linked to adverse effects can be used as surrogates for adverse outcomes, as seen in the use of steroidogenesis pathway function as a surrogate for reproductive effects. Linkages of pathways to outcome in AOPs then permit the use of pathways in a dose-response-based assessment risk framework. Benchmark dose (BMD) methods or no observed adverse effect levels (NOAELs) are used to calculate a point of departure (POD) of a biomarker or phenotype in an exposed animal group from that of a normal animal group. Regulatory agencies then use PODs, modified by uncertainty factors, to set safe levels for chemical exposures.[67]

A no observable transcriptional effect level (NOTEL) has been proposed as an alternative to NOAEL where any transcriptional effect is viewed as significant.[68,69] However, studies have found NOTEL may not be appropriate in all cases, especially where a U-shaped response curve is observed such as was found in gene expression induced by BPA exposure in fathead minnow and zebrafish.[70] Additionally, since these transcriptional effects were not typically based on pathways linked to adverse effects, reference dose values cannot be derived from NOTEL values.

If pathway changes are based upon transcriptional changes in pathways that have been mapped to mammalian orthologs, the POD could be estimated as the lowest dose at which significant pathway enrichment is seen. Dose-response relationships can also be used to determine BMD values, NOAEL, and lowest observed adverse effect levels (LOAEL). Physiological or biochemical NOAEL and LOAEL levels have long been used as a measure of what concentrations of a chemical will have no, or minimal, toxic effect on an animal. Similar criteria can also be used with toxicological pathways within an AOP. Where toxicological pathways are conserved, either as indicated by conservation of function or ortholog genes, impacts on these pathways in alternative species can be translated to mammalian species. For example, Hermsen et al.[71] exposed zebrafish embryos (0 hours post fertilization) to eight different concentrations of flusilazole, a known developmental toxicant, and compared transcriptional effects after 24 hours' exposure to morphological adverse effects after 24 and 72 hours' exposure. Flusilazole caused developmental delays and teratogenic effects such as pericardial edema and malformations of head and heart in the embryos. Microarray analysis identified four functionally enriched clusters that were dose-responsive including retinol metabolism (up), transcription and homeobox genes related to development (up), steroid biosynthesis (up), processes related to lipid metabolism and glycolysis/gluconeogenesis (down; Figure 7-5). Retinoic acid and upregulation of retinol-regulated genes have been demonstrated to result in skeletal deformities during development in zebrafish and mammals.[72-74] Since these pathways have been well characterized and linked to adverse outcomes, the impacts can be readily translated to mammals. A NOAEL could be derived as the lowest dose at which any pathway (here retinol metabolism) changed in response to flusilazole exposure. Since retinol metabolism is highly conserved with mammals, one could reasonably derive a NOAEL and LOAEL from the point at which the gene cluster enriched in retinol functions significantly departed from the control or 1.35 and 2.8 μM flusilazole, respectively (Figure 7-5).

NOAEL and LOAEL, directly based on experimental values, have the disadvantage that they are dramatically affected by the number of doses, spacing between doses, and sample size used in exposures. Benchmark dose (BMD) modeling was introduced to permit a more quantitative determination of the dose-response curve and the point at which chemically treated groups diverge from a control group.[75,76] BMD modeling has been applied to toxicogenomic-based pathway data in order to determine more quantitative hazard value for chemicals[77,78] and to use short *in vivo* exposures for assessing long-term impacts.[79,80] These approaches hold much promise in deriving hazard values from one species and translating it to another based on changes in pathways.

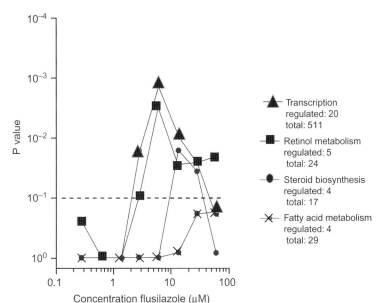

FIGURE 7-5 Pathway-based dose response of zebrafish embryos exposed to flusilazole. The level of significance of the change of expression from control is depicted by the dotted line. *Adapted from reference 71.*

A somewhat similar approach used by Judson et al.[81] applied pathway-based dose-response measurements to develop screening hazard levels through a combination of pathway point of departure (characterized as the pathway altering dose), reverse toxicokinetics, and modeling to incorporate uncertainty values and population variability. This framework has significant potential for translating pathway-level dose-response effects from alternative species, or nonanimal models like the zebrafish embryo test, to mammals—especially since mammalian-specific parameters such as toxicokinetics and toxicodynamics can be incorporated to make alternative species data more relevant to mammalian effects. Judson et al.[81] developed a framework in which dose-response data from human-oriented HTS assays organized in a pathway context can be used to derive a pathway-based NOAEL-like value for chemicals. First, a biological pathway altering concentration (BPAC) was defined in a simple context as the lowest chemical concentration at which an assay in the pathway has 50% of maximal activity. The *in vitro* BPAC value is then related to an *in vivo* dose using pharmacokinetic modeling to estimate the external dose required to achieve the internal dose of the BPAC. Judson et al.[81] used a simple reverse toxicokinetics approach suitable for chemicals mainly eliminated through metabolism and renal excretion to obtain a concentration to dose scaling factor in μM/(mg/kg/day).[41] The advantage of this approach is that simple assays exist for the rate of disappearance of parent chemical via hepatic metabolism and the fraction of chemical bound (or conversely unbound) to plasma proteins and allows one to tailor these parameters to specific species, e.g., human. The BPAC is converted to a biological altering pathway dose (BAPD), or the steady-state dose needed for a steady-state BPAC, by dividing the BPAC by the dose-scaling factor. The lower limit of the BAPD for a pathway would then be analogous to NOAEL. If one uses a pathway in an AOP framework, then the pathway is directly linked to an adverse outcome and can be used as a NOAEL-related reference dose for hazard assessments. Using either the dose at which a pathway effect is first seen for BPAC values or pathway-based BMD modeling to obtain a BMD lower limit, this general framework would permit conversion of pathway-based impacts on retinol pathways in zebrafish embryos to a more quantitatively accurate NOAEL-like reference dose in mg/kg/day. Such measures are likely to be more valuable in prioritizing chemical hazards and in a safety and risk framework both for mammals and for ecological health.

NETWORK INFERENCE AND MAPPING

Network analysis of gene expression and pathway data can enable integration of multiple data sets and help identify novel genes or functions in large data sets. A network can be constructed in two principal ways: using prior evidence (e.g., literature and manual or automatic curation) or *de novo* using statistical methods and experimental data. A number of recent reviews describe how one can construct and use biological networks.[82–86] Network inference methods can also be useful in identifying potential mechanisms underlying toxicity of chemicals. Perkins et al.[82] demonstrated that a transcriptional network analysis of plasma hormone levels and ovary gene expression data from fathead minnow exposed to different endocrine-disrupting chemicals was useful in identifying potential mechanisms underlying the effects of flutamide, an androgen receptor agonist. The potential to integrate a wide range of data types of interest to risk assessment was demonstrated by Williams et al.[87] where network inference was used to derive statistically significant relationships between different populations of fish collected from chemically contaminated sites and the sediment chemistry, microsatellite data, pathogen data, phenotypic data, gene expression data, and metabolomics data.

CROSS-SPECIES ANALYSIS USING NETWORKS

When comparing microarray data sets across multiple species, researchers face both technological and contextual challenges in comparing expression of divergent genomes. However, successful meta-analysis across species can be used to leverage information in one species with that of less-studied species, as well as to find common expression patterns that would reveal core gene functions. Analysis across species can be divided into two types (reviewed in reference 27): expression and co-expression meta-analyses. Expression meta-analysis studies the similarity between expression profiles of homologous genes in different species, while co-expression meta-analysis searches for conserved co-expressed gene clusters across species. One of the advantages of co-expression meta-analysis is that it allows the use of different conditions for the different species studied. Direct cross-species comparison of differentially expressed genes is particularly challenging when trying to compare distant species. Instead, it can be useful to look at enriched GO terms that are conserved across species.[27,47–51] Subramanian et al.[88] proposed a more sophisticated approach called gene set enrichment analysis (GSEA) to extract biological insight from expression data. GSEA focuses on groups of genes that share common biological function, chromosomal location, or regulation. This analysis can be extended to gene network enrichment analysis (GNEA), based upon the idea that the cell is associated with a protein-protein interaction network and each protein belongs to one or more gene sets associated with biological processes or molecular functions. When the cell is perturbed, some subset of the interaction network becomes affected; therefore, certain functional subnetworks may show significantly altered activity.[89] Other researchers have used network analysis to identify conserved genetic modules and conserved patterns of protein interactions.[90]

An example of how comparative toxicology and genomics can be used to assess conservation of adverse effects on different species can be found in Garcia-Reyero et al.[91] They used a genomics approach to compare and contrast the neurotoxic effects of RDX, a munitions constituent, among five phylogenetically disparate species: rat (Sprague Dawley), fathead minnow, earthworm (*Eisenia fetida*), Northern bobwhite quail, and coral (*Acropora formosa*). Their results showed that RDX accumulated into the brain of rat, Northern bobwhite quail, and fathead minnows and impacted neuronal function in rat, Northern bobwhite quail, and earthworm, but apparently not in fathead minnows. The comparison of Gene Ontology terms indicated several biological processes affected by RDX in all species, such as impacts on calcium signaling (involved in seizure response), effects on xenobiotic metabolism, electron transport, and cell signaling pathways. A transcriptional subnetwork focusing on neurotransmission pathways was created from the rat data set to complement the functional analysis describing the most common effects of RDX exposure among species (Figure 7-6). This was used for among-species comparisons to group differentially expressed genes on to common functionally enriched clusters on the rat transcriptional network.

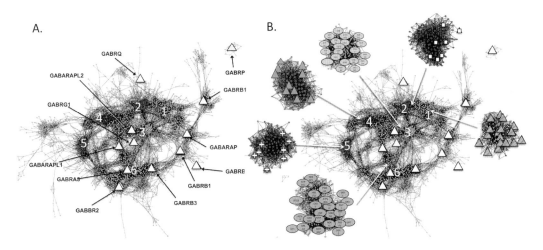

FIGURE 7-6 Mapping of genes differentially expressed in fathead minnow brain to Rat GABA$_A$R neurotransmission transcriptional network. A. Rat neurotransmission transcriptional network created filtering the source network using a list of genes related to neurotransmission and their first connected neighbors. The six most highly connected clusters are numbered 1–6. GABA$_A$R-related genes are highlighted with open triangles and gene names. First and second neighbor genes of GABA$_A$R-related genes are highlighted with dots. B. Genes differentially expressed in fathead minnow brain after exposure to RDX mapped to orthologous genes present in the six highly connected clusters connected to the Rat GABA$_A$R neurotransmission transcriptional network. *Adapted from reference 91.*

Highly connected clusters of genes in the rat neurotransmission network were enriched in pathways related to Parkinson's, Alzheimer's, and Huntington's disease; oxidative phosphorylation or glycolysis/gluconeogenesis, and others, in concordance with the results from the functional analysis (Figure 7-6). A significant number of differentially expressed genes for each species were mapped to the subnetwork, suggesting that many functional responses to RDX exposure were conserved from rat to coral. Overall, the meta-analysis using genomic data of the effects of RDX on several species suggested a common and conserved mode of action of the chemical throughout phylogenetically remote organisms.

TRANSLATING EFFECTS THROUGH COMPUTATIONAL MODELING AT THE SYSTEMS LEVEL

Modeling can generally be divided into statistical approaches that relate an experimental observation to an outcome. These approaches can be useful in translating effects on an individual to populations where molecular changes are mechanistically linked, or highly correlated to adverse outcomes such as fecundity can be used to create estimates of population-level impacts. Miller et al.[92] found that levels of the fish egg protein vitellogenin are highly correlated with fecundity and used this to predict the effects of endocrine-disrupting chemicals on the fathead minnow population. Field experiments exposing lake populations of fish to ethinylestradiol have validated that EDC perturbation of endocrine pathways does indeed crash populations.[93] Statistical models have been developed relating zebrafish embryos' toxicity to acute toxicity in adult fish[8] and to chemical toxicity in rats.[7]

More refined efforts use mechanistic models that simulate the key events of an AOP. For example, the inhibition of aromatase by fadrozole and its impact on estradiol has been well characterized.[94] Mechanistic simulations that model the HPG axis and steroidogenesis would be a useful tool in extrapolating effects from HTS assay data or short-term *in vivo* exposure results.[95,96] Such models are important for incorporating feedback and compensatory mechanisms that modulate the toxicity of chemicals. Efforts in dynamic energy budget modeling provide useful tools to extrapolate *in vitro* and pathway-level effects to population-level effects. Pathways affecting energy, growth, and function are incorporated into the dynamic energy budget and permit incorporation of chemical impacts in these areas.[97,98] Since the pathways underlying these models are highly conserved, species-specific data on sensitivities gained by HTS assays or testing can be used to adapt the models to a more relevant context.

FUTURE EFFORTS IN USE OF HIGH-THROUGHPUT SCREENING AND "OMICS" TECHNOLOGY AND COMPUTATIONAL TOOLS IN TRANSLATION OF NONMAMMALIAN SPECIES TO MAMMALIAN SPECIES TO MEET REACH AND NEXT GENERATION RISK ASSESSMENT NEEDS

The potential to translate alternative species and nonanimal models to relevant hazard values for mammals has greatly increased as our knowledge on gene conservation has increased. Recently, efforts in toxicology to focus on a pathway-based paradigm rather than a species-based focus have resulted in frameworks such as the adverse outcome pathway that facilitates linkage of molecular initiating events to pathways and to subsequent adverse effects relevant to risk assessment. This linkage is critical for identifying the potential to cause an adverse effect based on the impact or perturbation of a pathway. Once this is established, effects on pathways seen in alternative species and nonanimal models can be translated to mammalian species. Currently, pathway-based translation of effects is most robust due to the fact that pathways consist of a relatively small number of members; however, network-based approaches have demonstrated value in integrating diverse data and in understanding potential chemical modes of action. Development of a network-based point of departure is currently hampered by the lack of understanding of what a "normal" network is, since in reality biological networks are dynamic, constantly forming, and disassembling. Future studies on formation of stable networks or "attractor states" may prove useful in establishing a baseline from which a point of departure could be established. Ultimately, all of the information developed in screening, pathway analysis, and network analysis can be used in statistical models and mechanistic models that permit incorporation of species-specific parameters for chemical absorption, distribution, metabolism, excretion, and differential sensitivity. If AOPs can be created for significant classes of adverse effects, then this library may enable development of more sophisticated models. These models can be extended to understand how populations may respond to chemical exposure through agent-based models or in populations of models parameterized with different sensitivities. An even more realistic exposure-response modeling could include spatially explicit agent-based modeling to simulate environmental exposure routes and realistic responses.

References

[1] Dix DJ, Houck KA, Martin MT, Richard AM, Setzer RW, Kavlock RJ. The ToxCast program for prioritizing toxicity testing of environmental chemicals. Toxicol Sci 2006;95:5—12.
[2] Cote I, Anastas PT, Birnbaum LS, Clark RM, Dix DJ, Edwards SW, et al. Advancing the next generation of health risk assessment. Environ Health Perspect 2012;Aug 8 [Epub ahead of print].
[3] Rovida C, Longo F, Rabbit RR. How are reproductive toxicity and developmental toxicity addressed in REACH dossiers? ALTEX 2011;28:273—94.
[4] Schiffelers M-JWA, Blaauboer BJ, Hendriksen CFM, Bakker WE. Regulatory acceptance and use of 3R models: a multilevel perspective. ALTEX 2012;29:287—300.
[5] European Parliament. Directive 2010/63/EU of the European Parliament and of the council of 22 September 2010 on the protection of animals used for scientific purpose. Official J European Union L. 2010; 33—79.
[6] Embry MR, Belanger SE, Braunbeck TA, Galay-Burgos M, Halder M, Hinton DE, et al. The fish embryo toxicity test as an animal alternative method in hazard and risk assessment and scientific research. Aquat Toxicol 2010;97(2):79—87.
[7] Ali S, van Mil HG, Richardson MK. Large-scale assessment of the zebrafish embryo as a possible predictive model in toxicity testing. PLoS ONE 2011;6:e21076.
[8] Knöbel M, Busser FJM, Rico Rico A, Kramer NI, Hermens JLM, Hafner C, et al. Predicting adult fish acute lethality with the zebrafish embryo: relevance of test duration, endpoints, compound properties, and exposure concentration analysis. Environ Sci Technol 2012;46:9690—700.
[9] Judson RS, Houck KA, Kavlock RJ, Knudsen TB, Martin MT, Mortensen HM, et al. *In vitro* screening of environmental chemicals for targeted testing prioritization: the ToxCast project. Environ Health Perspect 2010;118:485—92.

[10] Wetmore BA, Wambaugh JF, Ferguson SS, Sochaski MA, Rotroff DM, Freeman K, et al. Integration of dosimetry, exposure, and high-throughput screening data in chemical toxicity assessment. Toxicol Sci 2011;125:157−74.

[11] Rotroff DM, Wetmore BA, Dix DJ, Ferguson SS, Clewell HJ, Houck KA, et al. Incorporating human dosimetry and exposure into high-throughput *in vitro* toxicity screening. Toxicol Sci 2010;117:348−58.

[12] Villeneuve DL, Garcia-Reyero N. Vision and strategy: predictive ecotoxicology in the 21st century. Environ Toxicol Chem 2010;30:1−8.

[13] Edwards SW, Preston RJ. Systems biology and mode of action based risk assessment. Toxicol Sci 2008;106:312−8.

[14] Vavouri T, Walter K, Gilks WR, Lehner B, Elgar G. Parallel evolution of conserved non-coding elements that target a common set of developmental regulatory genes from worms to humans. Genome Biol 2007;8:R15.

[15] Pan D. The hippo signaling pathway in development and cancer. Dev Cell 2010;19:491−505.

[16] Emoto K. The growing role of the Hippo-NDR kinase signalling in neuronal development and disease. J Biochem 2011;150:133−41.

[17] Vallee M, Aiba K, Piao Y, Palin MF, Ko MSH, Sirard MA. Comparative analysis of oocyte transcript profiles reveals a high degree of conservation among species. Reprod 2008;135:439−48.

[18] Ankley GT, Bennett RS, Erickson RJ, Hoff DJ, Hornung MW, Johnson RD, et al. Adverse outcome pathways: a conceptual framework to support ecotoxicology research and risk assessment. Environ Toxicol Chem 2010;29:730−41.

[19] Wilson VS, Cardon MC, Thornton J, Korte JJ, Ankley GT, Welch J, et al. Cloning and *in vitro* expression and characterization of the androgen receptor and isolation of estrogen receptor alpha from the fathead minnow (*Pimephales promelas*). Environ Sci Technol 2004;38:6314−21.

[20] Rider CV, Hartig PC, Cardon MC, Lambright CR, Bobseine KL, Guillette Jr LJ, et al. Differences in sensitivity but not selectivity of xenoestrogen binding to alligator versus human estrogen receptor alpha. Environ Toxicol Chem 2010;29:2064−71.

[21] Ankley GT, Grey L. 2013. Cross-species conservation of endocrine pathways: A critical analysis of tier 1 fish and rat screening assays with 12 model chemicals. Environ Toxicol Chem. doi: 10.1002/etc.2151. [Epub ahead of print].

[22] Raldúa D, Babin PJ. Simple, rapid zebrafish larva bioassay for assessing the potential of chemical pollutants and drugs to disrupt thyroid gland function. Environ Sci Technol 2009;43:6844−50.

[23] Thienpont B, Tingaud-Sequeira A, Prats E, Barata C, Babin PJ, Raldúa D. Zebrafish eleutheroembryos provide a suitable vertebrate model for screening chemicals that impair thyroid hormone synthesis. Environ Sci Technol 2011;45:7525−32.

[24] Hill AJ. Zebrafish as a model vertebrate for investigating chemical toxicity. Toxicol Sci 2005;86:6−19.

[25] Burgess-Herbert SL, Euling SY. Use of comparative genomics approaches to characterize interspecies differences in response to environmental chemicals: challenges, opportunities, and research needs. Toxicol Appl Pharmacol 2011; [In press].

[26] Bolser DM, Chibon PY, Palopoli N, Gong S, Jacob D, Angel VDD, et al. MetaBase—the wiki-database of biological databases. Nucleic Acids Res 2011;40:D1250−4.

[27] Lu Y, Huggins P, Bar-Joseph Z. Cross species analysis of microarray expression data. Bioinformatics 2009;25:1476−83.

[28] Kostich MS, Lazorchak JM. Risks to aquatic organisms posed by human pharmaceutical use. Sci Tot Environ 2008;389:329−39.

[29] Gunnarsson L, Jauhiainen A, Kristiansson E, Nerman O, Larsson DGJ. Evolutionary conservation of human drug targets in organisms used for environmental risk assessments. Environ Sci Technol 2008;42:5807−13.

[30] Ueda HR, Hayashi S, Matsuyama S, Yomo T, Hashimoto S, Kay SA, et al. Universality and flexibility in gene expression from bacteria to human. Proc Natl Acad Sci USA 2004;101:3765−9.

[31] Chen X, Burgoyne RD. Identification of common genetic modifiers of neurodegenerative diseases from an integrative analysis of diverse genetic screens in model organisms. BMC Genomics 2012;13:71.

[32] Shaye DD, Greenwald I. OrthoList: a compendium of *C. elegans* genes with human orthologs. PLoS ONE 2011;6:e20085.

[33] Brown PO, Botstein D. Exploring the new world of the genome with DNA microarrays. Nat Genet 1999;21:33−7.

REFERENCES

[34] Fan J-B, Gunderson KL, Bibikova M, Yeakley JM, Chen J, Wickham Garcia E, et al. Illumina universal bead arrays. Meth Enzymol 2006;410:57—73.

[35] Wang Z, Gerstein M, Snyder M. RNA-Seq: a revolutionary tool for transcriptomics. Nat Rev Genet 2009;10:57—63.

[36] Gust KA, Pirooznia M, Quinn MJ, Johnson MS, Escalon L, Indest KJ, et al. Neurotoxicogenomic investigations to assess mechanisms of action of the munitions constituents RDX and 2,6-DNT in Northern bobwhite (*Colinus virginianus*). Toxicol Sci 2009;110:168—80.

[37] Gong P, Guan X, Inouye LS, Pirooznia M, Indest KJ, Athow RS, et al. Toxicogenomic analysis provides new insights into molecular mechanisms of the sublethal toxicity of 2,4,6-trinitrotoluene in *Eisenia fetida*. Environ Sci Technol 2007;41:8195—202.

[38] Denslow ND, Garcia-Reyero NL, Barber DS. Fish 'n' chips: the use of microarrays for aquatic toxicology. Mol BioSyst 2007;3:172.

[39] Garcia-Reyero N, Griffitt RJ, Liu L, Kroll KJ, Farmerie WG, Barber DS, et al. Construction of a robust microarray from a non-model species (largemouth bass) using pyrosequencing technology. J Fish Biol 2008;72:2354—76.

[40] Garcia-Reyero N, Escalon BL, Loh PR, Laird JG, Kennedy AJ, Berger B, et al. Assessment of chemical mixtures and groundwater effects on *Daphnia magna* transcriptomics. Environ Sci Technol 2012;46:42—50.

[41] Ye J, McGinnis S, Madden TL. BLAST: improvements for better sequence analysis. Nucleic Acids Res 2006;34:W6—9.

[42] Ashburner M, Ball CA, Blake JA, Botstein D, Butler H, Cherry JM, et al. Gene ontology: tool for the unification of biology. The gene ontology consortium. Nat Genet 2000;25:25—9.

[43] Arbeitman MN, Furlong E, Imam F, Johnson E. Gene expression during the life cycle of *Drosophila melanogaster*. Science 2002;297:2270—5.

[44] Dennis G, Sherman BT, Hosack DA, Yang J, Gao W, Lane HC, et al. DAVID: Database for annotation, visualization, and integrated discovery. Genome Biol 2003;4:P3.

[45] Da Wei Huang BTS, Lempicki RA. Systematic and integrative analysis of large gene lists using DAVID bioinformatics resources. Nat Protoc 2008;4:44—57.

[46] Maere S, Heymans K, Kuiper M. BiNGO: a cytoscape plugin to assess overrepresentation of gene ontology categories in biological networks. Bioinformatics 2005;21:3448—9.

[47] Li Y, Agarwal P, Rajagopalan D. A global pathway crosstalk network. Bioinformatics 2008;24:1442—7.

[48] Viswanathan GA, Seto J, Patil S, Nudelman G, Sealfon SC. Getting started in biological pathway construction and analysis. PLoS Comp Biol 2008;4:e16.

[49] Werner T. Bioinformatics applications for pathway analysis of microarray data. Curr Opin Biotechnol 2008;19:50—4.

[50] Bauer-Mehren A, Furlong LI, Sanz F. Pathway databases and tools for their exploitation: benefits, current limitations and challenges. Mol Syst Biol 2009;5:290—303.

[51] Karp PD, Paley SM, Krummenacker M, Latendresse M, Dale JM, Lee TJ, et al. Pathway Tools version 13.0: integrated software for pathway/genome informatics and systems biology. Brief Bioinformatics 2010;11:40—79.

[52] Tong W, Cao X, Harris S, Sun H, Fang H, Fuscoe J, et al. Arraytrack—supporting toxicogenomic research at the U.S. Food and Drug Administration National Center for Toxicological Research. Environ Health Perspect 2002;110:1041—6.

[53] Salomonis N, Hanspers K, Zambon AC, Vranizan K, Lawlor SC, Dahlquist KD, et al. GenMAPP 2: new features and resources for pathway analysis. BMC Bioinformatics 2007;8:217.

[54] Cerami EG, Gross BE, Demir E, Rodchenkov I, Babur O, Anwar N, et al. Pathway commons, a web resource for biological pathway data. Nucleic Acids Res 2010;39:D685—90.

[55] Dunlap KL. Nitrobenzene and nitrotoluenes. In: Grayson M, Eckroth D, editors. Kirk-Othmer encyclopedia of chemical technology. 3rd ed. New York, NY: John Wiley and Sons; 1978. p. 930—1.

[56] Wintz H, Yoo LJ, Loguinov A, Wu YY, Steevens JA, Holland RD, et al. Gene expression profiles in fathead minnow exposed to 2,4-DNT: correlation with toxicity in mammals. Toxicol Sci 2006;94:71—82.

[57] ATSDR, Agency for Toxic Substances and Disease Registry. Toxicological profile for 2,4- and 2,6-dinitrotoluene. Atlanta, Georgia: Agency for Toxic Substances and Disease Registry, Division of Toxicology/Toxicology Information Branch; 1998.

[58] Quinn MJ, Bazar MA, McFarland CA, Perkins EJ, Gust KA, Gogal RM, et al. Effects of subchronic exposure to 2,6-dinitrotoluene in the northern bobwhite (Colinus virginianus). Environ Toxicol Chem 2007;26:2202–7.

[59] Johnson MS, Quinn MJ, Bazar MA, Gust KA, Escalon BL, Perkins EJ. Subacute toxicity of oral 2,6-dinitrotoluene and 1,3,5-trinitro-1,3,5-triazine (RDX) exposure to the northern bobwhite (Colinus virginianus). Environ Toxicol Chem 2009;26:1481–7.

[60] Rawat A, Gust KA, Deng Y, Garcia-Reyero N, Quinn MJ, Johnson MS, et al. From raw materials to validated system: the construction of a genomic library and microarray to interpret systemic perturbations in Northern bobwhite. Physiolog Genome 2010;42:219–35.

[61] Huang X, Madan A. CAP3: a DNA sequence assembly program. Genome Res 1999;9:868–77.

[62] Ye J, Fang L, Zheng H, Zhang Y, Chen J, Zhang Z, et al. WEGO: a web tool for plotting GO annotations. Nucleic Acids Res 2006;34:W293–7.

[63] Wu J, Mao X, Cai T, Luo J, Wei L. KOBAS server: a web-based platform for automated annotation and pathway identification. Nucleic Acids Res 2006;34:W720–4.

[64] Jones JR, Barrick C, Kim KA, Lindner J, Blondeau B, Fujimoto Y, et al. Deletion of PPAR-γ in adipose tissues of mice protects against high fat diet-induced obesity and insulin resistance. Proc Natl Acad Sci USA 2005;102:6207–12.

[65] Liang H, Ward WF. PGC-1: a key regulator of energy metabolism. Adv Physiol Ed 2006;30:145–51.

[66] Deng Y, Meyer SA, Guan X, Escalon BL, Ai J, Wilbanks MS, et al. Analysis of common and specific mechanisms of liver function affected by nitrotoluene compounds. PLoS ONE 2011;6:e14662.

[67] Gaylor DW, Aylward LL. An evaluation of benchmark dose methodology for non-cancer continuous-data health effects in animals due to exposures to dioxin (TCDD). Regul Toxicol Pharmacol 2004;40:9–17.

[68] Ankley GT, Daston GP, Degitz SJ, Denslow ND, Hoke RA, Kennedy SW, et al. Toxicogenomics in regulatory ecotoxicology. Environ Sci Technol 2006;40:4055–65.

[69] Poynton HC, Loguinov AV, Varshavsky JR, Chan S, Perkins EJ, Vulpe CD. Gene expression profiling in Daphnia magna part I: concentration-dependent profiles provide support for the no observed transcriptional effect level. Environ Sci Technol 2008;42:6250–6.

[70] Villeneuve DL, Garcia-Reyero N, Escalon BL, Jensen KM, Cavallin JE, Makynen EA, et al. Ecotoxicogenomics to support ecological risk assessment: a case study with bisphenol a in fish. Environ Sci Technol 2012;46:51–9.

[71] Hermsen SAB, Pronk TE, van den Brandhof EJ, van der Ven LTM, Piersma AH. Concentration-response analysis of differential gene expression in the zebrafish embryotoxicity test following flusilazole exposure. Toxicol Sci 2012;127:303–12.

[72] Yamaguchi M, Nakamoto M, Honda H, Nakagawa T, Fujita H, Nakamura T, et al. Retardation of skeletal development and cervical abnormalities in transgenic mice expressing a dominant-negative retinoic acid receptor in chondrogenic cells. Proc Natl Acad Sci USA 1998;95:7491–6.

[73] Bohnsack BL, Kasprick DS, Kish PE, Goldman D, Kahana A. A zebrafish model of Axenfeld–Rieger Syndrome reveals that pitx2 regulation by retinoic acid is essential for ocular and craniofacial development. Investig Ophthal Vis Sci 2012;53:7–22.

[74] Laue K, Pogoda H-M, Daniel PB, van Haeringen A, Alanay Y, Ameln von S, et al. Craniosynostosis and multiple skeletal anomalies in humans and zebrafish result from a defect in the localized degradation of retinoic acid. Am J Human Gen 2011;89:595–606.

[75] Crump KS. A new method for determining allowable daily intakes. Fundam Appl Toxicol 1984;4:854–71.

[76] Crump KS. Calculation of benchmark doses from continuous data. Risk Anal 1995;15:79–89.

[77] Burgoon LD, Zacharewski TR. Automated quantitative dose-response modeling and point of departure determination for large toxicogenomic and high-throughput screening data sets. Toxicol Sci 2008;104:412–8.

[78] Thomas RS, Allen BC, Nong A, Yang L, Bermudez E, Clewell HJ, et al. A method to integrate benchmark dose estimates with genomic data to assess the functional effects of chemical exposure. Toxicol Sci 2007;98:240–8.

[79] Thomas RS, Clewell HJ, Allen BC, Wesselkamper SC, Wang NCY, Lambert JC, et al. Application of transcriptional benchmark dose values in quantitative cancer and noncancer risk assessment. Toxicol Sci 2011;120:194–205.

[80] Thomas RS, Black MB, Li L, Healy E, Chu T-M, Bao W, et al. A comprehensive statistical analysis of predicting in vivo hazard using high-throughput in vitro screening. Toxicol Sci 2012;128:398–417.

[81] Judson RS, Kavlock RJ, Setzer RW, Cohen Hubal EA, Martin MT, Knudsen TB, et al. Estimating toxicity-related biological pathway altering doses for high-throughput chemical risk assessment. Chem Res Toxicol 2011;24:451–62.

[82] Perkins EJ, Chipman JK, Edwards SW, Habib T, Falciani F, Taylor R, et al. Reverse engineering adverse outcome pathways. Environ Toxicol Chem 2010;30:22–38.
[83] Emmert-Streib F, Glazko GV, Altay G, de Matos Simoes R. Statistical inference and reverse engineering of gene regulatory networks from observational expression data. Front Genet 2012;3:8.
[84] Ma'ayan A. Introduction to network analysis in systems biology. Sci Signal 2011;4 [tr5].
[85] Madhamshettiwar PB, Maetschke SR, Davis MJ, Reverter A, Ragan MA. Gene regulatory network inference: evaluation and application to ovarian cancer allows the prioritization of drug targets. Genome Med 2012;4:41.
[86] Hecker M, Lambeck S, Toepfer S, van Someren E, Guthke R. Gene regulatory network inference: data integration in dynamic models—a review. Biosystems 2009;96:86–103.
[87] Williams TD, Turan N, Diab AM, Wu H, Mackenzie C, Bartie KL, et al. Towards a system level understanding of non-model organisms sampled from the environment: a network biology approach. PLoS Comp Biol 2011;7:e1002126.
[88] Subramanian A, Tamayo P, Mootha VK, Mukherjee S, Ebert BL, Gillette MA, et al. Gene set enrichment analysis: a knowledge-based approach for interpreting genome-wide expression profiles. Proc Natl Acad Sci USA 2005;43:15545–50.
[89] Liu M, Liberzon A, Kong SW, Lai WR, Park PJ, Kohane IS, et al. Network-based analysis of affected biological processes in type 2 diabetes models. PLoS Genet 2007;3:e96.
[90] Sharan R, Ideker T, Kelley B, Shamir R, Karp RM. Identification of protein complexes by comparative analysis of yeast and bacterial protein interaction data. J Computat Biol 2005;12:835–46.
[91] Garcia-Reyero N, Habib T, Pirooznia M, Gust KA, Gong P, Warner C, et al. Conserved toxic responses across divergent phylogenetic lineages: a meta-analysis of the neurotoxic effects of RDX among multiple species using toxicogenomics. Ecotoxicol 2011;20:580–94.
[92] Miller DH, Jensen KM, Villeneuve DL, Kahl MD, Makynen EA, Durhan EJ, et al. Linkage of biochemical responses to population-level effects: a case study with vitellogenin in the fathead minnow (*Pimephales promelas*). Environ Toxicol Chem 2007;26:521–7.
[93] Kidd KA, Blanchfield PJ, Mills KH, Palace VP, Evans RE, Lazorchak JM, et al. Collapse of a fish population after exposure to a synthetic estrogen. Proc Natl Acad Sci USA 2007;104:8897–901.
[94] Villeneuve DL, Mueller ND, Martinović D, Makynen EA, Kahl MD, Jensen KM, et al. Direct effects, compensation, and recovery in female fathead minnows exposed to a model aromatase inhibitor. Environ Health Perspect 2009;117:624–31.
[95] Shoemaker JE, Gayen K, Garcia-Reyero N, Perkins EJ, Villeneuve DL, Liu L, et al. Fathead minnow steroidogenesis: *in silico* analyses reveals tradeoffs between nominal target efficacy and robustness to cross-talk. BMC Systems Biology 2010;4:89–106.
[96] Breen M, Breen MS, Terasaki N, Yamazaki M, Lloyd AL, Conolly RB. Mechanistic computational model of steroidogenesis in H295R cells: role of oxysterols and cell proliferation to improve predictability of biochemical response to endocrine active chemical—metyrapone. Toxicol Sci 2011;123:80–93.
[97] Soetaert A, Vandenbrouck T, van der Ven K, Maras M, van Remortel P, Blust R, et al. Molecular responses during cadmium-induced stress in *Daphnia magna*: Integration of differential gene expression with higher-level effects. Aquat Toxicol 2007;83:212–22.
[98] Swain S, Wren JF, Sturzenbaum SR, Kille P, Morgan AJ, Jager T, et al. Linking toxicant physiological mode of action with induced gene expression changes in *Caenorhabditis elegans*. BMC Syst Biol 2010;4:32–62.
[99] Salgado D, Gimenez G, Coulier F, Marcelle C. COMPARE, a multi-organism system for cross-species data comparison and transfer of information. Bioinformatics 2008;24:447–9.
[100] Hu Y, Flockhart I, Vinayagam A, Bergwitz C, Berger B, Perrimon N, et al. An integrative approach to ortholog prediction for disease-focused and other functional studies. BMC Bioinformatics 2011;12:357.
[101] Powell S, Szklarczyk D, Trachana K, Roth A, Kuhn M, Muller J, et al. EggNOG v3.0: orthologous groups covering 1133 organisms at 41 different taxonomic ranges. Nucleic Acids Res 2011;40:D284–9.
[102] Ostlund G, Schmitt T, Forslund K, Kostler T, Messina DN, Roopra S, et al. InParanoid 7: new algorithms and tools for eukaryotic orthology analysis. Nucleic Acids Res 2009;38:D196–203.
[103] Linard B, Thompson JD, Poch O, Lecompte O. OrthoInspector: comprehensive orthology analysis and visual exploration. BMC Bioinformatics 2011;12:11.

[104] Yu C, Desai V, Cheng L, Reifman J. QuartetS-DB: a large-scale orthology database for prokaryotes and eukaryotes inferred by evolutionary evidence. BMC Bioinformatics 2012;13:143.

[105] Croft D, O'Kelly G, Wu G, Haw R, Gillespie M, Matthews L, et al. Reactome: a database of reactions, pathways and biological processes. Nucleic Acids Res 2011;39:D691−7.

[106] Korcsmáros T, Szalay MS, Rovó P, Palotai R, Fazekas D, Lenti K, et al. Signalogs: orthology-based identification of novel signaling pathway components in three metazoans. PLoS ONE 2011;6:e19240.

[107] Kelder T, van Iersel MP, Hanspers K, Kutmon M, Conklin BR, Evelo CT, et al. WikiPathways: building research communities on biological pathways. Nucleic Acids Res 2011;40:D1301−7.

CHAPTER 8

Interpretation of Human Biological Monitoring Data Using a Newly Developed Generic Physiological-Based Toxicokinetic Model
Examples of Simulations with Carbofuran and Methyl Ethyl Ketone

Frans Jongeneelen[1], Wil ten Berge[2], and Peter J. Boogaard[3]

[1]IndusTox Consult, Universitair Bedrijven Centrum, Nijmegen, The Netherlands, [2]Santoxar, Westervoort, The Netherlands, [3]Shell Health, Shell International, The Hague, The Netherlands

INTRODUCTION

Human biological monitoring (HBM) assesses exposure by measuring chemicals or their metabolites in body fluids such as blood and urine. This method allows quantification of the amount of a chemical that has been absorbed into the body through all potential routes of exposure (inhalation, oral, dermal). It is a powerful tool for health risk assessment and can also be used to assess trends in exposure over time or evaluate the effectiveness of exposure reduction policies. Lin et al.[1] showed that the variance in the biological measurements is less likely to skew results than variance in external individual exposures, at least in the case of the inhalation exposure pathway.

In recent years, the wider availability of advanced analytical technologies has made HBM techniques more accessible. In addition, improvements in the analytical sensitivity of

HBM techniques, often in order of magnitudes, have enabled their capacity to detect not only the presence of a wider range of materials but also at much lower levels than previously possible. These developments have many potential benefits, most notably in the ability to consistently apply biomonitoring as a tool for the widespread evaluation of exposures and eventually for health risk assessments.

For the proper interpretation of HBM data, not only are toxicological data on the hazard and dose-response essential, but also data on the toxicokinetics.[2,3] However, these data are often lacking. The European strategy for deriving health-based biological limit values were reviewed by Bolt and Thier.[4] Only a limited number of compounds had enough data to be assigned with a health-based biological limit value. Since HBM integrates all routes of exposure, it makes sense to explore the totality of exposures by employing a top-down approach based on HBM (e.g., blood or urine sampling) rather than a bottom-up approach that samples air, water, food, and so on to reconstruct an estimate of exposure.[5] A challenge in implementing this approach is based on the limited availability of toxicokinetic information for many substances.

In the absence of a complete set of data, *in silico*/computational tools can help fill the data gaps. *In silico* modeling employs mathematical know-how and computer sciences to assist in the evaluation of exposures and may help to predict risk posed by chemicals. *In silico* models can also be applied to track the movement of chemicals through the environment and through the human body. Physiologically based toxicokinetic (PBTK) models comprise a family of such tools; their potential applications in human health risk assessment have stirred considerable interest. The salient feature of PBTK models is that through simulation they can approximate the kinetic behavior of chemicals in the human body. The models are actually designed to integrate the physical and biological characteristics of a chemical with the physiological processes in the body to estimate internal dose in target tissues/organs. A PBTK model can integrate available knowledge on route-specific absorption, distribution, metabolism, and excretion into one physiologically based mathematical routine. An optimized PBTK model can be used to predict tissue or blood concentrations of a specific chemical after different scenarios of exposure. Depending on the available data, more or less sophisticated kinetic models can be used to predict a provisional biomonitoring equivalent (BE). Biomonitoring equivalents (BEs) are defined as the concentration of a chemical (or metabolite) in a biological medium (blood, urine, human milk, etc.) consistent with defined exposure guidance values or toxicity criteria including reference doses and reference concentrations (RfD and RfCs), minimal risk levels (MRLs), tolerable daily intakes (TDIs), or external exposure limits such as the occupational exposure limits (OELs), threshold limit values (TLVs), or derived no effect levels (DNELs).[6]

However, the detailed toxicokinetic information for the organs, needed as input for such models, tends to be available primarily for widely used, high-volume substances and for substances of particular concern. Another drawback is that most PBPK models are still somewhat difficult to use and remain a tool primarily of specialists. To overcome this, generic, multichemical PBTK models are being developed that require a minimal amount of data on a substance. The recently developed generic PBTK model IndusChemFate is such a tool that was developed as a spreadsheet application.[7,8] One of the advantages of the spreadsheet approach is the easy overview of the model parameters. Essential parameters are estimated with internal algorithms based on quantitative structure-property relations (QSPR). The model is available as freeware.

In this chapter, two examples of predicting biomonitoring equivalents are presented using the spreadsheet application of the PBTK model IndusChemFate. Version 2.00 of the model holds the physiology of various subsets of the human population, which allows model studies of interindividual differences, related to characteristics such as sex, body fat, physical activity, and age. The aim of the present study was to test whether the generic PBTK model IndusChemFate can be used to predict provisional biomonitoring equivalents in humans when no biological limit value is available.

THE GENERIC PBTK MODEL INDUSCHEMFATE

IndusChemFate is a physiologically based toxicokinetic model (PBTK model) used to estimate blood and urine concentration of multiple chemicals, given a certain exposure scenario. It was developed as a software application in Microsoft Excel. Three uptake routes are considered (inhalation, dermal, and/or oral) as well as two built-in exercise levels (rest and light work). The layout of the PBTK model is presented in Figure 8-1. The model contains 11 body compartments: lung, heart, brain, skin, adipose tissue, muscles, bone, bone marrow, stomach and intestines, liver, and kidney.

The model holds the physiology of typical human subpopulations (adult males, adult females, children) and of two experimental animal species (rat and mouse). Standardized physiology parameters of humans with normal weight and obese individuals are used to dimension organ and tissue volumes and blood flows through these tissues. These parameters are scaled relative to the total body weight. The level of exercise can be set on two levels of exercise (at rest and at light work). The model contains published and in-house–developed quantitative structure-property relationship (QSPR) algorithms for blood:air partitioning, tissue:blood partitioning, and renal excretion. Dermal uptake is estimated by the use of a novel dermal physiologically based module that considers dermal deposition rate and duration of deposition. Moreover, dermal absorption includes evaporation during skin contact and is related to the volatility of the substance. Michaelis–Menten saturable metabolism is incorporated in the model. Metabolism can be modeled in any of 11 body compartments or in the liver only. Tubular resorption is considered optionally based on either user input or a built-in QSPR, dependent on the octanol:water partition coefficient. Figure 8-2 shows the graph of the sigmoidal relation of tubular resorption as dependent from log (K_{ow}).

Enterohepatic circulation is optional at a user-defined rate.

The model IndusChemFate is programmed in Visual Basic and runs in MS Excel. A description of version 1.6 has been published elsewhere.[7,8] The extended version of the model is now published and available as IndusChemFate version 2.00. Version 2.00

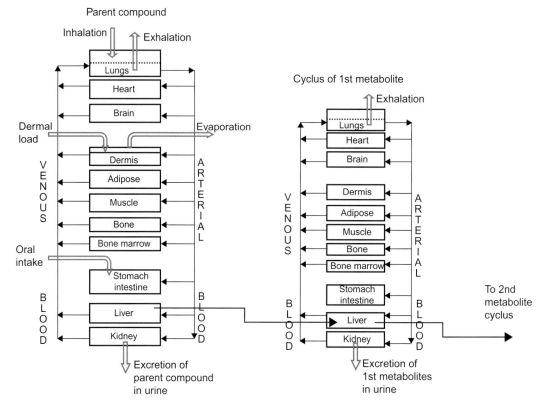

FIGURE 8-1 The IndusChemFate PBTK model structure.

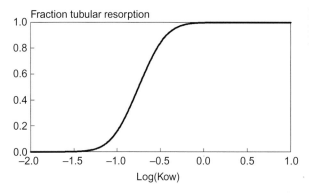

FIGURE 8-2 The sigmoidal relation of the fraction of tubular resorption as dependent from $\log(K_{ow})$.

holds additional physiological data to simulate various situations (male, female; normal or obese; at rest or light activity; rat and mouse; see Table 8-1) and improved rules for renal excretion of compounds and metabolites. Version 2.00 of IndusChemFate has been used for the calculations. The program is provided as freeware[9] and has open source code.

TABLE 8-1 Standardized Subjects That Can Be Selected in the Generic PBTK Model IndusChemFate

Number of Standardized Subject

1. Normal man (at rest)
2. Obese man (at rest)
3. Normal man (light work)
4. Obese man (light work)
5. Normal woman (at rest)
6. Obese woman (at rest)
7. Normal woman (light work)
8. Obese woman (light work)
9. Normal child (at rest)
10. Obese child (at rest)
11. Normal child (playing)
12. Obese child (playing)
13. Mouse (experimental animal)
14. Rat (experimental animal)

EXAMPLES

Example 1: Biological Monitoring of Carbofuran in Pesticide Production—A Provisional Guideline Value of 3-Hydroxy-Carbofuran in Urine

Carbofuran (2,3-dihydro-2,2-dimethyl-7-benzofuranyl-N-methylcarbamate) is a broad spectrum N-methyl carbamate insecticide widely used in agriculture for the control of insects, mites, and nematodes in soil or for protection of fruit, vegetables, and forest crops. Carbofuran and its metabolite, 3-hydroxycarbofuran, exert their toxicity by reversibly inhibiting acetylcholinesterase (AChE)[10] leading to the persistent action of the otherwise hydrolyzed neurotransmitter, acetylcholine, on its postsynaptic receptors. This example reports on biological monitoring of carbofuran and explores various approaches to propose a Biomonitoring Guidance Value of 3-OH-carbofuran in urine. A pilot study of biological monitoring of production workers in a carbofuran formulating plant was performed in 2001. Preshift and postshift samples of 7 production workers were repeatedly sampled during the cleaning activities after production of granular and liquid formulations of carbofuran. It was expected that cleaning activities after production would be a "worst case" exposure situation. As controls, spot samples of 4 nonexposed office workers were collected. The urine samples were analyzed without acid or enzymatic hydrolysis for the free primary metabolite 3-OH-carbofuran with HPLC-MS by HSL Laboratory (Sheffield, UK). The detection limit was 1.0 µg/L; precision was 10%. Results are presented in Table 8-2.

The level in the urine specimens of the control persons was for all samples <1.0 µg/L. The urinalysis made clear that in some workers, after certain activities, some absorption had occurred; the urine level after work was increased compared to the individual background and compared to the level in controls. However, a limit or guideline level of the biological indicator 3-OH-carbofuran in urine was not available. Thus, a direct health

TABLE 8-2 Measured Concentrations of 3-OH-Carbufuran in Urine Samples of Workers of a Pesticide Formulation Plant

Worker Code	Task	3-OH-Carbufuran in Urine (µg/L)	
		Preshift	Postshift
1	Cleaning after production	1.0	<1.0
1	Filling granulate line	–	<1.0
2	Cleaning after production	<1.0	<1.0
3	Cleaning after production	<1.0	3
3	Filling granulate line	–	2
3	Cleaning after production	<1.0	<1.0
4	Cleaning after production	<1.0	<1.0
4	Cleaning after production	<1.0	4
4	Cleaning after production	–	<1.0
5	Cleaning after production	<1.0	<1.0
5	Cleaning after production	<1.0	<1.0
6	Cleaning mixer	<1.0	<1.0
6	Cleaning after production	–	<1.0
6	Filters replaced + cleaning	–	2
6	Packing product	–	<1.0
6	1st day after holidays	<1.0	–
6	Cleaning mixer	<1.0	<1.0
6	Cleaning mixer	–	<1.0
7	Cleaning after production	<1.0	<1.0
7	Packing product	<1.0	<1.0
7	Cleaning after production	<1.0	<1.0

risk assessment was not possible, and the question was raised how the available information on the toxicological hazard of carbofuran can be used to derive a biomonitoring equivalent (BE).

TLV of Carbofuran

A TLV-TWA of $0.1\ mg/m^3$ (as inhalable vapor and aerosol) of carbofuran has been recommended for occupational exposure by ACGIH. This limit is intended to prevent the occurrence of cholinergic symptoms and is defined as a time-weighted-average concentration over 8 hours.[11]

Estimates of a Biological Exposure Guidance Value of Carbofuran

BY APPLICATION OF SIMPLE PHARMACOKINETICS

In the paper by Leung and Paustenbach,[12] concepts of pharmacokinetics were applied to establish a biological exposure guideline value in urine equivalent to the TLV. They derived the formula

$$\text{BEI in urine} = 6.7 * F * f_{ur} * TLV * [1 - 0.18(t^{1/2})(1 - e^{-5.5/t_{1/2}})] * MW_{metabolite}/MW_{parent} \quad (8\text{-}1)$$

where

F = bioavailability
f_{ur} = fraction of the absorbed chemical excreted in urine
TLV = threshold limit value [mg/m^3]
$t_{1/2}$ = biological half life [h]
MW $_{metabolite}$ = Molecular weight metabolite [g/mole]
MW $_{parent}$ = Molecular weight parent compound [g/mole]

Assuming that for carbofuran MW = 221, F = 1.0, f_{ur} = 0.05, TLV = 0.1 mg/m^3, and $t_{1/2}$ = 20 hours, the estimated biological exposure guideline value of 3-OH-carbofuran (MW = 236) in urine = 4.8 µg/L = 0.022 µmol/L.

BY APPLICATION OF THE GENERIC PBTK MODEL INDUSCHEMFATE

Another approach for establishing the biological exposure guideline value is to predict the concentration in urine with a generic physiologically based toxicokinetic model (PBTK model). The IndusChemFate PBTK model[7] was used to predict the urinary concentration of 3-OH-carbofuran after exposure during an 8-hour shift to carbofuran at the level of the TLV of 0.1 mg/m^3. Information on the metabolic pathways of carbofuran was derived from Zhang et al.[10] and Abass et al.[13] The metabolic pathway to 3-hydroxycarbofuran seems a major pathway in human metabolism. In the PBTK model, the metabolism is postulated as shown in Figure 8-3.

Kinetic parameters of carbofuran metabolism obtained with 10 human liver microsomes are from Abass et al.[14] It is assumed that the stoichiometric yield of 3-OH-carbofuran is 50%. Also some enterohepatic circulation of 3-OH-carbofuran-glucuronide was assumed.

FIGURE 8-3 Metabolism pathways of carbofuran as used in the PBTK model simulation.

TABLE 8-3 Physical-Chemical and Biological Properties of Carbofuran and Metabolites That Are Used for the Simulation with PBTK Model IndusChemFate

Compound Property	Carbofuran	3-OH-Carbofuran	3-OH-Carbofuran-glucuronide
CAS	1,563-66-2	16,655-82-6	53,305-32-1
Density (mg/cm^3 or grams/liter)	1,138	1,200	1,545
Molecular weight [g/mole]	221.26	237.26	413.38
Vapor pressure (Pa)	0.141	0.000656	$1*10^{-10}$
Log (K_{ow}) at skin pH 5.5	2.32		
Log (K_{ow}) at blood pH 7.4	2.32	0.76	−3.5
Water solubility (mg/liter)	320	6,200	1,000,000
Resorption tubuli (?/estimated fraction)	?	?	?
Enterohepatic removal (relative to liver venous blood)	0	0	0.5
V_{max} liver (parent[total] μmol/kg tissue/h)	10,000	1,200	
k_M liver (parent[total] μmol/liter)	34	20	
V_{max} liver (parent[specif] μmol/kg tissue/h)	2,500	300	
k_M liver (parent[specif] μmol/liter)	17	10	

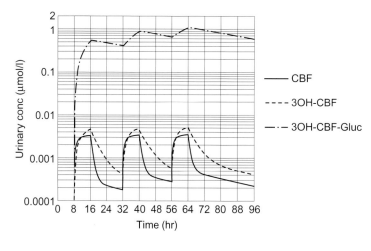

FIGURE 8-4 Simulated concentrations of carbofuran and two metabolites in urine after three 8-hour shifts with exposure at the TLV of 0.1 mg/m^3.

Table 8-3 shows the entry data used in the simulation. Physical-chemical properties are taken from the Chemspider chemical database.[15]

In Figure 8-4, the time course of the PBTK model–predicted concentration of carbofuran, 3-OH-carbofuran, and 3-OH-carbofuran-glucuronide in urine is presented for a man with light work in a period with three consecutive shifts of 8 hours when he is exposed to an airborne concentration at the level of the TLV of 0.1 mg/m^3. The urine sample voided directly at the end of the third shift was produced during the last 2 hours. The model-predicted concentration of free 3-OH-carbofuran is 0.005 μmol/L = 1.1 μg/L.

Provisional Biomonitoring Equivalent of Carbofuran

Due to uncertainties of the preceding two approaches, the most protective estimate is preferred. Therefore, a biomonitoring equivalent for 3-hydroxy-carbofuran in urine of 1.1 μg/L is proposed (as a free, nonconjugated compound). This is regarded as a tentative biomonitoring guidance value that should be updated when new scientific information becomes available.

Example 2: Differentiation of Biomonitoring Equivalents of Methyl Ethyl Ketone for Men and Women

ACGIH[16,17] proposed a Biological Exposure Index (BEI) for methyl ethyl ketone (MEK) in urine of 2 mg/L (28 μmol/L). The documentation shows that this value was derived from field studies in which it was shown that 8 hours of exposure to levels of 200 ppm resulted in end-of-shift urine concentrations of 2–5 mg/L (=28–69 μmol/L). The lower bound was adopted by ACGIH as the BEI level. There is no separate guideline for exposure of women, since this guideline is valid for humans. We used the PBTK model IndusChemFate to study the variation in the predicted concentration in urine of various human subgroups at the level of the TLV of MEK.

Entries for the PBTK Simulation of MEK and Metabolite 2,3-Butanediol in Blood and Urine

For the simulation, using the IndusChemFate model 2.00, we assumed that 2,3-butanediol is a major metabolite of MEK and that 2,3-butanediol is further metabolized to unidentified metabolites. The kinetic parameters and stoichiometric balance were derived as follows: the metabolic elimination rate of MEK was estimated by Liira et al.[18,19] (V_{max} was estimated to be 900 μmol/kg liver/h, k_M = 2 uM.) In our PBTK model, we applied the same metabolic clearance with slightly altered V_{max} = 1,800 μmol/kg liver/h and k_M = 4 uM. The stoichiometric ratio of the formation of metabolite 2,3-butanediol was unknown. The V_{max} and k_M were estimated by fitting the experimental urine concentration of 2,3-butanediol to the simulated levels (V_{max} = 90 μmol/kg liver/h, k_M = 4 uM). The metabolic elimination of 2,3-butanediol was assumed to have a V_{max} = 300 μmol/kg liver/h and k_M = 50 uM. All entry values of the parameters that are needed for the modeling are presented in Table 8-4.

Comparison of Observed with Simulated Levels of MEK in Urine of Males and Females

Recently, an experimental inhalation study of MEK with human volunteers became available.[20] In this study, the gender difference on the indicator MEK in urine was explored. A controlled experimental exposure to MEK was carried out with three subgroups of volunteers:

- 10 women with hormonal contraceptive (+HC),
- 5 women without hormonal contraceptive (−HC)
- 10 men

TABLE 8-4 Physical-Chemical and Biological Properties of MEK and the Metabolite 2,3-Butanediol That Are Used for the Simulation with PBTK Model IndusChemFate

Compound Property	MEK	2,3-Butanediol
CAS nr	78-93-3	513-85-9
Density (g/L)	805	984
MW	72.1	90.12
Vapor pressure (Pa)	12,100	73.3
Log (K_{ow}) (octanol-water) at pH 5.5	0.29	
Log (K_{ow}) (octanol-water) at pH 7.0	0.29	−0.92
Water solubility (mg/L)	223,000	984,000
Resorption tubuli (?/estimated fraction)	0.99	?
Enterohepatic removal (relative to venous blood)	0	0
V_{max} liver of removal (μmol/(kg tissue/h)	1,800	300
k_M liver of removal (μmol/L)	4	50
V_{max} liver of metabolite production (μmol/(kg tissue/h)	90	
k_M liver of metabolite production (μmol/L)	4	

TABLE 8-5 Observed and Predicted Level of MEK in Urine (AM ± SD) of Three SubGroups of Volunteers Before Exposure and After 6 Hours of Exposure to 298 mg/m^3 MEK

		Time (h)	
Observation	Group	0 (=before exposure)	6 (=end of exposure period)
Experimental[20]	Men ($n = 10$)	0.7 ± 0.7	13.8 ± 1.8
	Women with HC ($n = 10$)	0.7 ± 0.7	13.5 ± 3.2
	Women without HC ($n = 5$)	0.7 ± 0.7	20.0 ± 10.6
PBTK-predicted	Men at rest	0	34.6
	Women at rest	0	14.8

The volunteers were exposed to 99.2 ppm MEK (=298 mg/m^3) for 6 hours. Urine was collected at every 2 hours to test for urinary MEK. Table 8-5 shows the concentrations of MEK before and at the end of the 6-hour exposure period. Using the model IndusChemFate, we predicted the concentration at 6 hours for males and females at rest. The experimental results are also presented in Table 8-5. Metabolism kinetics of males was assumed to be equal to that of females.

The observed and simulated concentrations of MEK in urine are also presented as a function of time (Figure 8-5). The PBTK model simulations predicted an MEK level in

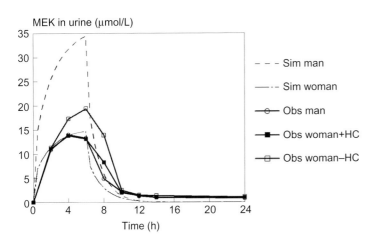

FIGURE 8-5 Observed (Obs) and simulated (Sim) MEK concentration in urine as a function of time of the three groups of volunteers at resting conditions after 6 hours' constant exposure to 298 mg/m^3 MEK.

urine of women that was in the same range as the observed values, but the simulated level in males was higher; however, the group size was small and the differences were small.

Variability of Biomonitoring Equivalents of MEK in Urine for Human Subgroups with Different Gender, Body Constitution, and Activity Level

ACGIH[16] proposed a Biological Exposure Index (BEI) for MEK in urine as 2 mg/L. The documentation from ACGIH shows that this value was derived from field studies in which it was shown that 8 hours of exposure to levels of 200 ppm (590 mg/m^3) resulted in end-of-shift urine concentrations of 2–5 mg/L (=28–69 μmol/L). The lower bound was adopted by ACGIH as the BEI level.

Using the IndusChemFate PBTK model for MEK, we predicted the level of MEK in urine for various human subgroups, as the end of the 8-hour shift concentration and as a concentration of the last 2 hours of exposure, assuming inhalation of MEK at a concentration of 590 mg/m^3. The reader should realize that PBTK modeling results in estimates of instantaneous concentrations. However, in real life a momentary urine concentration cannot be measured. Spot urine sampling will give an integration over a period of some hours, resulting in a time-weighted-average concentration that is a little lower.

Table 8-6 shows the results. It is predicted that males have higher urinary levels of MEK after exposure, and physical activity increases the level dramatically. In general, it seems that the model predicts higher levels of MEK in urine compared to data of the BEI documentation, especially for men. The estimates vary over various human subjects from 39 to 730 μmol/L. Liira et al.[19] used a PBTK model for MEK concentrations in blood. ACGIH's BEI committee[17] assumed a urine-blood partitioning coefficient equal to 1 and an instant equilibrium when they predicted that MEK concentrations in urine would be between 140 and 350 μmol/L (10–25 mg/L) at an exposure of 200 ppm, based on the findings of Liira et al.[19] This is within the range of our estimate for various human subjects.

TABLE 8-6 PBTK Model-Predicted Concentration of MEK in Urine of Various Human Subjects after 8 Hours of Exposure to 200 ppm MEK

Human Subject	Concentration of MEK in Urine after 8 hours of Exposure (in μmol/L)	Concentration of MEK in a Spot Urine Sample Voided over the Last 2 hours of 8-Hour Exposure Period (in μmol/L)
Male at rest (model subject 1)	161	140
Male with light work (model subject 3)	814	730
Obese male at rest (model subject 2)	182	160
Obese male with light work (model subject 4)	793	700
Female at rest (model subject 5)	44	39
Female with light work (model subject 7)	337	295
Obese female at rest (model subject 6)	57	50
Obese female with light work (model subject 8)	374	329

DISCUSSION

The present PBTK model IndusChemFate can be regarded as a tool for first-tier estimation of the fate of new and data-poor substances in the human body. The advantage of this tool is the ease with which the spreadsheet template may be changed due to the powerful software editor. The initial validations of this model indicate that predictions made using this approach are within an order of magnitude of the actual value.[7,8] This allows assessment of provisional biomonitoring equivalents as a first step in a tiered approach to determine whether or not there is a reason for concern.[21]

For some chemicals, when detailed toxicokinetic information is available, chemical-specific PBPK models have been developed. However, the detailed toxicokinetic information needed as input for such models tends to be available primarily for widely used, high-volume substances and for substances of particular concern. It is acknowledged that chemical-specific PBPK models may provide more accurate predictions because of their extensive calibration to a particular chemical, but a general model has the advantage of being applicable to a wide variety of chemicals. This general applicability of the model is useful for screening studies where a compound-specific PBTK model is not available and detailed data on partitioning over organs and fluids in the body are missing. The advantage of a generic model is that it can be consistently applied to different chemicals and therefore allows multiple chemicals to be compared under standardized conditions.

Like all models, the PBTK model IndusChemFate described here has limitations. It is designed primarily for neutral and predominantly ionic organic compounds. As a consequence, inorganic cations, which bind to specific tissues like bone tissue, cannot be modeled because of their unusual partitioning properties. Currently, it is assumed that typical occupational and/or environmental exposures are too low to cause enzyme induction, which may result in a conservative overestimate of tissue concentration by underestimating degradation rates. Another limitation of the current model is its simplicity. As the model lacks detailed descriptions for specific protein binding, intestinal transport, interaction with intestinal flora, and excretion by feces, it does not accurately describe the toxicokinetics for certain classes of chemicals where these factors play an important role. However, inclusion of detailed process descriptions increases the data demands for model applications. Moreover, a simple model will be sufficient for assessing the relative importance of the different elimination processes (urinary, respiratory, and metabolic elimination). The model outcomes have a satisfying accuracy for early testing, in so-called first-tier simulations where a differentiation is made between substances for which there is concern and prioritization required and substances for which there is no concern. It can be also used to optimize the sampling time and urine sampling frequency of a biomonitoring program and to support the interpretation of results in mixed populations.

In conclusion, the PBTK model IndusChemFate can be used to help predict blood and urine concentrations of data-poor substances in man. After exposure to an established limit value, such as an occupational exposure limit, predicted blood and/or urine concentrations might be used as a provisional guideline, indicated as a biomonitoring equivalent. In addition, the model allows prediction of biomonitoring equivalents for specific human subgroups by differentiation in sex, age, exercise level, and obesity.

SUPPLEMENTARY INFORMATION

The Excel spreadsheet file and manual for the PBTK model IndusChemFate are available free of charge from the CEFIC-LRI website (http://www.cefic-lri.org/lri-toolbox/induschemfate).

References

[1] Lin YS, Kupper LL, Rappaport SM. Air samples versus biomarkers for epidemiology. Occup Environ Med 2005;62:750—60.
[2] Boogaard PJ. Biomonitoring of the workplace and environment. In: Ballentyne B, Marrs T, Syversen T, editors. General and applied toxicology. 3rd ed. Chichester, UK: Wiley; 2009. p. 2559—89.
[3] Boogaard PJ, Money CD. A proposed framework for the interpretation of biomonitoring data. Environ Health 2008;7(S12):1—6.
[4] Bolt HM, Their R. Biological monitoring and Biological Limit Values (BLV): the strategy of the European Union. Mini-review. Toxicol Lett 2006;162:119—24.
[5] Rappaport SM. Implications of the exposome for exposure science. J Expo Sci Environ Epidemiol 2011;21:5—9.

[6] Hays SM, Becker RA, Leung HW, Aylward LL, Pyatt DW. Biomonitoring equivalents: a screening approach for interpreting biomonitoring results from a public health risk perspective. Regul Toxicol Pharmacol 2007;47:96−109.

[7] Jongeneelen FJ, ten Berge WF. A generic, cross-chemical predictive PBTK model with multiple entry routes running as application in MS Excel; design of the model and comparison of predictions with experimental results. Ann Occup Hyg 2011;55:841−64.

[8] Jongeneelen FJ, ten Berge WF. Simulation of urinary excretion of 1-hydroxypyrene in various scenarios of exposure to PAH with a generic, cross-chemical predictive PBTK-model. Int Arch Occup Environ Health 2012;85:689−702.

[9] CEFIC-LRI. Webpage of PBTK-Model IndusChemFate, <http://www.cefic-lri.org/lri-toolbox/induschemfate>; 2012. Last entered November 2012.

[10] Zhang X, Tsang AM, Okino MS, Power FW, Knaak JB, Harrison LS, et al. A physiologically based pharmacokinetic/pharmacodynamic model for carbofuran in Sprague-Dawley rats using the exposure-related dose estimating model. Toxicol Sci 2007;100:345−59.

[11] ACGIH. 2004 Documentation of TLVs and BEIs. Cincinnati, Ohio: Carbofuran.

[12] Leung HW, Paustenbach DJ. Application of pharmacokinetics to derive biological exposure indexes from threshold limit values. Am Ind Hyg Assoc J 1988;49:445−50.

[13] Abass K, Reponen P, Mattila S, Pelkonen O. Metabolism of carbosulfan. I. Species differences in the *in vitro* biotransformation by mammalian hepatic microsomes including human. Chem Biol Interact 2009;181:210−9.

[14] Abass K, Reponen P, Mattila S, Pelkonen O. Metabolism of carbosulfan. II. Human interindividual variability in its *in vitro* hepatic biotransformation and the identification of the cytochrome P450 isoforms involved. Chem Biol Interact 2010;185:163−73.

[15] Chemspider. Chemspider chemical database, <http://www.chemspider.com/>; 2012. Last entered November 2012.

[16] ACGIH 2001. Documentation of TLV and BEI. Methyl ethyl ketone 2001. Cincinnati, Ohio.

[17] ACGIH 2012. Documentation of TLV and BEI. Methyl ethyl ketone 2012 (draft). Cincinnati, Ohio.

[18] Liira J, Riihimäki V, Pfäffli P. Kinetics of methyl ethyl ketone in man: absorption, distribution and elimination in inhalation exposure. Int Arch Occup Environ Health 1988;60:195−200.

[19] Liira J, Johanson G, Riihimäki V. Dose-dependent kinetics of inhaled methylethylketone in man. Toxicol Lett 1990;50:195−201.

[20] Tomicic C, Berode M, Oppliger A, Castella V, Leyvraz F, Praz-Christinaz SM, et al. Sex differences in urinary levels of several biological indicators of exposure: a human volunteer study. Toxicol Lett 2011;202:218−25.

[21] LaKind JS, Aylward LL, Brunk C, DiZio S, Dourson M, Goldstein DA, et al. Guidelines for the communication of biomonitoring equivalents: report from the biomonitoring equivalents expert workshop. Regul Toxicol Pharmacol 2008;51:S16−26.

CHAPTER

9

Uses of Publicly Available Data in Risk Assessment

Isaac Warren[1] and Sorina Eftim[2]

[1]ICF International, Fairfax, VA, USA, [2]Department of Environmental and Occupational Health, George Washington University School of Public Health and Health Services, Washington, DC, USA

INTRODUCTION

Central to computational-based sciences, publicly available data sets are used in numerous fields and are essential in the process of scientific discovery. The free exchange of public data has increased with the advent of the Internet and advances in computing more generally. A key reason for investing in and creating publicly available data sets is that by sharing data, scientists will use the data to generate more hypotheses, methods, and conclusions that will benefit both researchers (i.e., those performing the analyses) and stakeholders (i.e., regulators and the public). While businesses may find ways to commercialize publicly available data, the benefits of free data analysis to scientists and public officials outweigh the restriction of these data. In the field of toxicology, multiple publicly available data sources reporting human exposure, traditional *in vivo* animal toxicity data, and newer computational toxicology data present a wealth of data to risk scientists.

As computational toxicology blossoms, new technologies are being used to collect and organize toxicological data in a high-throughput manner that promises to expand or reshape the process of chemical risk assessment. While developing new computational approaches and methods aims to add to our understanding of chemical risk, reduce the costs and the uses of animals associated with toxicology testing, and provide a more targeted and representative data landscape,[1] precisely how new data can be used in risk assessment has not yet been well defined. Nevertheless, scientists and regulators are coordinating efforts to purposefully collect and make use of new data and to make these data publicly available.[2] As more data are being generated, publicly released, and analyzed, new hypotheses and observations show how these data may be used in risk assessment (i.e., problem formulation, hazard assessment, exposure characterization, and risk characterization).

Ultimately, risk assessors will be able to use publicly available data at every stage in the risk assessment paradigm to address important toxicological questions, whether from the top-down, the bottom-up, or to devote minimal attention to a given chemical or endpoint. While the methods to mine new computational toxicology data and assess their predictive power and compatibility with more traditional toxicology data are still under development, this chapter presents publicly available data sets that currently can be mined to support risk assessment (see Table 9-1). Top-down approaches considered here include using publicly available data on human exposure and health outcome to examine associations between exposure to a certain chemical and specific health effects. Here we focus on the Fourth National Health and Nutrition Examination Survey (NHANES IV) from the National Center for Health Statistics (NCHS) and the Centers for Disease Control and Prevention (CDC). Bottom-up approaches using high-throughput *in vitro* and *in silico* methods are also reviewed, focusing on the ToxCast™ program of the U.S. Environmental Protection Agency (EPA) National Center for Computational Toxicology (NCCT).

In this chapter, we also assess the overlap between these publicly available data sets to suggest where evaluating multiple data sets can be useful in risk assessment. Notably, some of the chemicals assessed for both hazard and exposure using NHANES and ToxCast™ have already been thoroughly assessed using traditional methods. Currently, efforts are underway to compare new data and suggest how they can be used to predict toxic effects using mostly computational, data-focused methods. However, in order to make comparisons between traditional and emerging exposure and computational hazard data, both old and new toxicological data require careful databasing, organization, and curation in conjunction with expert analysis. Therefore, here we also review chemoinformatic and bioinformatic approaches to computational toxicology data management, while emphasizing the challenges in compiling emerging data sets for specific chemicals and comparing existing data sets across many chemicals.

Risk assessors are faced with adapting to toxicity testing in the 21st century and soon may be required to consider new types of data sets for use in risk assessment. In the meantime, further work is required to quantify the uncertainty associated with high-throughput screening (HTS) data, collect and analyze further exposure data, and organize and evaluate new data from different sources across the existing risk assessment paradigm or as components of a new risk assessment paradigm.

PUBLICLY AVAILABLE DATA SETS WITH USES IN RISK ASSESSMENT

In this section, we present examples of well-known publicly available data sets to illustrate the benefits and challenges of their uses in risk assessment. Table 9-1 presents data that are publicly available across a range of biological, chemical, and toxicological space. However, this chapter focuses specifically on data sets that lie at either extreme of the risk

TABLE 9-1 Publicly Available Computational Toxicology Data with Uses in Risk Assessment

Data Set Name	Data Set URL	Reference(s)
ACToR (contains the following data sets):	http://actor.epa.gov	37
ToxCast	http://www.epa.gov/ncct/toxcast/	16, 17, 19, 30, 32, 38
ToxRefDB	http://www.epa.gov/ncct/	19, 32
ExpoCast	http://www.epa.gov/ncct/expocast/	39–41
DSSTox	http://www.epa.gov/ncct/dsstox	37, 39, 42, 43
HPVIS	http://www.epa.gov/hpvis	
ECOTOX	http://cfpub.epa.gov/ecotox	
National Health and Nutrition Examination Survey (NHANES)	http://www.cdc.gov/nchs/nhanes.htm	
National Morbidity, Mortality, and Air Pollution Study (NMAPS)	Data security concerns caused the health data to be removed from public availability in 2011.	
Chemical Effects in Biological Systems (CEBS)	http://cebs.niehs.nih.gov	44–47
TOXNET	http://toxnet.nlm.nih.gov/index.html	
PubChem	http://pubchem.ncbi.nlm.nih.gov	
NCBI Online Mendelian Inheritance in Man (OMIM)	http://www.ncbi.nlm.nih.gov/omim/	48
ArrayExpress	http://www.ebi.ac.uk/arrayexpress/	49, 50
DrugBank	http://www.drugbank.ca	51
Carcinogenic Potency Database (CPDB)	http://potency.berkeley.edu	
Open Biological and Biomedical Ontologies (OBO) Foundry	http://www.obofoundry.org	24
GOPubMed	http://www.gopubmed.org/web/gopubmed	24
Protein Data Bank	http://www.pdb.org	
Kegg Database	http://www.genome.jp/kegg/kegg1.html	52
Pathway Commons	http://www.pathwaycommons.org/pc/home.do	53
Reactome	http://www.reactome.org/ReactomeGWT/entrypoint.html	54
Comparative Toxicogenomics Database (CTD)	http://ctdbase.org/	54, 55
PharmGKB	http://www.pharmgkb.org/	48
OpenTox	http://www.opentox.org	

assessment paradigm, that is, top-down uses of human exposure data and bottom-up uses of computational hazard data.

Human Biomonitoring Chemical Exposure Data

Current analytical methods allow for the detection of environmental chemicals and their metabolites at low concentrations in several media including human serum, blood, and urine. Recent analyses of environmental chemicals in human samples suggest that, within specific cohorts or populations, some health effects can be associated with exposure to some chemicals. Most often, risk-prone chemicals are selected for biomonitoring in human populations due to their known toxicity. For such chemicals, studies in large cohorts and smaller epidemiological studies provide valuable exposure and health outcome data. However, owing to data privacy, ownership, and other concerns, limited human exposure data are publicly available.

Some data are publicly available, mostly from larger cohorts such as the CDC's NHANES IV. While both large and small epidemiological studies each provide valuable exposure and effects data to risk scientists, larger representative studies that combine demographic data and health and exposure measurements are most useful for testing for associations between exposure and health effects. Publicly available data have been widely used to examine associations between exposures to environmental chemicals and health effects. Such analyses highlight the uses of publicly available exposure data in problem formulation via exposure-based chemical selection and prioritization, hazard characterization via association-based health outcome hypothesis generation, and risk characterization through detailing the likelihood and extent of exposure.

The National Health and Nutrition Examination Survey (NHANES)

The NHANES data set provides a series of cross-sectional surveys conducted by the CDC and the NCHS to assess the health and nutritional status of civilian, noninstitutionalized adults and children in the United States. Each survey is based on a complex, multistage probability design that is used to select a nationally representative sample of about 5,000 persons each year. These persons are located in counties across the country, 15 of which are visited each year. The NHANES program combines both interviews and physical examinations at a mobile examination center, creating a comprehensive set of analytical data.

NHANES is probably best known for the prevalence data it provides on obesity. Researchers in academic institutions, government laboratories, and private industry are making increased use of data from the NHANES surveys. Their research includes cross-sectional and longitudinal studies of exposures, health status, disease morbidity, and mortality. Data from NHANES are also used to determine the prevalence of major diseases and risk factors for diseases and to assess nutritional status and its association with health promotion and disease prevention. NHANES findings are also the basis for national standards for such measurements as height, weight, and blood pressure. Data from this survey are used in epidemiological studies and health sciences research to inform sound public health policy, direct and design health programs and services, and expand knowledge on the health of the U.S. population.

More specifically, the NHANES data have been used in research evaluating risk factors for chronic diseases such as arthritis, cardiovascular and respiratory diseases, diabetes, gallbladder and kidney diseases, osteoporosis, and cancer.[3] NHANES has also been used to assess exposure of the U.S. population to certain environmental chemicals and associations with risk of disease or biomarkers of disease (e.g., perchlorate exposure and thyroid function).[3] Exposure to persistent organic pollutants has been associated with a variety of adverse health effects including cancer, immune system suppression, decrements in cognitive and neurobehavioral function, disruption of sex steroid and thyroid function, and in some cases, the risk of chronic diseases, such as hypertension, cardiovascular disease, and diabetes.[4] Biomonitoring data from NHANES have also been used to characterize exposure to individual and multiple chemicals and their metabolites in populations of interest (e.g., pregnant women).[5] Data derived from early NHANES provided the first concrete evidence that blood levels of lead among Americans were becoming dangerously high. As a result, the EPA called for a reduction in production and sales of consumer products containing relatively large amounts of lead, most notably gasoline and household paints.

Analyses of NHANES biomonitoring data on more than 200 chemicals outline several benefits of the biomonitoring program as well as the benefits of making NHANES data publicly available. One of the most evident advantages is the capability to observe and follow long-term trends in the general population. For example, the increase in blood and breast milk levels of polybrominated diphenyl ethers, or PBDEs,[6] prompted the voluntary phase-out of some PBDEs in consumer products. Biomonitoring data can also help in establishing the impact of a specific policy decision. For example, the marked decrease in blood lead levels in the United States after the elimination of lead as a gasoline additive; similarly, polychlorinated biphenyl or PCB levels have also declined in the past 20 years as they are being phased out by industry.[7]

NHANES IV includes a variety of chemicals and their metabolites, such as dioxins, polychlorinated biphenyls (PCBs), polybrominated biphenyl ethers (PBDEs), and phthalates. Metabolites for various polycyclic aromatic hydrocarbons (PAHs) (e.g., fluorene, pyrene, phenanthrene) are also included. NHANES IV chemical exposure data are available in report format[8] for 212 chemicals measured during 1999 through 2004. An update to this report presented additional data on similar chemicals for measurements performed during 2004 through 2010 when an additional 34 new chemicals were also assessed.[9] In studies of environmental chemical exposure, random one-third or one-half samples of the NHANES population aged 6 years and older were selected for analysis of urinary, blood, or serum concentrations of environmental chemicals. Mercury concentrations in hair were also assessed in 1999–2000. Wherever detection limit varies by assay, chemical concentration data are coded as at or above detection limits, such as for dioxins, furans, PCBs, organochlorine pesticides, and some pesticides. For other measurements, detection limits are constant for each specimen analyzed and are reported in the metadata documentation for each data file, with some exceptions.

The National Morbidity, Mortality, and Air Pollution Study (NMMAPS)

Another data set that focused on exposure to airborne pollutants was the National Morbidity, Mortality, and Air Pollution Study (NMMAPS), a collaboration between Johns Hopkins Bloomberg School of Public Health and the Health Effects Institute. NMMAPS was designed to examine the health effects of air pollution in the United States.[10] The NMMAPS database was composed of daily observations of mortality counts (aggregated and stripped of identifiers) for three age groups from 108 U.S. cities for the years 1987–2000, along with census data from the U.S. Census Bureau. In addition, NMMAPS included air pollution data from the EPA's AirData and weather data from the National Climatic Data Center. The data were made publicly available first via the Web (http://www.ihapss.jhsph.edu/data/data.htm), and then via R as NMMAPSdata R Package.[11] Data privacy concerns caused the health data to be removed from public availability in 2011.

NMMAPS was designed with the goal of researching whether particulate matter was responsible for the associations between air pollution and daily mortality. However, NMMAPS has since been used to answer many important questions in environmental epidemiology. A systematic review to find peer-reviewed papers or reports that used the NMMAPS data found at least 67 papers or reports.[12] NMMAPS has been used to answer questions on the health effects of air pollution and temperature in the United States and to answer methodological questions that have been used in other countries and research areas.[8,13] NMMAPS has also been used to examine deaths during heat waves[11] and to predict future heat-wave deaths due to climate change. The EPA used NMMAPS for regulatory impact statements on particulates and ozone,[14] and for other reports on climate change[7] that have influenced policy outside the United States.

High-Throughput Computational Toxicology Data

Other emerging data being made available to risk scientists are high-throughput screening (HTS) data. HTS involves testing large numbers of chemicals *in vitro* and *in silico* against a variety of assays chosen for biological features that are suspected or known to contribute to apical toxic effects observed in whole organisms. Risk scientists are still assessing how predictive or accurate HTS data can be in assessing chemical risk in a screening, mechanistic, or risk-aversive manner. Nevertheless, HTS methods are being used to efficiently assess chemicals and provide putatively mechanistic data across a variety of biological endpoints. HTS methods provide some toxicogenomic data, through which "omic" (i.e., genomic, transcriptomic, proteomic, and metabolomic) techniques may increase our predictive and mechanistic understanding of toxicology.[15] However, despite the amount of detail each omic technology can produce, obtaining these data is still too costly to be performed for all chemicals.[16] Therefore, to perform toxicogenomic and other tests on chemicals, computational toxicologists carefully selected assays for high-throughput data collection. The resulting computational toxicology data can be mined along with existing biological and chemical data sets to provide useful insights into chemical hazards.

ToxCast™

The ToxCast™ program at the EPA National Center for Computational Toxicology has provided HTS data for a set of 309 ToxCast™ Phase I chemicals for which detailed traditional toxicology data are already available. This and other data sources published by the Tox21 community (http://ntp.niehs.nih.gov/go/tox21) such as the EPA's Aggregated Computational Toxicology Resource (ACToR) and NTP's Chemical Effects in Biological

Systems (CEBS) provide computational data that warrant further analysis into their applications for risk assessment. These and other publicly available data sources that can be useful in risk are presented in Table 9-1.

Researchers both from within and outside the Tox21 community have begun analyzing these publicly available HTS data to assess how predictive HTS data are of known effects *in vivo* and how these data can be interpreted to contribute to or reshape risk assessment. At present, these analyses suggest that HTS data will be useful to risk assessment. For example, HTS data can be used to inform problem formulation via improved chemical selection and prioritization, hazard characterization via increased mechanistic or computational understanding of potentiating or initiating toxic events, and risk characterization via reducing uncertainty associated with current or future risk assessments.

Kavlock et al.[17] review the assays selected for the EPA's ToxCast™ program, describe the chemical and biological space assessed, and highlight the public availability of ToxCast™ data. Dix et al.[18] also describe several priorities, decisions, and goals of the ToxCast™ program at its proof-of-concept stage. Several screening bioinformatic and chemoinformatic publications focusing on chemical risk were considered prior to the implementation of the ToxCast™ program.[18] High-throughput tests prior to ToxCast™ were largely focused on lead generation or screening for adverse effects in the pharmaceutical or agrochemical industries. These technologies have since been adapted for hazard assessment to provide a wealth of *in vitro* and *in silico* computational toxicology data. Whereas pharmaceutical applications of HTS assess specific drug targets, frequently for closely related compounds, HTS methods for risk assessment seek to provide information on a variety of endpoints and across a broader range of chemicals. Despite these and other differences, faced with the lack of hazard data, HTS methods were identified as a more practical and cost- and animal-conscious approach to hazard assessment.

Dix et al.[18] describe the chemical selection process used to select which chemicals would be tested during Phase I of ToxCast™. A total of 826 conventional active pesticide chemicals were considered for further study *in vitro*. These included a set of at least 270 food-use pesticides with extensive testing requirements and available test data. Chemicals with extensive data were selected and vetted by physico-chemical properties for practicality of HTS testing. Traditional *in vivo* hazard data availability was considered a key requirement when selecting chemicals for detailed study *in vitro*. Specific hazard data that were available to Dix et al.[18] have also been publicly released to enable further analysis of how HTS data compares to traditional animal toxicology test data. For example, data from Data Evaluation Records (DERs) from the EPA Office of Pesticide Programs and other conventional bioassay data from the EPA High Production Volume Challenge Program and the NTP High Throughput Screening Initiative were also considered when selecting chemicals for high-throughput analysis.[18] Some of these data are publicly available in various formats for use in validation of ToxCast™ data. For example, Martin et al.[19] describe EPA's Toxicity Reference Database (ToxRefDB), which stores submitted study data and serves as a primary source for validation of the predictive ability of ToxCast™ data. Altogether, these data sources provide a wealth of traditional *in vivo* toxicology data to scientists to evaluate new HTS computational toxicology data, following the completion of Phase I of ToxCast™.

The biological relevance of HTS data can be derived not only from the methods of an individual HTS assay, but also from how the results from one assay compare with biologically related assays or to *in vivo* effects. HTS data have not yet been fully characterized, unlike traditional *in vivo* data, which report effects across a series of validated "endpoints." Traditionally, "endpoint" is a broad term used to describe toxicological effects or specific changes to an organism's health, such as growth, reproduction, or any other insult to a specific organ or system. Within the realm of computational toxicology, endpoint can be defined to apply to a specific assay or *in vitro* response that is known or expected to be predictive of toxicity in the whole organism at an individual or population level. One goal in establishing the relevance of HTS data is to relate responses from specific chemicals for single or groups of HTS endpoints to more traditional endpoints.

By design, HTS data available thus far have been collected for chemicals with large, validated or "guideline" *in vivo* study data sets.[17] These chemicals include mostly pesticide active ingredients and some commercial chemicals. Therefore, publicly available human health and ecological risk assessments also exist for a number of these chemicals. One hypothesis for analyzing HTS data and comparing it to traditional toxicity data is that by analyzing the results from multiple biologically related, expert-selected assays, a single event or series of events observed *in vitro* may relate to apical effects observed *in vivo*. In theory, out of the over 600 HTS assays used in Phase I of ToxCast™, groups or single assays may be more predictive, more specific, or more sensitive in predicting adverse effects. If this hypothesis proves to be correct, a focused HTS test battery could be developed to screen more chemicals at a lower cost, albeit with a defined amount of uncertainty.

Within the ToxCast™ data set, half maximal activity (AC_{50}) or lowest effect concentration (LEC) are available for a series of 4 to 15 different concentrations across a variety of targets. These targets are diverse and include cell morphology, growth, cytotoxicity, DNA damage, oxidative stress, genotoxicity, stress kinase activation, nuclear and transmembrane receptor binding (i.e., agonism and antagonism), transcription factor response binding, protein-fragment complementation, effector-stimulated cocultures of primary human cells with quantitative digital imaging, and zebrafish embryonic exposures. Phase I and phase II xenobiotic metabolizing enzyme expression is assessed as a component to some assays, such as in tests using primary human hepatocytes. Limited to no pharmacokinetic and pharmacodynamic information is included in ToxCast™; however, complex PK and PD models are developed separately that have been proposed to be used in conjunction with ToxCast™ data when conducting risk assessments.[20]

Notably, dose-response data from *in vivo* studies require careful consideration before ruling out any association between positive HTS data and negative *in vivo* data, for example. For example, Shah et al.[21] present the case in which Rotenone was tested at only 3.75 ppm in its chronic study despite that at shorter durations it was shown to cause gastrointestinal injury at 150 ppm. The majority of HTS assays use a concentration-response format; however, uncertainty related to observing effects at the highest dose tested or at no tested doses remains a limitation. Judson et al.[20] propose that HTS methods can be used to measure dose-response at low concentrations to perhaps remove the need to extrapolate effects to low doses using benchmark dose modeling, for example.

As opposed to the traditional dose-response curve used in risk assessment, a newly proposed high-throughput risk assessment process[20] proposes overlaying or replacing this curve with a probability distribution based on computational data. In theory, cell lines representing sensitive endpoints or endpoints representing sensitive phenotypes could be used to generate distribution data using measured values. However, it is not clear whether such methods will provide greater sensitivity or predictive accuracy than pharmacodynamic models or other methods. Some recent studies assessing both publicly available exposure and genomic data suggest that data are available at the genetic level to generate population distributions and reduce uncertainty for genotoxicants, for example.[10,22] Other examples have shown how genetic variation, which can be assessed in a limited way using ToxCast™ data, may be associated with increased prevalence of adverse effects in exposed populations. For example, Scinicariello et al.[23] assessed how a polymorphism in the aminolevulinic acid dehydratase (ALAD) gene was associated with blood lead levels, prevalence of hypertension, and systolic and diastolic blood pressure. Scinicariello et al.[23] used NHANES III genomic data in their study.

COMPARISON OF THE NHANES IV AND TOXCAST™ DATA SETS

In an effort to assess overlap in the NHANES IV and ToxCast™ data sets, we selected measurements of environmental chemicals from the laboratory data sets in NHANES IV data from 1999 to 2010.[9] NHANES IV chemicals identified by their SAS variable labels were then compared to the ToxCast™ chemical inventory (see Table 9-2).

In order to establish more precise chemical identity, simplified molecular-input line-entry system (SMILES) notation and/or Chemical Abstract Services Registry Numbers (CASRNs) were gathered using PubChem or ChemSpider.com and then compared along with chemical name to assess overlap. Comparisons were performed only for chemicals for which there were NHANES measurements at or above detection limit in at least 500 individuals for at least one biennial study period. Urine, blood, or serum chemical concentrations were at or above detection in at least 500 individuals for approximately 176 unique chemicals during 1999 through 2010.

CASRNs and SMILES notation were not available for all metabolites included in the NHANES IV data set. Furthermore, ToxCast™ predominantly intends to test parent compounds only. Therefore, parent compounds were identified in the ToxCast™ inventory, where appropriate. Apart from data on metals (i.e., antimony, arsenic, barium, cadmium, cesium, lead, mercury, molybdenum, thallium, tungsten, and uranium), several chemicals have data in both NHANES and ToxCast™ (see Table 9-2). Where sufficient data are available from both sources, association studies can be performed to further characterize exposure by demographic group and health status. Furthermore, a number of the chemicals for which both NHANES IV and ToxCast™ data exist have already been assessed by the EPA's Integrated Risk and Information System (IRIS).

METHODS FOR COMPILING DATA FROM MULTIPLE SOURCES FOR RISK ASSESSMENT

In an effort to streamline the public release and analysis of 21st century toxicology data, bioinformatic and chemoinformatic techniques are used to index and organize data in a manner that will assist risk assessors in compiling data from different publicly available sources. Web-based and downloadable databases house and carefully match chemical and biological activity data to specific chemicals and targets.[24] These databases share

TABLE 9-2 Chemicals with NHANES IV Biomonitoring and ToxCast™ HTS Data

NHANES IV Chemical Name[a]	ToxCast™ Status[b]	IRIS Oral RfD Critical Effect[c]
2,4-D (2,4-Dichlorophenoxyacetic acid)	Phase I Version 1 Testing Completed	hematologic toxicity; liver toxicity; kidney toxicity
Bisphenol A	Phase I Version 1 Testing Completed	decreased mean body weight
Mono-(2-ethyl)-hexyl phthalate	Phase I Version 1 Testing Completed	
Perfluorooctane sulfonic acid	Phase I Version 1 Testing Completed	
Perfluorooctanoic acid	Phase I Version 1 Testing Completed	
Triclosan	Phase I Version 1 Testing Completed	
2,4,5-trichlorophenol	Phase II (in progress)	liver pathology; kidney pathology
Pentachlorophenol	Phase II (in progress)	liver pathology; kidney pathology
Dieldrin	Phase II (in progress)	liver lesions
p,p'-DDT	Phase II (in progress)	liver lesions
Mirex	Phase II (in progress)	liver cytomegaly; fatty metamorphosis; angiectasis; thyroid cystic follicles
Heptachlor epoxide	Phase II (in progress)	increased liver/body weight ratio—males and females
2,4-dichlorophenol	Phase II (in progress)	decreased delayed hypersensitivity response
1,4-dichlorobenzene	Phase II (in progress)	
2,4,6-trichlorophenol	Phase II (in progress)	
Beta-hexachlorocyclohexane	Phase II (in progress)	
p-bromodiphenyl ether (metabolites)	Phase II (in progress)	
Phenanthrene (metabolites)	Phase II (in progress)	
2,5-dichlorophenol	Phase II (in progress)	
3-hydroxyfluorene, fluorene (metabolites)	Phase II (in progress)	
3-phenoxybenzoic acid	Phase II (in progress)	
Butyl paraben	Phase II (in progress)	

(Continued)

TABLE 9-2 (Continued)

NHANES IV Chemical Name[a]	ToxCast™ Status[b]	IRIS Oral RfD Critical Effect[c]
Cotinine	Phase II (in progress)	
Daidzein	Phase II (in progress)	
Enterolactone	Phase II (in progress)	
Ethyl paraben	Phase II (in progress)	
Genistein	Phase II (in progress)	
Methyl paraben	Phase II (in progress)	
Naphthalene (metabolites)	Phase II (in progress)	
Perfluorodecanoic acid	Phase II (in progress)	
Perfluorononanoic acid	Phase II (in progress)	
Perfluorooctane sulfonamide	Phase II (in progress)	
Perfluoroundecanoic acid	Phase II (in progress)	
Propyl paraben	Phase II (in progress)	
Pyrene (metabolites)	Phase II (in progress)	

[a] Chemical names reported here are from the NHANES SAS 9.3 variable labels (SAS Institute Inc., Cary, NC, 2000), except for metabolites, for which the parent chemical name is reported matching the ToxCast Inventory. Source: http://www.epa.gov/ncct/dsstox/sdf_toxcst.html.
[b] As of March 20, 2012. Source: http://www.epa.gov/ncct/dsstox/sdf_toxcst.html.
[c] Where assessed, as of February 15, 2008. Source: http://www.epa.gov/ncct/dsstox/sdf_iristr.html.

at least some data formats, and generally are available in downloadable form to allow for more customized and detailed analysis. Indexing specific chemicals of interest involves matching complex formulaic names and other identifiers such as CASRNs to chemical structure. SMILES, structured data file (SDF), and International Chemical Identifier (InChI) are examples of common formats used to report chemical structure. While few standards are consistently adopted, by comparing chemicals on a structural basis instead of a name basis, there is less ambiguity in structure-activity relationships. Williams et al.[9] report on these and other data quality issues related to chemical identity in existing chemoinformatic databases.

In addition to correctly identifying and matching chemical identity between data sets, comparing dissimilar data sources presents a challenge for data management and analysis to risk scientists. Overall design, quality control, and indexing or linking capabilities deserve further attention before risk scientists can independently use and interpret data from several sources. The following paragraphs suggest how databasing and/or design of publicly available data sets may enable risk assessors to more easily compile and use these data sources in risk assessment.

DESIGNING PUBLICLY AVAILABLE TOXICOLOGICAL DATA SETS

In the absence of data standards or regulatory-endorsed workflows using publicly available data sets, several design considerations can assist in increasing the usefulness of these data sets to risk scientists and risk assessors. For example, adequate metadata and documentation are available for some, but not all, publicly available data sets. Furthermore, the specific toxicological vocabulary or nomenclature used as well as the units in which data are reported can vary significantly between data sources.

Various forms of both traditional and computational toxicological hazard data sets can be organized hierarchically to enable more relational or endpoint-specific comparisons. For example, when effect types (e.g., mortality, pathology, hematology), effect targets (e.g., liver, kidney), effect descriptions (e.g., hyperplasia, neoplasm), and effect type subclasses (e.g., alanine aminotransferase level) are defined, individual study effects reported in databases can be more readily analyzed. Other important data reflect variation in the species, strain, and age of test animals, as well as different descriptors for *in vitro* test systems or the number of animals, doses, or concentrations tested. Wherever risk assessors must compile data from various sources or studies, study reliability is also important to assess and report along with the data. One commonly accepted metric of study reliability is the Klimisch score.[25]

Given the complexity of toxicological information and varying methods for databasing this information, permitting stakeholders and outside researchers to review and offer comments on data organization or data records can provide the curators and publishers of publicly available data sets increased data integrity. Linking data sets can also help data integrity and foster easier communication and analysis between data sets. Once publicly available data sets enter the public domain, other private and public databases begin to reference the data sets or include them within their own data sets. While these new compiled data sets provide useful methods for analysis, they run the risk of being a snapshot of data sets that inevitably include some inaccuracies. Citing the importance of accurate physico-chemical and structural data, for example, Williams et al.[9] cite several related data sources that have used PubChem content wherefore erroneous name-structure synonyms or chemical names have proliferated across databases.

Recognizing the usefulness of comparing measured and modeled chemical property, hazard data, and risk management decisions across regulatory jurisdictions, the Organisation for Economic Cooperation and Development (OECD) has developed a publicly available QSAR Toolbox that searches several publicly available toxicological data sources in a workflow environment.[26] This toolbox is useful both to risk assessors and registrants seeking to fill data gaps on specific chemicals. However, at present, the use of curated toolboxes to compile data from validated sources cannot fully replace collecting or viewing data in the original data source. Nevertheless, as publicly available data sets receive more attention and further analyses suggest how they may inform risk assessment, designing new toolboxes and establishing regulatory workflows will assist in defining how computational data can be used to assess more chemicals.

ANALOGIES TO THE HUMAN GENOME PROJECT IN COMPUTATIONAL TOXICOLOGY

Similarities exist between the genome, transcriptome, and proteome data generation efforts of the Human Genome Project (HGP) and recent efforts to map the human toxome.[27] Indeed, the NRC[28] suggests that fully integrating toxicogenomic technologies into predictive toxicology will require a large coordinated effort similar to the HGP. Spurred by advances in analytical, *in vitro*, and *in silico* methods, toxicogenomics focuses more on data management, analysis, and interpretation as opposed to data generation, which was the initial focus of the HGP.[28] Indeed, the amount of toxicogenomic data collected over a course of a few days, weeks, months, or years today easily exceeds the amount of data collected during the first dozen or more years of the HGP. This is largely due to new, data-rich HTS technologies, many of which were originally designed to support the HGP or related efforts. Like the HGP, efforts to map the human toxome will rely on mostly government-sponsored, publicly available data sets. Researchers and companies have also shared data, both confidentially and openly. Given the vast amount of data available, however, useful public data sources must be carefully identified, compiled or linked, and ultimately assessed before they can meaningfully inform risk assessment.

Similarities exist in the publicly available data sets created for the HGP and those that can be used to map the human toxome or conduct other risk assessment–focused analyses. Each data set uses a specific alphabet, vocabulary, and syntax. For the HGP, nucleosides (e.g., adenosine, thymine, guanine, cysteine, etc.) provide a relatively simple alphabet. Sequence lengths are typically long, which creates a complex vocabulary. Genetic variation, random mutation, and manipulation of gene sequences produce a wide language of genes whose role in biology provides an elegant syntax. Understanding the biological function, syntax, and variation of genes and their transcription products is an area of continued research in both genomics and toxicogenomics. Genomics has already adopted standard data formats and methods that can be used to rapidly analyze sequence data across a variety of species (e.g., BLAST searches).

Indeed, similar methods can be developed to enable targeted searches of the human toxome, and perhaps the toxome of specific subpopulations or other sensitive organisms. Given that species share similar gene sequences and molecular pathways, toxicologists have begun to use toxicogenomics to develop data vocabularies for the toxome that build on and that can be used interchangeably with the vocabularies of the HGP. Toxicogenomics suggests that most if not all toxicological mechanisms of action (MOAs) can be explained at the genomic level at which a toxicant or novel molecule reacts with a living system to produce adverse effects.[29] Whereas risk assessors thus far are limited to using toxicogenomics as an *a posteriori* tool to assess the effects of specific chemicals, reverse pharmacologists use toxicogenomics to guide the discovery of safe biologics, therapeutic molecules, or drugs. Recently, reverse pharmacologists have provided risk assessors data on failed pharmaceuticals that can be used in the validation of predictive models based on computational toxicology data.[17,30] As computational toxicology matures, it is likely that connections between the genomic, pharmaceutical, and risk assessment data communities will strengthen, and further data exchanges will be useful across fields of study. However, the complexity of omic, chemical, and toxicological data underscores the importance of developing a more standard vocabulary and syntax prior to the use or sharing of data from multiple sources for risk assessment.

CHEMICAL DOMAIN AND LIMITATIONS TO DATA ANALYSIS OF TRADITIONAL AND COMPUTATIONAL TOXICOLOGY DATA

Chemical selection and prioritization techniques are central to obtaining good value and useful toxicological data that can be used to predict health hazards. Typically, both exposure estimates and hazard assessments are considered when selecting chemicals to assess in greater detail. Exposure estimates can be obtained from publicly available human exposure data and passive or active environmental sampling. Epidemiological studies arguably provide the highest level of detail through cross-sectional studies of the general population and through case-control studies of at-risk populations. In some cases, case-cohort studies can further contextualize the relative risk in at-risk populations compared to background levels of exposure in the general population. However, given the complexity and costs of conducting analyses of human samples, specific chemicals are selected for more focused analysis. The selection of which chemicals to study has been a complex topic. Knowledge of potential hazard and anecdotal or primary knowledge of exposure guide the selection of which chemicals to assess in these studies. For NHANES, the CDC considers factors such as evidence of exposure in the U.S. population; occurrence and significance of health effects at a given level of exposure; desire to track public health initiatives to reduce exposure to a given agent; existence of analytical methods to measure biologically relevant concentrations of the chemical; availability of sufficient tissue specimens, especially blood or urine; and cost-effectiveness.[31]

Similarly, given the expense, time, and use of animals needed to generate useful toxicological data, regulatory agencies adopt tiered approaches to selecting chemicals of concern for further hazard assessment. Tiered approaches are equally useful for exposure, hazard, and risk assessment; however, the resulting data sets do not always produce data on the same chemicals. There are many reasons for this, mostly relating to how scientists or regulators prioritize assessing hazard versus exposure. For example, in order to analyze positive results to assist hazard assessment and problem formulation, Phase I ToxCast™ and related toxicogenomic efforts have placed greater importance on hazard. NHANES and other exposure-based studies rely more on exposure but also take into account a margin of exposure (MOE) approach. The MOE approach suggests prioritizing chemicals for further assessment wherever hazard and exposure are expected to overlap in a population. Whether a MOE approach can be adopted for risk assessment purposes in light of emerging, publicly available hazard and exposure data sets requires further consideration. For instance, data are not always available from related data sources for the same chemical, and more generally, hazard or exposure data are not always available for a given chemical.

The types of conclusions that can be reached using computational toxicology data are limited by the extent and nature of the available data and what data are used. For example, Knudsen et al.[32] focused mostly on registered pesticides and reviewed their findings in comparison to those of Chernoff et al.,[33] who assessed more industrial chemicals, and Matthews et al.,[34] who assessed untested chemicals. Computational toxicology data has been assessed for chronic or cancer studies in rats and cancer studies in mice,[19] multi-generation reproductive studies in rats,[30] and prenatal developmental toxicity in rats and rabbits.[32]

Differences in the biology of test animals are known to impact the sensitivity of traditional toxicity testing to identify toxic effects. For example, in a database study characterizing developmental data now available in ToxRefDB, reduction in fetal weight was observed in both rats and rabbits[32] for pesticides selected for assessment by the EPA's Office of Pesticide Programs, whereas in pesticides selected for assessment by NTP, these reductions were observed less so in rabbits.[33] Trends in the traditional toxicity data landscape vary by species and endpoint, in addition to the specific chemicals targeted for assessment. These trends require further analysis prior to or in conjunction with new data analyses using computational data sets. For example, Knudsen et al.[32] suggest that an underlying biology may be observed that could identify specific apical endpoints that are most common, most sensitive, or that in some other way lend to easier clustering of chemicals or anchoring of *in vivo* data. Whether this underlying biology is a result of similarities between chemicals being assessed or an extension of traditional effects observed in animals, however, is not clear.

DATA SEMANTICS AND LIMITATIONS TO RELATING HTS DATA TO *IN VIVO* EFFECTS

Different methods have been proposed to analyze computational toxicology data using expert knowledge of specific *in vitro* assays[30,35] or using entirely statistical approaches.[35] The results from these analyses suggest that computational toxicology data may be useful in screening chemicals for further study and for identifying single or groups of related *in vitro* responses that can be likened to initiating molecular events that can be interpreted as risk factors for toxicity.

For example, Martin et al.[30] developed a predictive model for reproductive toxicity using *in silico* and *in vitro* ToxCast™ data and assessed the accuracy of its results using *in vivo* data from ToxRefDB. A set of 36 assays were selected as likely significant indicators of reproductive toxicity based on univariate association. Through the use of combinatorial techniques, several genes were identified as predictive targets of reproductive toxicity. The feature set, which included mostly nuclear receptors, was as follows: peroxisome proliferator activated receptor alpha and gamma (PPARA and PPARG), androgen receptor (AR), estrogen receptor alpha (ESR1), cytochrome P450 enzyme (CYP), G protein-coupled receptor (GPCR), pregnane X receptor (PXR, NR1L2), epidermal growth factor 1 (EGFR1), transforming growth factor beta 1 (TGFB1), and nuclear factor kappa light-chain enhancer of activated B cells (NFKB). Cross-validation balanced accuracies and other performance metrics were presented for the model. The model was also tested with an external validation set and compared to classification and labeling data for a subset of chemicals. Activities in individual assays or groups of related assays were identified as being associated with reproductive toxicity observed *in vivo*, with varying rates of accuracy.

Using statistic approaches and without prefiltering HTS data by endpoint using *t*-test or univariate associations, HTS data have been shown to be much less predictive of *in vivo* responses.[35] Nevertheless, by analyzing what types of data analysis are or are not predictive of observed *in vivo* responses, researchers may further understand how predictive new HTS data can be and further test the hypothesis that mechanistic or initiating events can be observed in these data for known toxicants for use in screening other compounds for toxicity. Various publicly available data sources provide researchers with additional methods to make these comparisons. However, owing to semantic differences across publicly available data sets, further standardization of these data sets and the development of new data sets reporting existing traditional toxicology knowledge will likely be required for further validation of HTS data.

CONCLUSIONS

In an effort to prevent unnecessary testing, reduce costs, and avoid the use of animals in chemical hazard assessment, computational toxicologists have begun using HTS technologies to generate *in vitro* effects data for a variety of chemicals. While HTS data are collected efficiently without the use of animals, whether using only HTS data will be sufficient for regulatory purposes is not clear and is perhaps unlikely.[36] However, when combined with large-scale biomonitoring data, publicly available data sets can be used to suggest where chemical exposure and hazard are expected to overlap across the risk assessment paradigm. New uses of these data sets may further suggest how MOE or other approaches can be used in risk assessment.

Regardless of the outcome of how new data sources will be used in risk assessment, this chapter describes which emerging data sources are publicly available and provides examples of how these data have been analyzed to date. These analyses suggest that top-down and bottom-up approaches can provide a more exposure-based or mechanistic understanding of toxicology. While using publicly available computational and population-based data sets may not fully address the epistemology (i.e., knowing that vs. knowing how) of toxicology, as the use and confidence in new publicly available data sets increases, toxicology testing in the 21st century is expected to lead to new methods in risk assessment. Through a commitment to data sharing for related chemicals and endpoints across local, national, and international regulatory boundaries, risk assessors may soon be able to use new publicly available data sets in an observational or semi-quantitative way to inform risk assessment.

References

[1] NRC, National Research Council. Toxicity testing in the 21st century: a vision and a strategy. Washington DC: The National Academies Press; 2007.

[2] NTP, National Toxiclogy Program. High throughput screening initiative. Durham, NC: U.S. Department of Health and Human Services; 2012.
[3] Mendez Jr. W, Eftim SE. Biomarkers of perchlorate exposure are correlated with circulating thyroid hormone levels in the 2007–2008 NHANES. Environ Res 2012;118:137–44.
[4] Carpenter DO. Health effects of persistent organic pollutants: the challenge for the Pacific Basin and for the world. Rev Environ Health 2011;26:61–9.
[5] Woodruff TJ, Zota AR, Schwartz JM. Environmental chemicals in pregnant women in the United States: NHANES 2003–2004. Environ Health Perspect 2011;119:878–85.
[6] CDC, Centers for Disease Control and Prevention. Fourth national report on human exposure to environmental chemicals. Atlanta, GA: Department of Health and Human Services; 2012.
[7] IOM, Institute of Medicine. Climate change, the indoor environment, and health (Committee on the Effect of Climate Change on Indoor Air Quality Public Health). Washington, DC: The National Academies Press; 2011.
[8] CDC, Centers for Disease Control and Prevention. Fourth national report on human exposure to environmental chemicals. Atlanta, GA: Department of Health and Human Services; 2009.
[9] CDC, Centers for Disease Control and Prevention. National Health and Nutrition Examination Survey (NHANES) data [1999–2010]. Hyattsville, MD: U.S. Department of Health and Human Services; 2012.
[10] Samet JM, Zeger SL, Dominici F, Curriero F, Coursac I, Dockery DW, et al. The national morbidity, mortality, and air pollution study. Part II: morbidity and mortality from air pollution in the United States. Res Rep 2000;94:5–70 [discussion 71-9].
[11] Peng RD, Welty LJ. The NMMAPSdata package. R News 2004;4:10–4.
[12] Barnett AG, Huang C, Turner L. Benefits of publicly available data. Epidemiology 2012;23:500–1.
[13] Dominici F, Peng RD, Ebisu K, Zeger SL, Samet JM, Bell ML. Does the effect of PM10 on mortality depend on PM nickel and vanadium content? A reanalysis of the NMMAPS data. Environ Health Perspect 2007;115:1701–3.
[14] U.S. EPA. Regulatory impact analysis for the federal implementation plans to reduce interstate transport of fine particulate matter and ozone in 27 states. Correction of SIP approvals for 22 states. Washington, DC: Office of Air and Radiation; 2011.
[15] Afshari CA, Hamadeh HK, Bushel PR. The evolution of bioinformatics in toxicology: advancing toxicogenomics. Toxicol Sci 2011;120(Suppl. 1):S225–37.
[16] Martin MT, Knudsen TB, Judson RS, Kavlock RJ, Dix DJ. Economic benefits of using adaptive predictive models of reproductive toxicity in the context of a tiered testing program. Syst Biol Reprod Med 2012;58:3–9.
[17] Kavlock R, Chandler K, Houck K, Hunter S, Judson R, Kleinstreuer N, et al. Update on EPA's ToxCast program: providing high throughput decision support tools for chemical risk management. Chem Res Toxicol 2012;25:1287–302.
[18] Dix DJ, Houck KA, Martin MT, Richard AM, Setzer RW, Kavlock RJ. The ToxCast program for prioritizing toxicity testing of environmental chemicals. Toxicol Sci 2007;95:5–12.
[19] Martin MT, Judson RS, Reif DM, Kavlock RJ, Dix DJ. Profiling chemicals based on chronic toxicity results from the U.S. EPA ToxRef database. Environ Health Perspect 2009;117:392–9.
[20] Judson RS, Kavlock RJ, Setzer RW, Hubal EA, Martin MT, Knudsen TB, et al. Estimating toxicity-related biological pathway altering doses for high-throughput chemical risk assessment. Chem Res Toxicol 2011;24:451–62.
[21] Shah I, Houck K, Judson RS, Kavlock RJ, Martin MT, Reif DM, et al. Using nuclear receptor activity to stratify hepatocarcinogens. PloS one 2011;6:e14584.
[22] Rusyn I, Daston GP. Computational toxicology: realizing the promise of the toxicity testing in the 21st century. *Environ Health Perspect* 2010;118:1047–50.
[23] Scinicariello F, Yesupriya A, Chang MH, Fowler BA. Modification by ALAD of the association between blood lead and blood pressure in the U.S. population: results from the Third National Health and Nutrition Examination Survey. Environ Health Perspect 2010;118:259–64.
[24] Judson R. Public databases supporting computational toxicology. J Toxicol Environ Health B Crit Rev 2010;13:218–31.
[25] Kanehisa M, Goto S, Hattori M, Aoki-Kinoshita KF, Itoh M, Kawashima S, et al. From genomics to chemical genomics: new developments in KEGG. Nucleic Acids Res 2006;34:D354–7.

[26] Cerami EG, Gross BE, Demir E, Rodchenkov I, Babur O, Anwar N, et al. Pathway commons, a web resource for biological pathway data. Nucleic Acids Res 2011;39:D685–90.
[27] Hartung T, McBride M. Food for thought ... on mapping the human toxome. ALTEX 2011;28:83–93.
[28] NRC, National Research Council. Applications of toxicogenomic technologies to predictive toxicology and risk assessment. Washington, DC: The National Academies Press; 2007.
[29] Boekelheide K, Campion SN. Toxicity testing in the 21st century: using the new toxicity testing paradigm to create a taxonomy of adverse effects. Toxicol Sci 2010;114:20–4.
[30] Martin MT, Knudsen TB, Reif DM, Houck KA, Judson RS, Kavlock RJ, et al. Predictive model of rat reproductive toxicity from ToxCast high throughput screening. Biol Reprod 2011;85:327–39.
[31] Gangwal S, Hubal EAC. Nanotoxicology: Gangwal et al. Respond. Environmental health perspectives 2012;120:a13–4.
[32] Knudsen TB, Martin MT, Kavlock RJ, Judson RS, Dix DJ, Singh AV. Profiling the activity of environmental chemicals in prenatal developmental toxicity studies using the U.S. EPA's ToxRefDB. Reprod Toxicol 2009;28:209–19.
[33] Chernoff N, Rogers EH, Gage MI, Francis BM. The relationship of maternal and fetal toxicity in developmental toxicology bioassays with notes on the biological significance of the "no observed adverse effect level." Reprod Toxicol 2008;25:192–202.
[34] Matthews EJ, Kruhlak NL, Daniel Benz R, Contrera JF. A comprehensive model for reproductive and developmental toxicity hazard identification: I. Development of a weight of evidence QSAR database. Regul Toxicol Pharmacol 2007;47:115–35.
[35] Thomas RS, Black M, Li L, Healy E, Chu TM, Bao W, et al. A comprehensive statistical analysis of predicting *in vivo* hazard using high-throughput *in vitro* screening. Toxicol Sci 2012;128:398–417.
[36] Bhattacharya S, Zhang Q, Carmichael PL, Boekelheide K, Andersen ME. Toxicity testing in the 21 century: defining new risk assessment approaches based on perturbation of intracellular toxicity pathways. PloS one 2011;6:e20887.
[37] Judson RS, Martin MT, Egeghy P, Gangwal S, Reif DM, Kothiya P, et al. Aggregating data for computational toxicology applications: the U.S. Environmental Protection Agency (EPA) Aggregated Computational Toxicology Resource (ACToR) system. Int J Mol Sci 2012;13:1805–31.
[38] Judson RS, Houck KA, Kavlock RJ, Knudsen TB, Martin MT, Mortensen HM, et al. *In vitro* screening of environmental chemicals for targeted testing prioritization: the ToxCast project. Environ Health Perspect 2010;118:485–92.
[39] Kavlock R, Dix D. Computational toxicology as implemented by the U.S. EPA: providing high throughput decision support tools for screening and assessing chemical exposure, hazard and risk. J Toxicol Environ Health B 2010;13:197–217.
[40] Egeghy PP, Judson R, Gangwal S, Mosher S, Smith D, Vail J, et al. The exposure data landscape for manufactured chemicals. Sci Total Environ 2012;414:159–66.
[41] Cohen Hubal EA, Richard A, Aylward L, Edwards S, Gallagher J, Goldsmith MR, et al. Advancing exposure characterization for chemical evaluation and risk assessment. J Toxicol Environ Health B Crit Rev 2010;13:299–313.
[42] Richard AM, Williams CLR. Public sources of mutagenicity and carcinogenicity data: use in structure-activity relationship models. In: Benigni R., editor. Quantitative Structure-Activity Relationship (QSAR) models of mutagens and carcinogens. Boca Raton, FL: CRC Press; 2003. p. 151-79.
[43] Richard AM, Williams CR. Distributed structure-searchable toxicity (DSSTox) public database network: a proposal. Mutat Res 2002;499:27–52.
[44] Fostel JM. Towards standards for data exchange and integration and their impact on a public database such as CEBS (Chemical Effects in Biological Systems). Toxicol Appl Pharmacol 2008;233:54–62.
[45] Waters M, Stasiewicz S, Merrick BA, Tomer K, Bushel P, Paules R, et al. CEBS—Chemical Effects in Biological Systems: a public data repository integrating study design and toxicity data with microarray and proteomics data. Nucleic Acids Res 2008;36:D892–900.
[46] Waters M, Jackson M. Databases applicable to quantitative hazard/risk assessment—towards a predictive systems toxicology. Toxicol Appl Pharmacol 2008;233:34–44.
[47] McHale CM, Zhang L, Hubbard AE, Smith MT. Toxicogenomic profiling of chemically exposed humans in risk assessment. Mutat Res 2010;705:172–83.

[48] NCBI, National Center for Biotechnology Information. NCBI online mendelian inheritance in man. Baltimore, MD: Johns Hopkins University, McKusick-Nathans Institute of Genetic Medicine; 2012.
[49] Parkinson H, Kapushesky M, Kolesnikov N, Rustici G, Shojatalab M, Abeygunawardena N, et al. ArrayExpress update—from an archive of functional genomics experiments to the atlas of gene expression. Nucleic Acids Res 2009;37:D868—72.
[50] Williams-Devane CR, Wolf MA, Richard AM. Toward a public toxicogenomics capability for supporting predictive toxicology: survey of current resources and chemical indexing of experiments in GEO and ArrayExpress. Toxicol Sci 2009;109:358—71.
[51] Southan C, Varkonyi P, Muresan S. Quantitative assessment of the expanding complementarity between public and commercial databases of bioactive compounds. J Cheminform 2009;1:10.
[52] Ankley GT, Bennett RS, Erickson RJ, Hoff DJ, Hornung MW, Johnson RD, et al. Adverse outcome pathways: a conceptual framework to support ecotoxicology research and risk assessment. Environ Toxicol Chem 2010;29:730—41.
[53] Davis AP, Murphy CG, Saraceni-Richards CA, Rosenstein MC, Wiegers TC, Mattingly CJ. Comparative toxicogenomics database: a knowledgebase and discovery tool for chemical-gene-disease networks. Nucleic Acids Res 2009;37:D786—92.
[54] Mattingly CJ, Rosenstein MC, Davis AP, Colby GT, Forrest Jr. JN, Boyer JL. The comparative toxicogenomics database: a cross-species resource for building chemical-gene interaction networks. Toxicol Sci 2006;92:587—95.
[55] Klein TE, Chang JT, Cho MK, Easton KL, Fergerson R, Hewett M, et al. Integrating genotype and phenotype information: an overview of the PharmGKB project. Pharmacogenetics research network and knowledge base. Pharmacogenomics J 2001;1:167—70.

CHAPTER

10

Computational Toxicology Experience and Applications for Risk Assessment in the Pharmaceutical Industry

Nigel Greene[1] and Mark Gosink[2]

[1]Compound Safety Prediction, Pfizer Inc., Groton, CT, USA, [2]Investigative Toxicology, Pfizer Inc., Groton, CT, USA

BACKGROUND

The costs to develop a new drug are staggeringly high. Cost estimates to take a new drug to market typically range between $1 and $2 billion.[1,2] However, *Forbes* magazine writer Matthew Herper said costs could be much higher, reaching an astounding $11 billion for some companies.[3] These costs are cumulative with a significant portion of the costs occurring in the late-stage clinical trials. For example, if a compound reaches phase III only to fail for safety issues, approximately 95% of the costs it would have taken to bring a safe compound to market have already been spent.[2] Even compounds that fail in late preclinical safety studies have consumed 45%–70% of the costs of a marketed compound.[2,4] The cost of drug development has continued to rise faster than the pace of new drugs reaching the marketplace.[5] In the long term, continuing the current drug development model is untenable. Among those recognizing this challenge was the U.S. Food and Drug Administration (FDA), which launched the Critical Path Initiative as an attempt to address this and other issues with the current drug development paradigm. As part of that initiative, the FDA identified the use of informatics techniques to streamline drug development.[6]

Informatics techniques offer a number of significant enhancements to the current drug discovery pipeline. Because these techniques take place in a virtual environment, they are both fast and low cost. Indeed, automated structural alerting has been utilized by the pharmaceutical industry for many years. The sophistication of these tools is regularly increasing our ability to detect certain types of toxicity with a high degree of certainty. As the numbers of structural models increases, it has also become significantly more difficult to compare all of these models to new chemical entities without the use of computers. More recently, genetic evaluations have begun to be considered in toxicological investigations. It has long been recognized that exaggerated pharmacology can lead to drug toxicities.[7,8] With the sequencing of the human genome, individual genetic variation has been recognized as the underlying mechanism behind a number of adverse drug reactions. An examination of the drug warnings at the FDA indicates that 10% of drugs have pharmacogenomic information in their labels, but this number is increasing exponentially.[9] New technologies have also allowed more thorough investigations of off-target activities by examining whole-cell or even whole-organism expression changes and thereby providing insights into the mechanisms driving these toxicological effects.

TWO MAIN CONSIDERATIONS

Relationship between the Chemical Structure of the Compound and the Adverse Biological Phenotype

The prediction of a compound's molecular properties based on an evaluation of its chemical structure, commonly referred to as the structure-activity relationship (SAR), has been commonplace in the chemical sciences for many decades. However, the early approaches developed by Hansch and Fujita and Free and Wilson using linear free energy formed the basic scientific framework for the quantitative correlation of chemical structure with biological activity and have spurred many developments in the field of quantitative structure-activity relationships (QSAR).[10,11] These QSAR techniques have been used in the modeling of biological properties of molecules, more specifically used to optimize the pharmacological activity within a series of chemically similar compounds.

More recently, SAR and QSAR techniques have been applied to the prediction and characterization of chemical toxicity.[12] These computational techniques have been applied to a wide variety of toxicological endpoints, from the prediction of LD50 (the dose at which 50% of a population exposed to the chemical will die) and maximum tolerated dose (MTD) values to *Salmonella typhimurium* (Ames) assay results, carcinogenic potential, and developmental toxicity effects. Generally, however, toxicological endpoints such as carcinogenicity, reproductive effects, and hepatotoxicity are mechanistically complex when compared to the prediction of pharmacological activity; hence, this leads to added difficulty when trying to predict these endpoints.[13]

Computational Prediction of Genotoxicity

In the late 1980s, large sets of compounds with Ames mutagenicity assay data became more widely accessible, helping to catalyze research into structure-based methods for predicting genotoxicity. A compound that has been shown to cause mutagenicity in the Ames assay is highly unlikely to be considered a viable candidate for further development as a new drug, so early identification of genotoxic compounds has been a major area of investment for the pharmaceutical industry.[14] This investment has naturally led to many commercial systems being developed, probably the best known of which are those developed by Lhasa Limited (http://www.lhasalimited.org), MultiCASE Inc. (http://www.multicase.com), and Leadscope Inc. (http://www.leadscope.com)—namely, Derek Nexus,[15] MC4PC,[16] and Leadscope Predictive Data Miner (LPDM), respectively.[16]

Many reviews and comparisons of these commercial systems have been published, and generally, the accuracy of these models ranges from 60% to 92% depending on the test set being used.[13,17–20] The practical application of these commercial systems in the pharmaceutical processes has focused primarily on the later stages of drug development where they are used to assess the potential hazards presented by low-level impurities and synthetic intermediates. The reason for this is not due to any limitation of the approaches, but more that medicinal chemistry practice is to avoid the incorporation of reactive chemical groups in order to minimize the potential for protein adducts and/or reactive metabolites, which can lead to a variety of adverse effects, not least of which is genotoxicity. In contrast, synthetic intermediates are designed to be somewhat reactive, and since these tend to be the most common source for drug impurities, these compounds are more of a concern for early assessment and control during the development process.

In more recent years, in parallel with the development of regulatory guidelines on the limits for genotoxic impurities (GTIs),[21,22] the pharmaceutical industry has developed strategies to help with the identification, monitoring, and control of these GTIs,[23] which have incorporated the use of these *in silico* predictions as part of a screening cascade. *In silico* prediction of genotoxicity has now become so widely accepted that the use of these predictive approaches has been incorporated into draft proposals for new regulatory guidelines addressing the assessment of GTIs in drug products.[18] While not yet universally accepted, these proposals recommend the use of two complementary methodologies (i.e., expert-derived SAR rules as well as a statistical or machine-learning approach) that have shown to have adequate coverage and sensitivity when used to predict the large quantity of public domain data[24] now available for both model development and testing.

Despite the relative maturity of the area of genotoxicity prediction, Ames mutagenicity prediction still remains an active area of research with new approaches and methods being developed every year; for example, see reference 25.

Carcinogenicity Prediction

It should not be forgotten that the original intent of the Ames assay was to provide an *in vitro* assessment of a chemical's carcinogenic potential, which is the actual adverse effect that we wish to avoid in the development of any new compound or drug. However, the underlying biological processes and mechanisms that can lead to carcinogenicity are far more complex than simply causing mutations in the DNA of a cell through direct or indirect chemical reactivity. Often these other mechanisms are not always fully understood, so the translation from genotoxic potential to causing cancer is not guaranteed. As a consequence, there have been numerous efforts to develop *in silico* models of carcinogenicity based primarily on data available from 2-year rodent carcinogenicity studies that are available in the public domain.[26–28]

Despite the complexity of the biological endpoint, these computational modeling efforts have had some degree of success in being able to distinguish carcinogens from noncarcinogens with sensitivity (number of positives identified/total number positives) in the range of 40%–50% and a specificity (number of negatives identified/total number of negatives) of 80%–90%.[16] However, when predictions were made prospectively for 30 chemicals that were to be tested by the National Toxicology Program (NTP), the accuracy of *in silico* systems ranged from as low as 25% to a high of 65% depending on the system used.[29]

With most *in silico* methods, there is a trade-off between sensitivity and specificity in that increased sensitivity often comes at the expense of a decrease in specificity. This is often as a result of multiple factors, including a general lack of knowledge of the multiple mechanisms involved in the event being modeled as well as inadequate representatives of each mechanism to be able to differentiate these in the model descriptor space. This is particularly true in the *in silico* prediction of carcinogenicity where the number of different biological mechanisms is vast and the number of chemicals with high-quality data is relatively few by comparison.

Prediction of the carcinogenic potential of pharmaceutical products has not been as extensively studied as environmental contaminants. This is partly due to the relative paucity of data available for marketed drugs and partly due to the fact that most drugs are screened much earlier in the development process for their genotoxic potential and so are rarely carried forward if deemed positive. In one evaluation, Snyder showed that Derek and MCASE had a positive predictivity (number of correct positive predictions/total number of positive predictions) of 57% and 67%, respectively, for a set of marketed pharmaceuticals. However, it should be noted that the number of positive predictions made by MCASE were considerably lower than Derek (33 versus 91, respectively), and therefore, the overall accuracy of each system would be considerably lower.[30] In the drug set reported by Snyder, 148 drugs (38%) were negative in the Ames assay but resulted in carcinogenic findings in one or both rodent species compared to just 24 drugs (6%) that were positive in both Ames and the 2-year rodent bioassay. It is therefore thought that most drugs cause carcinogenicity through nongenotoxic mechanisms or induce cancer as a result of their pharmacological mechanism; for example, the suppression of the immune system can lead to an increased risk of developing certain types of cancer such as lymphomas.[31] Since genotoxic carcinogens are of higher concern, use of *in silico* models has focused on identifying those compounds that cause genetic toxicity rather than attempting to predict the more complex and possibly species-specific carcinogenic effects of drug candidates.

Prediction of Specific Organ Toxicities

Drug-induced liver injury (DILI) is still a leading cause of the withdrawal of marketed drugs.[32,33] These hepatic effects can range from liver enzyme elevations to organ failure[34,35] and are often difficult to predict in the preclinical stages of drug development. For the pharmaceutical industry, hepatotoxicity discovered late in clinical trials or after the launch of the drug, leading to its withdrawal, has huge financial implications.[36] It is, therefore, important to try to identify these effects early in the drug development process. As a consequence, the prediction of this adverse organ toxicity based on the chemical structure of a compound would offer a valuable screening tool.[37] However, only recently have computational methods been more widely applied to the complex endpoint of hepatotoxicity.[38]

Despite the complexities of the toxicological endpoint, structural alerts for hepatotoxicity have been developed and incorporated into rule-based systems like Derek, and when evaluated against a set of marketed drugs and other compounds known to cause liver injury, showed a sensitivity of 46%, a specificity of 73%, with an overall accuracy of 56%.[39]

Investigators at the U.S. Food and Drug Administration used the Adverse Event Reporting System (AERS) data to evaluate the utility of structure-based QSAR approaches in predicting human hepatobiliary and urinary tract toxicities.[40–42] These approaches used multiple commercial tools including MC4PC, LPDM, and BioEpisteme to create predictive models for each of the AERS ontology terms for classifying adverse events. The predictive performance of these models was variable depending upon the tool and the endpoint, but sensitivity ranged from 28% to 52% with specificity ranging from 84% to 87%.[41]

Other computational models have used Bayesian approaches for predicting drug-induced liver injury, achieving similar levels of predictive performance with an overall accuracy of 60% and a sensitivity of 56% and specificity of 67%.[43] Lui et al. used a consensus modeling approach to improve upon the predictive performance for hepatotoxicity. They subdivided the DILI effects of compounds into 13 different side effects and then developed a separate computational model for each one. If a compound was predicted to be positive by three or more models, then an overall call of positive for DILI was assigned to the compound. Using this consensus approach, the authors claimed to achieve a sensitivity between 52% and 66%, a specificity of 67% to 93%, and accuracy of 66% to 70% depending on the data set used for testing. This approach gave an overall positive predictive value (correct positive predictions/total number positive predictions) of 91% across the three data sets used to test the system.[44]

Other sample areas of toxicology that have been studied with respect to structure-based computational modeling are ocular toxicity,[45] and the complex set of effects grouped under reproductive and developmental toxicity[46,47] where the authors claimed to achieve sensitivities ranging from 5% to 45% with specificities of between 65% and 95%. Similarly, researchers at the U.S. FDA looked at structure-based models for predicting the safe human clinical starting dose using the maximum recommended daily dose (MRDD).[48] The authors claimed to be able to predict the MRDD for 74% to 78% of the 160-compound test set to within 10-fold of the actual MRDD. However, on closer inspection of the test set, it is apparent that 75% of the compounds have an MRDD between 1 mg/kg and 100 mg/kg, so simply predicting everything to have an MRDD of 10 mg/kg would have achieved similar levels of success.

As with models for carcinogenicity, with most of these SAR or QSAR approaches, there tends to be a marked trade-off between sensitivity and specificity whereby as sensitivity increases, so specificity decreases. Many reasons for this observation can be rationalized in the same way that they can be for carcinogenic potential. One reason could be that the binary classification applied to the biological observation of the undesired effect is independent of the concentration or exposure at which it is observed; i.e., adverse effects at any concentration are classified as a positive. Similarly, absence of effect is classified as a negative but is not put into context with the maximum concentration tested in the studies. Another possible rationale for this observation is that while compounds may contain the necessary structural features to induce toxicity, they are not reaching the necessary site of action in sufficient concentrations to observe the effect. There are many variables that contribute to a lack of exposure, and only some of them can be predicted from the chemical structure alone.

The lack of sensitivity and/or specificity combined with an inability to predict the exposure that would elicit the adverse effect invariably makes these models difficult to use in practice as part of drug development. The majority of toxic effects can be tolerated with novel drug candidates provided there is an adequate margin of safety between the efficacious concentration, the concentration required to treat the disease, and the concentration where toxicity is observed. Although these models exist, they are not yet widely utilized in pharmaceutical development.

Relationship of Physical Properties to In Vivo Toxicity

In recent years there have been numerous efforts to study the relationship of physicochemical properties of compounds to their *in vivo* toxicity profiles. To get around the issue of concentration dependent effects identified in the previous section, Hughes et al. used a binary classification scheme to flag compounds that caused an *in vivo* effect at maximum total drug concentration in plasma (C_{max}) less than 10 μM as "toxic" and those that showed no *in vivo* effects at a 10 μM C_{max} concentration or greater being classified as "clean."[49] In an analysis of some 280 compounds, the authors observed that compounds with a calculated octanol/water partition coefficient (cLogP) greater than 3 and a calculated topological polar surface area (TPSA) of less than 75Å^2 are six times more likely to elicit *in vivo* effects at concentrations below 10 μM C_{max} (total drug) compared to compounds that have cLogP > 3 and TPSA > 75Å^2. The authors also observed that compounds with cLogP > 3 and TPSA < 75Å^2 are also more likely to have increased propensity for off-target pharmacology.

Luker et al.[50] studied the relationship between the physico-chemical properties of a set of 34 basic molecules and their success in being considered safe enough to be approved for dosing in man as part of a Phase I clinical trial. The analysis showed that planarity or flatness of a molecule led to an increased risk of preclinical attrition, along with increasing numbers of potential positively charged centers ("Poison centers"). The authors suggest that the relationships between properties are complex, but excessive numbers of aromatic atoms cannot be simply overcome with the introduction of polar acceptor atoms. However, the authors note that the relatively small data set makes any definitive rules or conclusions difficult.

Finally, Sutherland et al. studied the relationship of molecular properties and *in vitro* assays to *in vivo* disposition and toxicity.[51] In their analysis, they looked at a database of some 3,773 compounds, which they divided into 173 chemical series to statistically control for covariances within the data. In looking at molecular pairs within each series, they observed that cytotoxicity (LC_{50}) measured in rat primary hepatocytes (RPHs) along with the experimentally derived volume of distribution (V_d) for a compound. When the difference in V_d is compared to the change in the lowest observed adverse effect level (LOAEL), defined as the C_{max} (μM) where histologic changes were first observed, between molecular pairs, it was observed that a decrease in V_d would lead to an increase in the LOAEL concentration. Similarly, a decrease in the RPH LC_{50} would lead to a decrease in the LOAEL for a given compound. Other significant factors that affected the LOAEL of a compound were the plasma clearance in rats normalized to the predicted free fraction and the AUC measured in the *in vivo* study.

While these properties were not derived computationally per se, there are reports of computational models to predict the steady-state volume of distribution (VD_{ss}) of compounds in humans,[52] which may ultimately allow for the *in silico* prediction of the *in vivo* LOAEL of a novel structure. This approach shows that lipophilicity, polar surface area, and pKa are all important determinants of VD_{ss} and hence support the previous observations that cLogP and TPSA as well as ionization state are key predictors of the plasma concentration at which toxicity is observed.

Polypharmacology Approaches to Predicting Adverse Effects

Following on from the analysis by Hughes et al. described earlier, the relationship between off-target promiscuity and *in vivo* toxicity was further explored by Wang et al.,[53] who reported that a correlation could be observed between a computed measure of promiscuity that would differentiate toxic compounds from nontoxic compounds as defined using the same scheme published by Hughes et al. They also noted that a promiscuity measure based on a small number of *in vitro* pharmacology binding assays performs as well as one calculated from a much larger assay panel, suggesting that compounds can cause toxicity through a generic disruption of pharmacology as well as through a specific highly potent interaction with a unique receptor. However, the authors also state that there still remain a high number of false negatives giving rise to the conclusion that toxicity can result from multiple mechanisms, so it will be necessary to combine multiple approaches to measure and predict the *in vivo* effects of a compound.

Lounkine et al.[54] published a computational approach, using the SEA approach,[55–57] to predict the off-target effects of a set of 656 drug molecules across a panel of 73 receptors that have been linked to causing adverse effects in the clinic. After removing those interactions present in the training database of the algorithm, the authors claimed to accurately predict the off-target activity of 48% of the 1,042 compound-target interactions considered to be significant by the SEA methodology. Three hundred forty-eight interactions were confirmed through other sources of data not available to the SEA algorithm, but no comment was made on the potency of these interactions. However, subsequent testing of 694 predicted interactions yielded only 151 that were active at concentrations less than 30 µM, and of these, 48 were active below 1 µM. While the performance of this methodology is not at a level where it could replace compound-target experimental testing, it is considered to provide value in prioritizing potential off-target interactions for consideration.

Relationship between the Gene Function and the Adverse Biological Phenotype

When a new gene is being evaluated as a therapeutic target, it is common practice to consider not only what role that gene plays in disease efficacy, but also other cellular roles for that target. When this is done, potential risks directly related to target activities can either be avoided or at least have derisking strategies in place.

Most pharmaceutical companies maintain a list of proteins with well-characterized safety concerns. If a new drug is active against any of those proteins, it is often enough to flag that new therapeutic as a potential safety risk. For example, pharmaceuticals routinely screen new compounds for their activity against the human Ether-a-go-go Related Gene (hERG). Inhibition of this potassium channel is known to cause QT prolongation and can ultimately lead to heart failure.[58,59] Significant hERG activity has led to termination of otherwise promising compounds. This prescreening of new compounds is now so much of standard practice that companies such as Cerep (http://www.cerep.com/), Charles River Laboratories (http://www.criver.com/), and WuXi (http://www.wuxiapptec.com/) have arisen to sell gene-activity screening services to pharmaceutical organizations. The number of these providers has grown so great and their toxicity-related services so diverse that several organizations have developed websites such as the Assay Depot (http://www.assaydepot.com/) and Biocompare (http://www.biocompare.com/) to list and compare these vendors.

A significant difficulty in toxicology predictions arises from the complex interactive nature of the genome. For some genetic diseases to manifest into observable phenotypes, multiple genes may have to be mutated. Similarly, a number of drugs are known to have activity toward multiple gene targets. The adverse phenotypes associated with some of these compounds may require activity against multiple genes. Another consideration is that 90% of the human genome is believed to undergo alternate splicing.[60,61] These splice variants can have significantly different enzymatic/regulatory activities. For example, Lamba et al. reported on the splice-variant forms of the important drug metabolism regulators PXR and CAR and how alterations of these genes alter drug metabolism.[62] Similarly, Rodriguez-Paz demonstrated that splice-variant forms of the NMDA receptor can have significantly different responses to phencyclidine-like drugs.[63] Since splice variants are almost universally not assayed when considering off-target effects, a number of splice-variant-driven adverse events could be missed in these screening panels. Of particular concern in drug discovery is the splice variation seen in drug transport and metabolism.[64–66]

Another complexity to toxicity predictions arises from the interactive nature of the metabolic and regulatory pathways in which all genes lie. Pathways can converge on single genes or may regulate one or more cellular processes. Because of this, compounds active against any one gene in a pathway or cellular process may all exhibit similar phenotypes. For example, in mice, at least 29 gene mutations have been reported to induce an "increased systemic arterial diastolic blood pressure" phenotype (Table 10-1). These genes span a broad range of activities from membrane receptors, kinases, and signal transducers to hormones. It would be next to impossible to identify a common compound structure that had significant activities against all of these moieties. Similarly, in human genome-wide association studies, dozens of genes spanning diverse biological functions have been linked to hypertension.[67,68]

Literature Mining

A significant portion of toxicity knowledge is in the form of textual data consisting of journal articles, drug documents/labels, safety study reports, and even books. Additionally, the amount of text-based information continues to increase dramatically

TABLE 10-1 Mouse Gene Mutations Resulting in Increased Blood Pressure Phenotype

Gene—Description	Mutation PubMed Reference ID	Activity
Abcc9—ATP-binding cassette, sub-family C (CFTR/MRP), member 9	12122112	Potassium Channel
Add2—adducin 2 (beta)	10988280	Actin Binder
Agtr2—angiotensin II receptor, type 2	7477267	GPCR
Cacnb3—calcium channel, voltage-dependent, beta 3 subunit	12920136	Calcium Channel
Cd47—CD47 antigen	19284971	Cell Adhesion
Ctsa—cathepsin A	18391110	Peptidase
Cyp4a14—cytochrome P450, family 4, subfamily a, polypeptide 14	11320253	Cytochrome p450
Dll1—delta-like 1 (Drosophila)	19562077	Notch Binder
Drd2—dopamine receptor D2	11566895	GPCR
Drd3—dopamine receptor D3	9691085	GPCR
Drd5—dopamine receptor D5	15598876	GPCR
Ednrb—endothelin receptor type B	16868309	GPCR
Eln—elastin	14597767	Extracellular Matrix
Fkbp1b—FK506 binding protein 1b	11907581	Isomerase
Hras1—Harvey rat sarcoma virus oncogene 1	18483625	GTPase
Igf1—insulin-like growth factor 1	8958230	Hormone Activity
Irs1—insulin receptor substrate 1	9541510	Signal Transducer
Kcnn4—potassium intermediate/small conductance calcium-activated channel, subfamily N, member 4	16873714	Potassium Channel
Kras—v-Ki-ras2 Kirsten rat sarcoma viral oncogene homolog	15864294	GTPase
Nos3—nitric oxide synthase 3, endothelial cell	15350840	NO synthase
Nppa—natriuretic peptide type A	10100923	Hormone Activity
Npr1—natriuretic peptide receptor 1	7477288	Guanylyl cyclase
Pdc—phosducin	19959875	Signal Transducer
Prkg1—protein kinase, cGMP-dependent, type I	18448676	Kinase
Ptgs2—prostaglandin-endoperoxide synthase 2	19959875	Peroxidase
Thbs1—thrombospondin 1	19284971	Extracellular Matrix
Tlr5—toll-like receptor 5	20203013	Toll-like Receptor
Vdr—vitamin D receptor	12122115	Nuclear Hormone Receptor
Wnk1—WNK lysine deficient protein kinase 1	20921400	Kinase

every year. Because of the vast amount of information stored as text, literature mining may represent one of the most highly utilized but ultimately ineffectively exploited sources of predicting toxicity. Successfully mining this information can yield rapid insights into potential adverse activities for a new target or compound.

Search engines such as Google, Yahoo!, and Bing offer a powerful means to search literature quickly and easily. For example, while Google indexes approximately 50 billion documents,[69] it takes only a second or two to yield search results. The search engine behind Google is commercially available and is utilized by a number of companies to index their internal documents.[70] These search tools can sometimes be useful for general searches and can sometimes provide links to toxicity information about specific drugs. However, the sheer volume of information returned can make identification of toxicity information hard to find and the interpretation difficult. Because of this, a number of search engines dedicated to scientific reports have been developed. A list of many of these dedicated search engines is available from Wikipedia.[71] One example, the Google academic search site, Google Scholar, is dedicated to indexing scientific information ranging from journal articles to conference abstracts.[72] Microsoft and Elsevier also offer search engines dedicated to scientific literature called Microsoft Academic Search[73] and SCIRUS,[74] respectively. One important aspect provided by several scientific search engines is the translation and indexing of articles written in other languages. Another feature offered by Google Scholar is the digitizing of older scientific articles.[75] Since many articles written before the advent of computers are stored as digital images, direct searches for textual information is impossible. However, Google uses optical character recognition software to convert these images into text documents that can then be indexed in the search engines.

While scientific search engines, such as Google Scholar, offer an easy means of searching vast amounts of scientific literature, they are limited in their ability to filter and otherwise customize their searches. This limitation can lead to a lack of specificity in the search results returned.[76–78] This lack of specificity can be problematic when searching for specific articles about areas of toxicity. The complex search capabilities available at PubMed are tailored to the needs of the biomedical community. With over 21 million annotated articles, PubMed remains one of the most highly utilized scientific search engines.[79,80] The PubMed database is also integrated with the Entrez databases, thereby permitting easy links to compounds, genes, and other useful data sources.[81] Unlike the simple queries available at Google Scholar, PubMed searches can be enhanced through the use of numerous filter options and the ability to use complex Boolean expressions.[82,83] MedLine articles, which make up the majority of articles in PubMed, have also been manually annotated with information about the content of the articles that can be utilized as fields within a PubMed query.[84,85] One set of fields, Medical Subject Headings (MESH), is particularly useful for focusing searches to specific research topics. MESH is a hierarchically controlled vocabulary with over 26,000 terms.[86] Many of the MESH headings can further be restricted using subheadings. MESH headings can be extremely useful when searching for particular aspects of toxicity. For example, for a researcher interested in learning about the relatively rare thrombocytopenia observed with doxorubicin treatment, a simple search of PubMed for doxorubicin returns over 46,000 articles. Even searching for "doxorubicin AND thrombocytopenia" returns over 1,000 articles. However, searching with the MESH term "thrombocytopenia" (i.e., "doxorubicin AND thrombocytopenia[mesh]") returns a much more manageable 367 articles. The results can be further restricted by restricting the search to those articles where thrombocytopenia is the major topic by using the field "[major]" rather than "[mesh]." At the time this article was written, the results of the search yielded 61 articles. If one was further interested in preclinical rat studies, adding the MESH heading "rats[mesh]" to the search narrowed the results to two articles. Despite the significant advantages to the researcher in terms of specificity, only about 29% of PubMed queries contain a Boolean or a field of any kind, whereas about 71% contain neither (Majda Valjavec-Gratian, Ph.D., NCBI User Services, personal communication).

Beyond the publically available literature repositories, most pharmaceutical companies maintain large repositories of internal toxicity data in the form of documents and reports. To extract information from these sources, one must utilize natural-language processing (NLP) approaches. NLP can also be utilized on public textual repositories to extract additional information. One immediate issue that arises before any NLP efforts can take place is converting textual documents into computer-readable formats. If the documents to be processed are stored as images (as is the case with many older documents), this work can take a considerable amount of effort involving optical character recognition (OCR). OCR can further be complicated with documents that are handwritten or have handwritten modifications and/or annotations. A full discussion of OCR techniques and applications is beyond the scope of this review.

Once documents are in a machine-readable format, the next step is identification of entities (genes, compounds, phenotypes, etc.) within the document and then mapping those entities to the root term.[87] Such mapping to root terms is prone to error because of the lack of consistency in entity names and even in single names mapping to multiple entities.[87,88] At this stage, the simplest text-mining approach, co-occurrence, can be applied to the documents. Co-occurrence can be measured in a number of ways. In its simplest terms, co-occurrence is a count of the number of times two entities occur in a sentence, a paragraph, a paper, or even a set of documents. Some applications apply a statistical analysis to the counts to look for an overabundance of counts compared to the frequency of each entity.[87,89] While this approach is prone to a lack of specificity, it can be useful in some applications. It has been used successfully in a number of applications such as the GenSensor Suite and CoPub to identify pathways related to disease phenotypes and disease phenotypes related to genes.[90,91] The precision of text-mining can be increased through the use of machine-learning or statistical text-mining approaches.[87] Machine-learning approaches use word frequencies within documents and text corpora to determine the significance of co-occurrences.

Next, natural-language processing (NLP), sometimes called rules-based, approaches use semantic parsing to break document sentences into their components such as nouns (entities) and verbs/adverbs (interactions). NLP approaches allow computers to determine entity interactions with significantly better precision than do co-occurrence approaches.[87,92] For example, in this abstract sentence using the preceding co-occurrence approach, the sentence "2-Butoxyethanol induced cycle delay but did not induce either sister chromatid exchanges or chromosomal aberrations in ... (pmid12571679)" would associate "2-butoxyethanol" with "chromosomal aberrations." NLP approaches could detect the negation and therefore would not make a positive association.

Finally, there are a number of open source initiatives and libraries to assist researchers who wish to develop their own text-mining approaches. Two particularly useful resources are the U.S. National Library of Medicine's libraries of terms called the Medical Encyclopedia of Subject Headings (MESH) and the Unified Medical Language System (UMLS). MESH is a hierarchical controlled vocabulary of over 199,000 terms used to annotate MedLine articles.[93] The UMLS is a collection of over 900,000 biomedical terms that integrates millions of medical concepts into a controlled vocabulary.[94] The UMLS collection is useful for not only biomedical research articles but also health records and other such reports. A number of programming languages, such as PERL, also have built-in functions that simplify the extraction of words and sentence structures. There are also a number of text-mining initiatives that allow researchers to compare their methodologies against standardized literature repositories. One such group is BioCreAtIvE (Critical Assessment of Information Extraction systems in Biology), which was organized specifically for mining biomedical text sources.[95]

Genomic-Analysis Methods

Genomic analyses have been in use within the research community since the early 1990s. They have also become a mainstay of many toxicologic investigations and have been successfully utilized to deconvolute toxicity mechanisms.[96] Despite these successes, the techniques, tools, and analysis methodologies are still undergoing rapid changes and improvements. Because of the lack of standardization and the exploratory nature of genomic technologies, the FDA does not make use of genomic data when making regulatory decisions; however, to better understand the development of biomarkers, the FDA does allow for voluntary genomic data submissions (VGDS). Toward that end, the FDA and the EMEA issued the "Guidance for Industry: Pharmacogenomic Data Submissions (70 FR 14698)" and the "Guideline on Pharmacogenetics Briefing Meetings," respectively.[97,98]

MICROARRAYS

A number of technologies are available to simultaneously quantify the expression of large numbers of genes in tissues and cells. One of the most common and well-established methodologies is the use of microarrays. Microarray technology was developed in the 1990s, and a number of companies supply the instruments and reagents for microarray analysis. There are also companies that offer complete analysis from sample preparation to bioinformatic analyses of the results. Microarrays and off-shoot technologies such as bead arrays share the basic aspect of predefined oligonucleotides bound to a solid matrix. Labeled RNA or DNA is generated from the samples of interest and hybridized to the bound oligos. The nonspecific label is washed off and the RNA quantitated by the amount of label bound. The exact method that maps the label to the appropriate gene is technology specific. For example, with Affymetrix microarrays, the oligonucleotide probes are laid down in a defined location on a "chip." RNA expression is quantified based on the intensity of fluorescence in specific "spot" locations. However, with bead arrays, each oligonucleotide probe is bound to a different bead and each bead tagged with a nucleotide or fluorescent label. Beads are either mapped to individual genes after being stochastically bound to a microarray chip or are mapped to genes using fluorescence labeling by flow cytometry. In addition to RNA expression array, numerous array subtypes and methodologies have been developed to address specific research needs.

NEXT-GENERATION SEQUENCING

Recently, the costs of next-generation (or high-throughput) sequencing (NGS) have dropped significantly while sequence quality continues to improve.[99–101] (See Table 10-2.) NGS has several advantages over microarray technologies because microarrays require that the sequences to be assayed be known in advance so that probes can be developed. Therefore, tissues from nonstandard animals cannot be analyzed, nor can novel splice variants and micro-RNAs. (It should be noted that some companies such as Affymetrix and Illumina offer exon and/or miRNA arrays that can detect some splice-variant forms or miRNAs, respectively.) Because of these factors, NGS has begun to supplant microarrays for genomic analyses. For DNA samples, the DNA is analyzed directly; however, RNA samples are first retro-transcribed to DNA before sequencing. Three NGS technologies were recently reviewed by Loman et al.[102] The technologies reviewed all utilize some form of sequencing-by-synthesis approaches; however, several other technologies are available or are being developed.[103] One important aspect with NGS is the large amount of sequence data generated. For example, we found a single sample from an NGS experiment generated over 20 million sequence reads and over 5 gigabytes of data (unpublished data). Because of the amount of data generated, analysis can take hours or days to run on a

TABLE 10-2 Specialized NGS Techniques

Technique	Description	Comments
Whole Genome	Whole genome DNA is fragmented and then sequenced.	—
Target Region	One or more targeted regions are PCR amplified before sequencing.	Can save costs if only select region(s) is (are) of interest
Exome Sequencing	Most methods use DNA fragmentation followed by binding to matrix bound transcriptomic regions. (Several techniques are available.)	Focuses on protein coding regions where most Mendelian traits have been identified
DNA Methylation	DNA is chemically modified with bisulfite, converting cytosine to uracil. Bisulfite-treated DNA versus untreated DNA is compared to identify protected methylcytosine residues.	Used to identify epigenetic modifications
ChIP-Seq	DNA fragments are immunoprecipitated using antibodies against one or more transcription factors, histones, or other DNA-binding proteins.	Identifies promotor-binding or other DNA regulatory regions
RNA-Seq	Purified RNA is reverse transcribed to DNA before sequencing.	Used to measure differential gene expression and/or splice variation and miRNAs

desktop computer. Toward that end, a number of open source sequence analysis applications (such as TopHat, Cufflinks, and Bowtie) have been developed.[104,105] Like microarray technology, several derivative approaches have been developed to address specific scientific needs and to keep costs low.

PROTEOMICS/METABOLOMICS

Proteomics/metabolomics is at once one of the oldest and one of the newest technologies applied to medical disorder diagnosis. Tasting of urine was advocated by Hippocrates in 300 BCE for medical diagnosis, and the sweetness of urine in diabetics was noted even earlier.[106] However, because of the inherent technical difficulties, the use of protein-microarrays and mass-spectrometric analyses has only really been applied to toxicology since the early 1990s to 2000s. One such difficulty with MS proteomic analysis has been the identification of the protein in an MS peak. The use of MS/MS approaches has aided this effort by allowing individual peaks to be isolated, fragmented, and the fragments subjected to a further round of MS analysis. This second fragmentation pattern is compared to databases of fragmentation patterns and thereby de-convolutes the individual molecular species. Quantitation of molecular species between samples is often addressed either by "spiking" in isotopically labeled protein or compounds or in the case of proteomic analysis using post-extraction modification to isotopically tag proteins from two samples with heavy and light versions of a label. In the latter case, a number of isotope labeling techniques and compounds have been developed, such as ICAT, iTRAQ, and Tandem Mass Tags.[107] Many of these labels have been developed so that they react only with certain classes of proteins such as glycoproteins or protein with cysteine residues.[108]

Analysis Approaches

All of the preceding "omic" technologies can generate huge lists of genes, proteins, or metabolites and their differential expression values. While it may be possible to identify important biomarkers from expression change lists of samples exposed to toxicants, it's generally necessary to apply some sort of statistical analysis techniques to identify relevant changes. Toward this end, a large number of applications and methods can be utilized to identify these changes. In addition, many of the systems used to measure gene, protein, or metabolite values will have a custom set of included applications and methods to identify changes and may also have tools to interpret these changes. There are also a number of open source applications for change analysis. Perhaps the most utilized set of applications can be found at the Bioconductor project website (http://bioconductor.org/). The Bioconductor project is an open source, community-supported collection of R packages and data libraries. Bioconductor takes advantage of the open source statistics software environment known as R (http://www.r-project.org/). It is not uncommon for new computational biology analysis methodologies to be published as an R or Bioconductor package. Bioconductor contains not only numerous tools to identify gene, protein, and metabolite changes but also packages to interpret these changes. In addition to the Bioconductor tools, numerous other open source and commercial packages are available for omic analyses. We discuss some of the major categories of tools used in industry in the following sections.

GENE SETS, PATHWAYS, AND GENE SIGNATURES

Interpretation is a critical aspect, as many of these genomic technologies can generate lists of significantly changed elements that number in the hundreds to thousands. In computational toxicology, these large lists often need to be analyzed for patterns of expression in order to understand the nature of toxicity. Toward that end, a number of tools have been developed to identify these patterns. One popular method is gene set enrichment.[109] Gene set enrichment uses a statistical methodology to compare the list of changed genes, generated in one experiment, to collections of gene lists to identify significant overlap. Some gene set methodologies also utilize the directionality of gene changes.[109] Some of the earliest gene sets utilized in this approach were Gene Ontologies (http://www.geneontology.org/). Gene Ontologies (GOs) were originally developed as an informatics initiative to develop a controlled vocabulary for describing cellular functions, processes, and components in which a gene/protein participates. Using GO terms to analyze microarray data can help provide a better understanding of the mechanisms behind an observed toxicity. For example, troglitazone, an antidiabetic agent, was withdrawn from the market due to cases of idiosyncratic liver injury. One of the driving mechanisms behind this toxicity was later shown to be through mitochondrial injury.[110,111] Microarray analysis of troglitazone treated 3T3-L1 cells followed by gene set enrichment analysis using the DAVID tool reveals mitochondrion as the most significant GO cellular component (unpublished results). Another type of data that is commonly used with gene set analyses is metabolic and regulatory pathway data. In this case, each pathway is considered to be a collection of genes or metabolites, and analyses methodologies use these collections in the same way that they utilize GO category gene sets. However, one consideration the researcher needs to know is that there is no definite set of pathways. Numerous commercial and noncommercial pathway collections exist, and the genes identified as belonging to one pathway in one source (i.e., the "Citrate cycle [TCA cycle]" from KEGG) will likely not be the same as belonging to a similarly named pathway in a second source (i.e., the "Citrate Cycle" from Ingenuity). These gene differences will likely result in somewhat different pathways identified when using different tools. Some of the commonly used pathway sets are listed in Table 10-3. A more extensive list can be found at the Pathguide website (http://www.pathguide.org/).

However, even with these differences, pathway analysis has successfully been used to gain a deeper understanding of the mechanisms behind such diverse toxicities as acetaminophen in liver to diesel exhaust in lung.[112,113] Finally, a number of groups have developed gene expression signature databases. Some of the signature databases contain

TABLE 10-3 Commonly Used Pathway Collections

Pathway Set	Provider	Availability
BioCarta Pathway Diagrams	BioCarta	Free
KEGG	Kanehisa Laboratories	Free to academics
Pathway Studio	Elsevier	Commercial
Pathways Knowledge Base	Ingenuity	Commercial
Reactome KnowledgeBase	Reactome	Free

lists of genes that are differentially expressed in predefined cells upon treatment with drugs with known liabilities. Others are lists of genes known to be differentially expressed in established mechanisms of toxicity. In essence, these databases contain signature expression patterns of toxicity. One aspect of gene expression signatures is that since the number of genes comprising that signature is limited, focused arrays can be utilized rather than full microarray scans, which can provide significant cost savings. These lists are generally analyzed using the same methodologies as used with GO categories and pathways. A significant match between differentially expressed genes from an experimental compound treatment and one of these gene signatures suggests that the experimental compound may have the same toxicity profile as the matching drug's liabilities or that the experimental compound induces a known toxicity mechanism.

CAUSAL REASONING

Recently, some organizations have begun to utilize causal reasoning (CR) analysis to identify potential mechanisms behind expression data.[114] Like gene signatures and some forms of gene set enrichment (GSEA), CR analysis uses gene sets that change in response to some treatment. In the case of CR analysis, the direction of change (i.e., increased or decreased) is also considered in the calculation of significance. However, unlike the gene sets behind gene signature analysis and GSEA, much of the data behind CR analysis is based on expert manual annotation of the literature to identify gene changes. This approach allows CR analysis to have much greater flexibility than gene signature analysis. In CR analysis, the gene signatures (or hypotheses) are generally built by experts' manual read of the scientific literature in a particular area (i.e., hypoxic response) and their identification of which genes change and in which direction. Therefore large libraries of these hypotheses can be constructed around numerous toxicological mechanisms. This hypothesis library can then be used in CR analysis to identify hypotheses matching the gene changes seen. For example, in CR analysis, a hypothesis of cholestasis can be built from an examination of the literature for genes which increase (or decrease) under cholestatic conditions. If treatment with a particular compound induces many of the gene changes reported for the cholestasis hypothesis, then the treatment would be given a positive score for the induction of cholestasis.

CLASS DISCOVERY (UNSUPERVISED) AND CLASS PREDICTION (SUPERVISED)

Finally, a number of classical statistical approaches can be utilized to characterize expression data from genomic studies in the absence of pre-existing information. These approaches generally fall into unsupervised and supervised learning. In the absence of any additional information, unsupervised approaches are used to determine if the data naturally segregates into distinct classes or groups. Commonly used techniques include clustering (i.e., hierarchical, k-means, etc.), principal component analysis, and decision trees. This approach can be particularly useful if an observed toxicity is actually the result of multiple mechanisms. By clustering data both by sample and by expression values (2D clustering), one may be able to identify coregulated gene clusters, which can then be further analyzed using some of the methods described previously. Unsupervised clustering methods have also been applied to clinical samples to segregate patients into responders and nonresponders.[115,116] When nonresponders are precluded from a set of patients receiving a drug, the risk-benefit profile in a disease population can be significantly improved. These methods can sometimes provide insight into the underlying mechanisms in the various cluster groups. Similarly, a number of groups have applied unsupervised clustering approaches to toxicity studies to elucidate the underlying mechanisms.[117] For example, Waring et al. profiled a number of known hepatotoxicants and found that the compounds clustered into groups, which resulted in similar histopathology mechanisms.[118] Finally, once appropriate gene clusters have been identified, they can be utilized for predictive approaches or supervised methodologies.

In supervised or class prediction approaches, classes of mechanisms or compounds have already been established. New expression profiles are compared to the gene profiles in these class sets to predict into which known cluster a new profile fits. The sample from which the new profile is derived is then inferred to have similar effects. This approach is similar to the gene set approaches described previously; however, many gene set analysis methodologies do not consider the magnitude of change or even the directionality of change for some gene set approaches. A number of pharmaceutical companies are investigating these predictive approaches as a means of identifying potential toxic compounds early in development.[115,118,119] Some have even reported using these techniques to stop work on toxic molecules before they reach animal trials.[115] Early elimination of toxicants could save millions of dollars in drug development costs. There are numerous methodologies with which to do class prediction. Often these methodologies involve some sort of machine-learning approaches, which is a whole field of study, and an in-depth view is beyond the scope of this review. However, common approaches used in expression analysis include co-clustering, *t*-test, decision trees, regression, and support vector analysis. Many of these predictive methodologies are available through standard microarray analysis packages. One particularly useful set of tools is found in the open source BioConductor library for R.

SYSTEMS BIOLOGY/VIRTUAL LIVER OR HEART

In the past few years, several groups have begun using systems biology approaches to model and predict compound toxicity. There is some ambiguity in the literature around what is meant by "systems biology." Some authors use this term loosely to describe any type of omic data analysis. For the purposes of this review, we define "systems biology" as the development of mathematical models capable of predicting the appearance or disappearance of individual metabolites or the passage of a signal through a pathway. These approaches generally require detailed knowledge of the metabolic and regulatory pathways involved in a process or cell. Even simple models can be extremely complex using differential equations to follow the flow and levels of metabolites and/or enzyme states (i.e., amount phosphorylated versus unphosphorylated) through a pathway. To build this kind of pathway, one must have not only enzymatic linkages in a pathway but also their reaction rates. Consequently, most systems biology models are of smaller pathways and tend to focus on basic biology rather than predicting toxicity.[120–123] However, some groups are beginning to develop models to help predict some aspects of toxicity. For example, Acikgöz et al. developed a biotransformation model for diazepam in the liver.[124] Similarly, Stamatelos and colleagues developed a hepatocyte model of arsenic metabolism.[125] They were able to compare their model to a previously developed model for rats and suggested that differences in glutathione conjugation could explain some of the species differences in toxicity. Another group used a systems biology model to access how individual genetic variations would affect the metabolism of statins.[126] They showed that even a simple model could account for a significant amount of the observed individual variations. A few groups have begun to assemble more complex models involving hundreds of proteins and metabolites. These complex models require large amounts of transcriptomic or other omic data generated from multiple treatments and can require many man-years to assemble. For example, Entelos's Physiolab has been used to help companies predict drug responses to new compounds.[127,128] Other groups have also begun modeling the activities of the liver and heart, which may ultimately lead to a deeper understanding of how new drugs could affect these vital organs.[129–131]

SUMMARY

Computational toxicology has a proven track record of saving time, effort, and costs in the development of safe, new drugs. The ability of computational methods to predict genotoxicity and carcinogenicity has been utilized by the pharmaceutical industry for decades. The ability to predict organ-specific toxicities based on chemical structure alone is improving and is being explored to improve early compound selection. Target-mediated toxicity is utilized by industry today to screen for activity against known risky targets. More recently, companies have begun to utilize omic technologies to understand the mechanisms of toxicity. While these newer technologies are not yet included in the required filings to regulatory agencies, a number of companies have used these techniques to gain insight into mechanisms behind some of their drugs' adverse risks. Ultimately, omics technologies in conjunction with computational chemistry methodologies will result in getting safe and effective drugs to market in shorter timelines and with lower costs.

References

[1] Morgan S, Grootendorst P, Lexchine J, Cunninghama C, Greyson D. The cost of drug development: a systematic review. Health Policy 2011;100(1):4–17.
[2] Paul SM, Mytelka DS, Dunwiddie CT, Persinger CC, Munos BH, Linborg SR, et al. How to improve R&D productivity: the pharmaceutical industry's grand challenge. Nat Rev Drug Discov 2010;9(3):203–14.
[3] Herper M. The truly staggering cost of inventing new drugs. [Cited September 18, 2012]; Available from: <http://www.forbes.com/sites/matthewherper/2012/02/10/the-truly-staggering-cost-of-inventing-new-drugs/>; 2012.
[4] Rawlins MD. Cutting the cost of drug development? Nat Rev Drug Discov 2004;3(4):360–4.
[5] Dickson M, Gagnon JP. Key factors in the rising cost of new drug discovery and development. Nat Rev Drug Discov 2004;3(5):417–29.
[6] U.S. FDA. Critical path opportunities list. [Cited September 13, 2012]; Available from: <http://www.fda.gov/downloads/ScienceResearch/SpecialTopics/CriticalPathInitiative/CriticalPathOpportunitiesReports/UCM077254.pdf>; 2006.
[7] Gillespie MN, et al. Exaggerated nicotine-induced norepinephrine release from atherosclerotic rabbit hearts. Toxicology 1985;37(1-2):147–57.
[8] Edwards JG. Adverse effects of antianxiety drugs. Drugs 1981;22(6):495–514.

[9] Frueh FW, et al. Pharmacogenomic biomarker information in drug labels approved by the United States Food and Drug Administration: prevalence of related drug use. Pharmacotherapy 2008;28(8):992–8.
[10] Franke R. Theoretical drug design methods, vol. 7. Amsterdam: Elsevier; 1984.
[11] Fujita T. The extrathermodynamic approach in drug design. In: Hansch C, et al., editors. Comprehensive medicinal chemistry. New York: Pergamon; 1990. p. 497.
[12] Benigni R, Richard AM. Quantitative structure-based modeling applied to characterization and prediction of chemical toxicity. Methods 1998;14(3):264–76.
[13] Greene N. Computer systems for the prediction of toxicity: an update. Adv Drug Deliv Rev 2002;54(3):417–31.
[14] Muster W, et al. Computational toxicology in drug development. Drug Discov Today 2008;13(7-8):303–10.
[15] Marchant CA. Computational toxicology: a tool for all industries. Comput Mol Sci 2012;2(3):424–34.
[16] Matthews EJ, et al. Combined use of MC4PC, MDL-QSAR, BioEpisteme, leadscope PDM, and Derek for Windows software to achieve high-performance, high-confidence, mode of action-based predictions of chemical carcinogenesis in rodents. Toxicol Mech Methods 2008;18(2-3):189–206.
[17] Hillebrecht A, et al. Comparative evaluation of *in silico* systems for Ames test mutagenicity prediction: scope and limitations. Chem Res Toxicol 2011;24(6):843–54.
[18] Naven RT, Greene N, Williams RV. Latest advances in computational genotoxicity prediction. Expert Opin Drug Metab Toxicol 2012;8(12):1579–87.
[19] Snyder RD. Assessment of atypical DNA intercalating agents in biological and *in silico* systems. Mutat Res 2007;623(1-2):72–82.
[20] White AC, et al. A multiple *in silico* program approach for the prediction of mutagenicity from chemical structure. Mutat Res 2003;539(1-2):77–89.
[21] U.S. FDA. Guidance for industry, genotoxic and carcinogenic impurities in drug substance and drug products: recommended approaches. December 2008; Available from: <http://www.fda.gov/downloads/Drugs/GuidanceComplianceRegulatoryInformation/Guidances/ucm079235.pdf>.
[22] European Medicines Agency. Guideline on the limits of genotoxic impurities. June 28, 2006; Available from: <http://www.emea.europa.eu/docs/en_GB/document_library/Scientific_guideline/2009/09/WC500002903.pdf>.
[23] Muller L, et al. A rationale for determining, testing, and controlling specific impurities in pharmaceuticals that possess potential for genotoxicity. Regul Toxicol Pharmacol 2006;44(3):198–211.
[24] Hansen K, et al. Benchmark data set for *in silico* prediction of Ames mutagenicity. J Chem Inf Model 2009;49(9):2077–81.
[25] Xu C, et al. *In silico* prediction of chemical Ames mutagenicity. J Chem Inf Model 2012;52(11):2840–7.
[26] Gold LS, et al. Supplement to the Carcinogenic Potency Database (CPDB): results of animal bioassays published in the general literature in 1993 to 1994 and by the National Toxicology Program in 1995 to 1996. Environ Health Perspect 1999;107(Suppl. 4):527–600.
[27] Gold LS, et al. Supplement to the Carcinogenic Potency Database (CPDB): results of animal bioassays published in the general literature through 1997 and by the National Toxicology Program in 1997–1998. Toxicol Sci 2005;85(2):747–808.
[28] Haseman JK. Using the NTP database to assess the value of rodent carcinogenicity studies for determining human cancer risk. Drug Metab Rev 2000;32(2):169–86.
[29] Benigni R, Zito R. The second National Toxicology Program comparative exercise on the prediction of rodent carcinogenicity: definitive results. Mutat Res 2004;566(1):49–63.
[30] Snyder RD. An update on the genotoxicity and carcinogenicity of marketed pharmaceuticals with reference to *in silico* predictivity. Environ Mol Mutagen 2009;50(6):435–50.
[31] Vial T, Descotes J. Immunosuppressive drugs and cancer. Toxicology 2003;185(3):229–40.
[32] Holt MP, Ju C. Mechanisms of drug-induced liver injury. AAPS J 2006;8(1):E48–54.
[33] Kaplowitz N. Idiosyncratic drug hepatotoxicity. Nat Rev Drug Discov 2005;4(6):489–99.
[34] Williams DP. Toxicophores: investigations in drug safety. Toxicolog 2006;226(1):1–11.
[35] Zimmermann HJ, editor. The adverse effects of drugs and other chemicals on the liver. Philadelphia: Lippincott Williams & Wilkins; 1999.
[36] Kola I, Landis J. Can the pharmaceutical indstry reduce attrition rates? Nat Rev Drug Discov 2004;3:711–6.
[37] Greene N, Naven R. Early toxicity screening strategies. Curr Opin Drug Discov Devel 2009;12(1):90–7.

[38] Li AP. An integrated, multidisciplinary approach for drug safety assessment. Drug Discov Today 2004; 9(16):687–93.
[39] Greene N, et al. Developing structure-activity relationships for the prediction of hepatotoxicity. Chem Res Toxicol 2010;23(7):1215–22.
[40] Ursem CJ, et al. Identification of structure activity relationships for adverse effects of pharmaceuticals in humans: A. Use of FDA post-market reports to create a database of hepatobiliary and urinary tract toxicities. Regul Toxicol Pharmacol 2009;54(1):1–22.
[41] Matthews EJ, et al. Identification of structure-activity relationships for adverse effects of pharmaceuticals in humans: B. Use of (Q)SAR systems for early detection of drug-induced hepatobiliary and urinary tract toxicities. Regul Toxicol Pharmacol 2009;54(1):23–42.
[42] Matthews EJ, et al. Identification of structure-activity relationships for adverse effects of pharmaceuticals in humans: part C: use of QSAR and an expert system for the estimation of the mechanism of action of drug-induced hepatobiliary and urinary tract toxicities. Regul Toxicol Pharmacol 2009;54(1):4–65.
[43] Ekins S, Williams AJ, Xu JJ. A predictive ligand-based Bayesian model for human drug-induced liver injury. Drug Metab Dispos 2010;38(12):2302–8.
[44] Liu Z, et al. Translating clinical findings into knowledge in drug safety evaluation—drug induced liver injury prediction system (DILIps). PLoS Comput Biol 2011;7(12):e1002310.
[45] Somps CJ, et al. A current practice for predicting ocular toxicity of systemically delivered drugs. Cutan Ocul Toxicol 2009;28(1):1–18.
[46] Matthews EJ, et al. A comprehensive model for reproductive and developmental toxicity hazard identification: I. Development of a weight of evidence QSAR database. Regul Toxicol Pharmacol 2007; 47(2):115–35.
[47] Matthews EJ, et al. A comprehensive model for reproductive and developmental toxicity hazard identification: II. Construction of QSAR models to predict activities of untested chemicals. Regul Toxicol Pharmacol 2007;47(2):136–55.
[48] Contrera JF, et al. Estimating the safe starting dose in phase I clinical trials and no observed effect level based on QSAR modeling of the human maximum recommended daily dose. Regul Toxicol Pharmacol 2004; 40(3):185–206.
[49] Hughes JD, et al. Physiochemical drug properties associated with *in vivo* toxicological outcomes. Bioorg Med Chem Lett 2008;18(17):4872–5.
[50] Luker T, et al. Strategies to improve *in vivo* toxicology outcomes for basic candidate drug molecules. Bioorg Med Chem Lett 2011;21(19):5673–9.
[51] Sutherland JJ, et al. Relating molecular properties and *in vitro* assay results to *in vivo* drug disposition and toxicity outcomes. J Med Chem 2012;55(14):6455–66.
[52] Lombardo F, et al. A hybrid mixture discriminant analysis-random forest computational model for the prediction of volume of distribution of drugs in human. J Med Chem 2006;49(7):2262–7.
[53] Wang X, Greene N. Comparing measures of promiscuity and exploring their relationship to toxicity. Mol Inform 2012;31(2):145–59.
[54] Lounkine E, et al. Large-scale prediction and testing of drug activity on side-effect targets. Nature 2012; 486(7403):361–7.
[55] Hert J, et al. Quantifying the relationships among drug classes. J Chem Inf Model 2008;48(4):755–65.
[56] Keiser MJ, et al. Relating protein pharmacology by ligand chemistry. Nat Biotechnol 2007;25(2):197–206.
[57] Keiser MJ, et al. Predicting new molecular targets for known drugs. Nature 2009;462(7270):175–81.
[58] Raschi E, et al. hERG-related drug toxicity and models for predicting hERG liability and QT prolongation. Expert Opin Drug Metab Toxicol 2009;5(9):1005–21.
[59] Priest BT, Bell IM, Garcia ML. Role of hERG potassium channel assays in drug development. Channels (Austin) 2008;2(2):87–93.
[60] Pan Q, et al. Deep surveying of alternative splicing complexity in the human transcriptome by high-throughput sequencing. Nat Genet 2008;40(12):1413–5.
[61] Croft L, et al. ISIS, the intron information system, reveals the high frequency of alternative splicing in the human genome. Nat Genet 2000;24(4):340–1.
[62] Lamba J, Lamba V, Schuetz E. Genetic variants of PXR (NR1I2) and CAR (NR1I3) and their implications in drug metabolism and pharmacogenetics. Curr Drug Metab 2005;6(4):369–83.

[63] Rodriguez-Paz JM, Anantharam V, Treistman SN. Block of the N-methyl-D-aspartate receptor by phencyclidine-like drugs is influenced by alternative splicing. Neurosci Lett 1995;190(3):147–50.
[64] Zaharieva E, Chipman JK, Soller M. Alternative splicing interference by xenobiotics. Toxicology 2012;296(1-3):1–12.
[65] Srinivasan S, Bingham JL, Johnson D. The ABCs of human alternative splicing: areview of ATP-binding cassette transporter splicing. Curr Opin Drug Discov Devel 2009;12(1):149–58.
[66] Turman CM, et al. Alternative splicing within the human cytochrome P450 superfamily with an emphasis on the brain: the convolution continues. Expert Opin Drug Metab Toxicol 2006;2(3):399–418.
[67] Wain LV, et al. Genome-wide association study identifies six new loci influencing pulse pressure and mean arterial pressure. Nat Genet 2011;43(10):1005–11.
[68] Johnson AD, et al. Association of hypertension drug target genes with blood pressure and hypertension in 86,588 individuals. Hypertension 2011;57(5):903–10.
[69] de Kunder M. The size of the World Wide Web (The Internet). [May 10, 2012]; Available from: <http://www.worldwidewebsize.com/>.
[70] Google Enterprise Search. [May 10, 2012]; Available from: <http://www.google.com/enterprise/search/>.
[71] List of academic databases and search engines. [May 10, 2012]; Available from: <http://en.wikipedia.org/wiki/Academic_databases_and_search_engines>.
[72] Scholar Help. [May 10, 2012]; Available from: <http://scholar.google.com/intl/en/scholar/help.html>.
[73] Microsoft Academic Search. [May 10, 2012]; Available from: <http://academic.research.microsoft.com/>.
[74] Scirus—for scientific information. [May 10, 2012]; Available from: <http://www.scirus.com/srsapp/>.
[75] Quint B. Changes at Google Scholar: A conversation with Anurag Acharya. [May 10, 2012]; Available from: <http://newsbreaks.infotoday.com/nbReader.asp?ArticleId=37309>; 2007.
[76] Anders ME, Evans DP. Comparison of PubMed and Google Scholar literature Sarches. Respir Care 2010;55(5):578–83.
[77] Freeman MK, et al. Google Scholar versus PubMed in locating primary literature to aswer drug-related questions. Ann Pharmacother 2009;43(3):478–84.
[78] Shultz M. Comparing test searches in PubMed and Google Scholar. J Med Libr Assoc 2007;95(4):442–5.
[79] Lu Z. PubMed and beyond: a survey of web tools for searching biomedical literature. Database (Oxford) 2011;2011:baq036.
[80] Islamaj Dogan R, et al. Understanding PubMed user search behavior through log analysis. Database (Oxford) 2009;2009:bap018.
[81] Sayers EW, et al. Database resources of the National Center for Biotechnology Information. Nucleic Acids Res 2011;39(Database issue):D38–51.
[82] Golder S, Loke YK. Sensitivity and precision of adverse effects search filters in MEDLINE and EMBASE: a case study of fractures with thiazolidinediones. Health Info Libr J 2012;29(1):28–38.
[83] PubMed Help. August 8, 2012 [cited October 15, 2012]; Available from: <http://www.ncbi.nlm.nih.gov/books/NBK3827/>.
[84] Chang AA, Heskett KM, Davidson TM. Searching the literature using medical subject headings versus text word with PubMed. Laryngoscope 2006;116(2):336–40.
[85] Nelson SJ, et al. The MeSH translation maintenance system: structure, interface design, and implementation. Stud Health Technol Inform 2004;107(Pt 1):67–9.
[86] Fact Sheet: Medical Subject Headings (MeSH®). August 29, 2012 [cited October 15, 2012]; Available from: <http://www.nlm.nih.gov/pubs/factsheets/mesh.html>.
[87] Krallinger M, Leitner F, Valencia A. Analysis of biological processes and diseases using text mining approaches. Methods Mol Biol 2010;593:341–82.
[88] Winnenburg R, et al. Facts from text: can text mining help to scale-up high-quality manual curation of gene products with ontologies? Brief Bioinform 2008;9(6):466–78.
[89] Krallinger M, Valencia A, Hirschman L. Linking genes to literature: text mining, information extraction, and retrieval applications for biology. Genome Biol 2008;9(Suppl. 2):S8.
[90] Gosink M, et al. Gensensor suite: a web-based tool for the analysis of gene and protein interactions, pathways, and regulation. Adv Bioinformatics 2011;2011:271563.
[91] Fleuren WW, et al. CoPub update: CoPub 5.0 a text mining system to answer biological questions. Nucleic Acids Res 2011;39(Web Server issue):W450–4.

[92] Ananiadou S, et al. Event extraction for systems biology by text mining the literature. Trends Biotechnol 2010;28(7):381–90.

[93] U.S. National Library of Medicine. Fact Sheet Medical Subject Headings (MeSH®). [Cited October 22, 2012]; Available from: <http://www.nlm.nih.gov/pubs/factsheets/mesh.html>; 2012.

[94] Bodenreider O. The Unified Medical Language System (UMLS): integrating bomedical terminology. Nucleic Acids Res 2004;32(Database issue):D267–70.

[95] Arighi CN, et al. BioCreative III interactive task: an overview. BMC Bioinformatics 2011;12(Suppl. 8):S4.

[96] Laifenfeld D, et al. The role of hypoxia in 2-butoxyethanol-induced hemangiosarcoma. Toxicol Sci 2010;113 (1):254–66.

[97] U.S. FDA. Pharmacogenomic data submissions. Guidance for industry [PDF]. [Cited October 21, 2012]; Available from: <http://www.fda.gov/downloads/Drugs/GuidanceComplianceRegulatoryInformation/Guidances/UCM079849.pdf>.

[98] European Medicines Agency. Guideline on pharmacogenetics briefing meetings [PDF]. [Cited October 21, 2012]; Available from: <http://www.ema.europa.eu/docs/en_GB/document_library/Scientific_guideline/2009/09/WC500003886.pdf>; 2006.

[99] Su Z, et al. Comparing next-generation sequencing and microarray technologies in a toxicological study of the effects of aristolochic acid on rat kidneys. Chem Res Toxicol 2011;24(9):1486–93.

[100] Su Z, et al. Next-generation sequencing and its applications in molecular diagnostics. Expert Rev Mol Diagn 2011;11(3):333–43.

[101] Mane SP, et al. Transcriptome sequencing of the Microarray Quality Control (MAQC) RNA reference samples using next generation sequencing. BMC Genomics 2009;10:264.

[102] Loman NJ, et al. Performance comparison of benchtop high-throughput sequencing platforms. Nat Biotechnol 2012;30(5):434–9.

[103] Pickrell WO, Rees MI, Chung SK. Next generation sequencing methodologies—an overview. Adv Protein Chem Struct Biol 2012;89:1–26.

[104] Langmead B, Salzberg SL. Fast gapped-read alignment with Bowtie 2. Nat Methods 2012;9(4):357–9.

[105] Trapnell C, et al. Differential gene and transcript expression analysis of RNA-seq experiments with TopHat and Cufflinks. Nat Protoc 2012;7(3):562–78.

[106] Robertson DG, Watkins PB, Reily MD. Metabolomics in toxicology: preclinical and clinical applications. Toxicol Sci 2011;120(Suppl. 1):S146–70.

[107] Schneider LV, Hall MP. Stable isotope methods for high-precision proteomics. Drug Discov Today 2005; 10(5):353–63.

[108] Tao WA, Aebersold R. Advances in quantitative proteomics via stable isotope tagging and mass spectrometry. Curr Opin Biotechnol 2003;14(1):110–8.

[109] Huang da W, Sherman BT, Lempicki RA. Bioinformatics enrichment tools: paths toward the comprehensive functional analysis of large gene lists. Nucleic Acids Res 2009;37(1):1–13.

[110] Rachek LI, et al. Troglitazone, but not rosiglitazone, damages mitochondrial DNA and induces mitochondrial dysfunction and cell death in human hepatocytes. Toxicol Appl Pharmacol 2009;240(3):348–54.

[111] Lee YH, et al. Troglitazone-induced hepatic mitochondrial proteome expression dynamics in heterozygous Sod2(+/−) mice: two-stage oxidative injury. Toxicol Appl Pharmacol 2008;231(1):43–51.

[112] Stamper BD, et al. Differential regulation of mitogen-activated protein kinase pathways by acetaminophen and its nonhepatotoxic regioisomer 3′-hydroxyacetanilide in TAMH cells. Toxicol Sci 2010;116(1):164–73.

[113] Stevens T, et al. Increased transcription of immune and metabolic pathways in naive and allergic mice exposed to diesel exhaust. Toxicol Sci 2008;102(2):359–70.

[114] Chindelevitch L, et al. Causal reasoning on biological networks: interpreting transcriptional changes. Bioinformatics 2012;28(8):1114–21.

[115] Roth A, et al. Gene expression-based *in vivo* and *in vitro* prediction of liver toxicity allows compound selection at an early stage of drug development. J Biochem Mol Toxicol 2011;25(3):183–94.

[116] Golub TR, et al. Molecular classification of cancer: class discovery and class prediction by gene expression monitoring. Science 1999;286(5439):531–7.

[117] Jeffrey F, Waring RGU. In: Burczynsk ME, editor. Unsupervised hierarchical clustering of toxicants, in an introduction to toxicogenomics. Danvers, MA: CRC Press; 2003.

[118] Waring JF, et al. Clustering of hepatotoxins based on mechanism of toxicity using gene expression profiles. Toxicol Appl Pharmacol 2001;175(1):28–42.
[119] Elferink MG, et al. Microarray analysis in rat liver slices correctly predicts *in vivo* hepatotoxicity. Toxicol Appl Pharmacol 2008;229(3):300–9.
[120] Li C, et al. BioModels database: an enhanced, curated and annotated resource for published quantitative kinetic models. BMC Syst Biol 2010;4:92.
[121] Lloyd CM, et al. The CellML model repository. Bioinformatics 2008;24(18):2122–3.
[122] Lloyd CM, Halstead MD, Nielsen PF. CellML: its future, present and past. Prog Biophys Mol Biol 2004;85(2-3):433–50.
[123] Finney A, Hucka M. Systems biology markup language: level 2 and beyond. Biochem Soc Trans 2003;31(Pt 6):1472–3.
[124] Acikgöz A, et al. Two compartment model of diazepam biotransformation in an organotypical culture of primary human hepatocytes. Toxicol Appl Pharmacol 2009;234(2):179–91.
[125] Stamatelos SK, et al. Mathematical model of uptake and metabolism of arsenic(III) in human hepatocytes—incorporation of cellular antioxidant response and threshold-dependent behavior. BMC Syst Biol 2011;5:16.
[126] Bucher J, et al. A systems biology approach to dynamic modeling and inter-subject variability of statin pharmacokinetics in human hepatocytes. BMC Syst Biol 2011;5:66.
[127] Mamchak AA, et al. Preexisting autoantibodies predict efficacy of oral insulin to cure autoimmune diabetes in combination with anti-CD3. Diabetes 2012;61(6):1490–9.
[128] Natsoulis G, et al. The liver pharmacological and xenobiotic gene response repertoire. Mol Syst Biol 2008;4:175.
[129] Real Time Embedded Systems Lab. Penn Virtual Heart Model (VHM). [Cited October 21, 2012]; Available from: <http://mlab.seas.upenn.edu/index.php/research/medical/mdvv/vhm/>; 2012.
[130] Shah I, Wambaugh J. Virtual tissues in toxicology. J Toxicol Environ Health B Crit Rev 2010;13(2-4):314–28.
[131] U.S. EPA. The Virtual Liver Project (v-Liver). [Cited October 21, 2012]; Available from: <http://www.epa.gov/ncct/virtual_liver/>; 2012.

CHAPTER 11

Omics Biomarkers in Risk Assessment
A Bioinformatics Perspective

Hong Fang[1], Huixiao Hong[2], Zhichao Liu[2], Roger Perkins[2], Reagan Kelly[2], John Beresney[3], Weida Tong[2], and Bruce A. Fowler[3]

[1]Office of Scientific Coordination, National Center for Toxicological Research, U.S. Food and Drug Administration, Jefferson, AR, USA, [2]Division of Bioinformatics and Biostatistics, National Center for Toxicological Research, U.S. Food and Drug Administration, Jefferson, AR, USA, [3]ICF International, Fairfax, VA, USA

ABBREVIATIONS AND GLOSSARIES

- ALL—acute lymphoblastic leukemia
- AML—acute myeloid leukemia
- DF—decision forest
- GOFFA—Gene Ontology for Functional Analysis
- NCTR—National Center for Toxicological Research
- Omics—genomics, proteomics, and metabolomics
- SELDI-TOF MS—surface enhanced laser deposition/ionization time-of-flight mass spectrometry
- SNP—single nucleotide polymorphisms
- SVM—support vector machine
- SRBCTs—small round blue-cell tumors

INTRODUCTION

High-throughput and high-content "omics" technologies including genomics, proteomics, and metabolomics have revolutionized toxicological and biological studies.[1] These technologies provide a precise and detailed measurement of the response of an organism, tissue, or cell—for example, a disease, a natural biological process such as aging, or exposure to a toxic compound. Omics assays allow researchers to simultaneously investigate hundreds or thousands of genotypes, transcripts, proteins, or metabolites. This comprehensive profile not only provides much more information than traditional assays, but also allows both data-driven and hypothesis-driven approaches to uncover mechanisms underlying biological processes.[2] Because of the rapid advancement and development of biotechnologies, omics technologies have had and will continue to have an impact on clinical practice, toxicology, and safety assessment.

One of the most promising applications of omics technologies is the development of biomarkers. Biomarkers are molecular tools that can be interrogated to understand a system's physiological responses to environmental stress, chemical insult, and drug treatment at the molecular, cellular, or tissue levels. Molecular biomarkers are routinely encountered in the FDA regulatory process and are increasingly common in clinical applications[3,4] with pharmacogenomics[3,5] and safety/risk assessment using toxicogenomics.[6,7] Biomarkers are also useful tools for monitoring the biological effects of pollutants and environmental stress.[8] However, few biomarkers for risk assessment using omics technologies have been reported in the last decade.[9–11] Nevertheless, the potential application for omics technologies is tremendous. A more thorough understanding of these emerging molecular technologies and their applications and uses can provide researchers an opportunity to devise more sensitive and more powerful screens for risk assessment.

Omics biomarkers are the biomarkers derived from omics technologies. Unlike traditional biomarkers that normally involve either a single molecule or a few molecules, an omics biomarker can and often does consist of a panel of molecules (e.g., mRNA, SNPs, proteins, metabolites). This makes omics biomarker discovery a different process. Researchers must understand the source of their data, the information and opportunities that the data present, and the best methods for developing an omics biomarker with those data. A robust integrative bioinformatics framework is often necessary in order to take full advantage of the potential offered by omics technologies.

Discovering omics biomarkers from high-dimensional and diverse data requires state-of-the-art analytical methods and algorithms from multidisciplinary fields including bioinformatics, biostatistics, machine learning, chemoinformatics, and artificial intelligence. An omics biomarker discovery workflow is typically composed of a number of approaches, but primarily depends on the omics biomarkers being looked for and the data used. A successful workflow (Figure 11-1) must provide a way to integrate different types of data (sample, phenotype, and genotype information), identify potential omics biomarker candidates, and test the predictive power of a candidate omics biomarker. The challenges for the whole process of omics biomarker discovery and qualification regard standard sampling procedures, standard data quality control, and integrated data analysis and

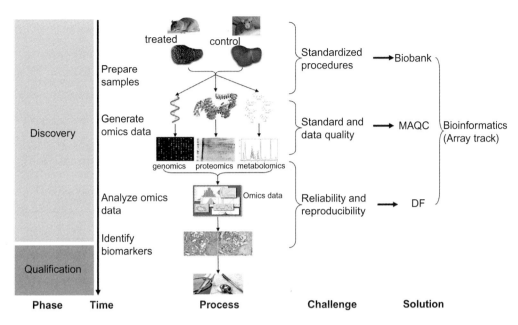

FIGURE 11-1   A depiction of the different phases, timelines, processes, challenges, and solutions for the discovery and qualification of biomarkers based on omics data.

bioinformatics tools. Applications that are capable of these tasks are available commercially, as well as from the open source community. The FDA has built and freely distributes a software system, ArrayTrack™, which offers a suite of analysis tools and a set of federated libraries with data relating to gene and protein functions and pathways. ArrayTrack™ addresses the FDA's need for an integrated bioinformatics system with the capacity to manage, analyze, and interpret pharmacogenomic data submitted by regulated product sponsors.[12] To support the FDA for effective review of biomarkers based on omics data for regulatory purposes, the National Center for Toxicological Research (NCTR) of the FDA initiated the MicroArray Quality Control (MAQC) project[13,14] to broadly address reliability and technical performance related to quality and data analysis for use of DNA microarrays for gene expression measurements in discovery of transcriptomic and SNP biomarkers.

This chapter is an exploration of using integrative bioinformatics approaches to discover omics biomarkers for improved risk assessment. We begin with a brief discussion about biomarkers with specific emphasis on the development of omics biomarkers and their prospective use in risk assessment. Next, we introduce various bioinformatics tools that have been successfully used in omics biomarker development. The chapter emphasizes proper approaches to avoid pitfalls when developing predictive models based on omics biomarkers. We present the use of the decision forest (DF) ensemble methodology as an example for omics biomarker discovery with specific application to transcriptomic, proteomic, and genotypic data and summarize our lessons learned using the classification method in discovery and identification of omics biomarkers.

BIOMARKERS

Omics Biomarkers and Risk Assessment

A biomarker is an indicator of normal biological processes, pathogenic/disease processes, or response to chemical exposure that can be objectively measured and evaluated.[15] Biomarkers can serve as indicators of disease progression (diagnosis and prognosis), therapeutic responses (efficacy), and adverse side effects (safety) in organisms, organs, and cells.[16–18] Biomarkers are widely used in medicine and toxicology.[19] However, biomarker application in risk assessment has been limited but is undergoing rapid growth[20,21] (http://www.inchem.org/documents/ehc/ehc/ehc222.htm).

Biomarkers are capable of linking chemical exposure to biological effects and so are relevant to the risk assessment process. In addition, omics biomarkers (e.g., SNPs) are also capable of providing useful risk assessment information with regard to susceptibility to toxicity for agents such as lead.[22] Current biomarker discovery increasingly relies on emerging molecular technologies because these measurements not only can be used as a way to identify exposure, effect, or response, but also carry the additional benefit of elucidating causal and mechanistic relationships of the markers with environmental and human clinical endpoints. Data from high-throughput molecular technologies such as transcriptomic profiles, protein expression, and metabolite patterns can all represent the state of a biological system and can be a suite of potentially omics biomarkers. The representative omics biomarker types and their associated technologies are summarized in Table 11-1.[23] Sources of possible omics biomarkers include SNP genotyping arrays, gene expression from microarrays, or next generation sequencing, protein expression from mass spectrometry experiments, and metabolite profiles. Each type of omics data conveys a different type of information to characterize conditions in a tissue, cell, or organ, and so has a different utility. All of the different omics data types, however, do require a complex workflow with numerous analysis steps to allow biomarker discovery (Figure 11-2). A detailed process is discussed in the "Bioinformatics Approaches: Challenges and Solutions in Omics Biomarker Discovery" section.

Discovering and developing omics biomarkers for assessing risk to human health from exposure to chemical agents requires special attention to criteria for selecting statistically informative (causative) molecular markers and for assuring validity, reliability, and limitations for use. Omics biomarkers suitable for safety assessment sometimes

TABLE 11-1 Molecular Biomarkers at Different Biological Levels that may be Used for Toxicological Risk Assessment

Biological Levels	Representative Molecular Technologies	Data Type	Biomarkers
DNA	SNP microarray or next generation sequencing	Genotype	Genetic
RNA	Genomics: microarrays; next generation sequencing	Gene expression	Genomic
Protein	Proteomics: 1D/2D gel coupled with MS or MS/MS; High-resolution NMR	Protein production	Protein
Metabolism	Metabolomics: NMR and MS	Metabolite	Metabolomic

require development for individual organ systems, particularly the hepatic, renal, hematological, immune, pulmonary, reproductive/developmental, and nervous systems (http://www.inchem.org/documents/ehc/ehc/ehc155.htm). Omics biomarkers for carcinogenicity risk are particularly advanced owing to the fact that a large effort has been focused on understanding the mechanisms of cancers.[2,16]

The effect of chemical exposure on a human or animal can be measured using gene expression, protein expression, and metabolite profiling. Shifts in expression patterns due to a toxic effect or exposure can tell researchers the mode of action of a particular chemical or the particular deleterious effect exposure may cause. While we normally think of biomarkers as tools for detecting a biological effect, separate biomarkers for detecting exposure are also important. Exposure biomarkers are most important where there are concerns of contamination, such as environmental quality assessments, food chain management, and workplace safety, especially for chemicals known or suspected of causing adverse health effects. Omics technologies such as genome-wide genotyping (either via SNP genotyping microarray or next generation sequencing) also allow the development of "biomarkers of sensitivity," through which human or animal subpopulations that are more susceptible to harm from a particular exposure can be identified. Thus, omics biomarkers based on understanding of the genetic associations of toxicity mechanisms can be anticipated to lead to improved risk assessment, as well as the possibility of intervention to mitigate exposure.

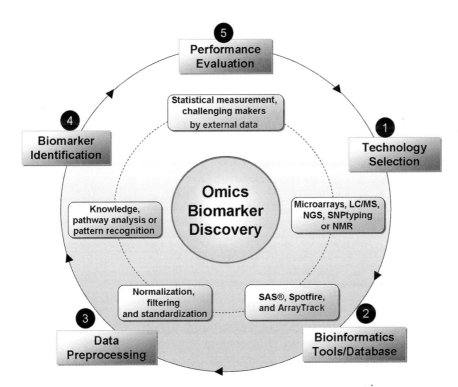

FIGURE 11-2   Key steps and methods for omics biomarker discovery.

Omics Biomarkers and Development of Predictive Models

An ideal biomarker should be a single and quantitative molecular species with high sensitivity and specificity to treatment effects and completely understood mechanisms. Unfortunately, omics biomarkers generally lack these characteristics. Rather, they normally comprise a panel of molecular species and have moderate sensitivity and specificity; also, their biological relevance is often ambiguous.[2] To confound matters further, the particular composition of a biomarker can vary with the choice of analytical methods employed to analyze the omics data in the first place. Thus, caution is warranted to avoid bias in feature selection, avoid overfitting, and carefully validate omics biomarkers for use in the domain or population intended.

Discovery of omics biomarkers via development of predictive models has become widespread in biomedical product development and in clinical research for diagnostics, prognostics, and treatment selection, with the central goal being the development of predictive models sophisticated enough to accurately predict the class of untested samples.[24] An iterative approach for biomarker validation and identification can be used to enhance the confidence of the models by adding new samples to experimental design, as depicted in Figure 11-3. The process starts with an initial set of omics data followed by data preprocessing including quality inspection and noise filtering, and then model development. The internal validation tests the statistical significance of the model and the external validation assesses its validity of biomarker identification. It is highly recommended that this should be viewed as an iterative process which includes new or additional model training that will further improve model accuracy.

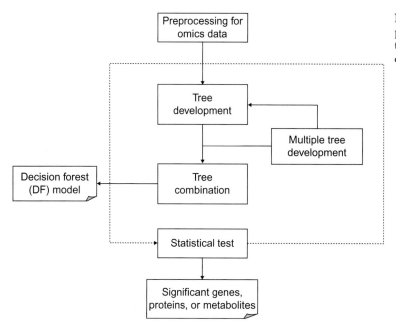

FIGURE 11-3 The iterative process for biomarker validation and identification in predictive model building.

The iterative nature in model development requires a collaborative environment between the bench scientists and modelers. Several benefits accrue from collaborative integration of the experimental and modeling efforts. Immediate feedback can be given to the experimentalists so that suspected problems (such as mislabeled samples, apparent batch effects) can be rapidly investigated and corrected. Each new assay data point directly from the field and lab becomes a challenge to the evolving model; the result is either further confirmation of its validity, identification of a limitation, or an outlier prediction. Failure of the model also provides important information, such as a biomarker is not sufficiently robust to represent the real-world application. A research hypothesis is spawned in each iteration, which can lead to new data, an improved training set, and an improvement to the living model. In the case of pharmacogenomics, estimates can be made by inclusion of additional samples to attain needed statistical power and accuracy. In the case of toxicogenomics, there is a need for inclusion of additional chemicals to span the chemistry domain space in order to reach firmer conclusions. This process could lead to omics biomarkers with robust discriminatory power to distinguish chemical classes, toxicity risk, the effect of exposure level, etc., to guide risk assessment.

BIOINFORMATICS APPROACHES: CHALLENGES AND SOLUTIONS IN OMICS BIOMARKER DISCOVERY

As noted previously, although there is tremendous interest in the discovery of omics biomarkers, few omics biomarkers are currently in use for risk assessment. There are a number of challenges in developing omics biomarkers and translating them into practical use. Omics biomarker discovery requires integrated bioinformatics approaches due to the need to process large and complex data sets. Figure 11-2 illustrates an omics biomarker discovery process with essential bioinformatics steps; starting with proper sample and omics technology selection, followed by bioinformatics tools and databases that manage massive amounts of omics data and conjunct with sample information; a good data QA/QC process and a robust statistical analysis and modeling technique for biomarker identification; and finally to evaluation of biomarker performance by external data and real-work application. The process can be iteratively improved for more accurate biomarkers and better predictions.

Choosing Samples and Technologies

The first step in the discovery process is the choice of samples and the choice of a technology to interrogate the samples. Table 11-1 lists the types of omics data, DNA/mRNA, proteins, and metabolites that are collected from animal or human samples. The choice of sample and technology is dependent on the question being asked. For example, if a researcher was attempting to build a biomarker to assay the exposure of a person to a known liver toxicant, then it would be reasonable to use primary human hepatocytes or blood samples. Blood samples, because of their minimally invasive nature, are often recommended for the study of liver toxicity.[25] These are the cells that would be affected in a person exposed to the liver toxicant, so the reaction of these cells offers an opportunity to discover a biomarker. Because the goal is to discover a biomarker of exposure, a typical approach would be to assay the gene expression level of both control cells and cells treated with the chemical of interest. This could be done using either a microarray (a mature omics technology that provides expression measurements for a predetermined number of genes or next generation sequencing), or a newer omics technology that sequences the RNA expressed and gives a more complete picture of the state of the cell but requires more sophisticated analysis than a microarray.

Developing a biomarker for a different use would necessitate the use of a different strategy. For example, if a researcher was attempting to develop a biomarker of susceptibility (i.e., a biomarker that identifies individuals at increased risk of harm from exposure to a particular chemical), then the use of gene expression may not be the best choice. Because the use case for a biomarker of susceptibility is to identify susceptible individuals prior to exposure the likely source of the biomarker would be a genetic component that does not vary depending on an individual's environment. For this purpose single nucleotide polymorphisms (SNPs) are an excellent choice. These variations are a static part of an individual's genome and can be easily measured using a sample of DNA. The source of the sample is less important in this case because DNA does not vary by tissue. As with gene expression, there are two choices for the technology used to measure the SNPs. The first is an SNP genotyping array, which is analogous to a microarray for gene expression and provides the genotypes of 1 million or more SNPs. The second option is next generation sequencing, which reads the full genome and will identify any differences between a subject's genome and a reference genome. The extra information from the next generation sequencing, however, is offset by the additional complexity of the analysis workflow necessary.

Bioinformatics Tools and Databases

Analyses of high-dimensional genomics, proteomics, metabolomics, and SNP data require integrated bioinformatics tools. Normally, workflows rely on multiple methods and tools drawn from molecular biology, systems biology, mathematics, statistics, computational science, and information science. Bioinformatics skills are required in areas of data analyses, mining, and management, and biomarker and predictive model development.

Bioinformatics is involved in essentially all steps in the discovery of biomarkers from omics data. Efficient bioinformatics tools and workflows can overcome obstacles that would otherwise delay discovery or diminish the value of omics biomarkers. For example, correct analysis of epigenetic variation helps selection of candidate epigenetic biomarkers that have a small variation among healthy individuals, thus reducing the risk of false classification due to population heterogeneity.[26]

Molecular biomarker discovery becomes more efficient and cost-effective when the proper software and computing infrastructure is available. Ideally, data and tools to integrate different data become rapidly apparent. The combining of omics data for the purpose of identifying and confirming relevant biological processes and biomarkers, such as pathways and gene or protein functions, is crucial to realize the benefits of these high-throughput molecular technologies in drug development and public health.

Research to identify genes and proteins affected by experimental conditions commonly uses information on gene and protein function and ontology information from data repositories such as KEGG, Gene Ontology, and biological pathway software such as GeneGo and Ingenuity Pathway Analysis to aid in determining the biological relevancy of prospective biomarkers. Numerous software tools [commercial sources (Partek, Spotfire, SAS, JMP, MATLAB, JMP Genomics, GeneData, and GeneSpring) and publicly available and open sources (TMeV, Bioconductor, BRB-ArrayTools, SAM, etc.)] have been used to integrate this information into the biomarker discovery process.

As described earlier, the FDA has developed and made available ArrayTrack™, a free tool for the analysis of omics data. ArrayTrack™ is routinely used in the review of pharmacogenomic data submitted to the FDA. It provides a rich collection of functional information about genes, proteins, and pathways drawn from various public biological databases for facilitating data interpretation. Many data analysis and visualization tools are available within ArrayTrack™ for analysis of individual platform data and integrated analysis of omics data with toxicological and clinic study data. Users can select an analysis method from the ArrayTrack™ toolbox, apply it to selected omics data, and the results can be directly linked to individual genes, pathways, and Gene Ontology results. The FDA has also developed and released two companion software systems: SNPTrack,[27] an integrated system for storage, analysis, and interpretation of genotype data; and atBioNet,[28] a network analysis tool that can be used to uncover the underlying mechanisms of toxicology and diseases. These tools are routinely used to identify genomic signatures, biomarkers, and predictive models for disease and toxicity.[25,29]

The current challenges, as well as opportunities in biomarker discovery, lie in the integration of data of multiple types together with diverse data in the public domain to assess disease and toxicity at the systems level (systems biology). Omics technology in chemical risk assessment requires an integrated knowledge base and a better understanding of the relationships between specific responses and biomarkers to adverse toxic events. The knowledge-based approach takes advantage of current advances in computer technology and bioinformatics to integrate diverse data sets into a content-centric resource for evaluation of molecular biomarkers in determining toxin exposure to animals and humans. Such a knowledge base will spawn hypotheses to develop new studies to address the current gaps and lead to further improvements in biomarker discovery.

Ensuring Data Quality

The next step in the process of biomarker discovery is preprocessing and filtering the data. The amount and complexity of the required processing vary depending on the technology used to generate the data. Despite that, however, it is absolutely essential to ensure that the data used are of the highest possible quality. As with the previous steps, the specific methods and techniques depend largely on the type of data being used for discovery. With microarrays, it is necessary to filter out low expressed genes and to normalize the data appropriately, while for SNP genotyping arrays, it is necessary to filter out low-quality genotype calls. Next generation sequencing must first be mapped to a reference genome and then used to derive either expression values or variant genotypes as required. Mass spectrometry data such as metabolomics and proteomics must be matched against a database to identify the peptide or metabolite observed, and further questions of data filtering and normalization must be addressed.

It is beyond the scope of this chapter to provide comprehensive instructions on the way in which omics data must be processed; however, it is an important step in biomarker discovery and must be given the appropriate attention. The quality of data used in discovery is a prerequisite to successful omics biomarker development.

Biomarker Identification

Once the data have been collected and processed, they are ready to be used for biomarker discovery. As with the previous steps, there is not one single correct way to approach this problem. The particular statistical tools used will depend heavily on the type of data, the knowledge level of the researcher, and the particular application.

Some biomarker research is done using fairly simple unsupervised learning tools such as hierarchical clustering analysis (HCA) and principal components analysis (PCA). Unsupervised learning investigates the natural grouping without regard to the class that the sample belongs to (e.g., exposed or not exposed to a particular compound). HCA, which is routinely used in microarray analysis, computes the distance between two samples by comparing the expression values (or some other quantitative omics measurement) for a set of genes (or proteins or other molecular species) and arranging a set of samples so that the most similar are the closest together. If the clustering works well and accurately splits the samples into two groups that correspond with the classes of the samples, then the gene expression measurements used to cluster them can be investigated as a possible biomarker. PCA reduces the dimensionality of the data by estimating loading factors for each of the variables being used, and can thus capture information from a data set with 30,000 or more measurements with only a handful of variables. The resulting first two or three principal components, which capture most of the variation in the data, can be visualized in a 2D or 3D plot and often effectively separates the sample classes. Caution should be taken when PCA-based biomarkers are developed, as they may require the measurement of a potentially very large number of variables, possibly making it impractical for real-world use.

Supervised learning techniques are the predominant tools for omics biomarker development. In these methods a set of training samples is first selected for use in the learning algorithm. The algorithm then uses components of the candidate biomarker (often called features in the machine learning field) to build a classifier, or a model that is able to make a prediction about the class of an unknown sample. In some methods the features are selected before training, and in others, selection is integrated into the model training procedure. The accuracy of a derived classifier can be estimated by internal cross-validation using the training set, but should be validated by challenging the model with known subjects excluded from the training (a test set).[30,31] A number of classification methods have been applied to high-throughput molecular data[24,32] including total principal component regression (TPCR),[33] discriminant partial least squares,[34] artificial neural networks (ANNs),[35] K-nearest neighbor (KNN),[36] decision tree (DT),[37] support vector machines (SVMs),[38] logistic regression (LR),[39] Fisher's linear discriminant (FLD),[40] and soft independent modeling of class analogy (SIMCA).[41] For example, based on gene expression profiling using an Affymetrix GeneChip of 6,817 genes, Golub et al.[42] were able to classify acute myeloid leukemia (AML) and acute lymphocytic leukemia (ALL) samples using a variant of the linear discriminant analysis method; the same data have been successfully analyzed by other investigators using various classification methods. In another example, Yeoh et al.[43] successfully constructed multiclass prediction models using SVMs for the diagnosis and risk stratification of pediatric ALL based on gene expression profiling. In proteomics, Petricoin et al.,[44] using a genetic algorithm (GA)-based SVM, developed classification models for early detection of ovarian and prostate cancers based on SELDI-TOF MS data.

Whatever method is used to identify possible elements of a biomarker, researchers must next understand the roles of those elements in the biological process being investigated and the relationships among those elements. In the past this investigation was typically done by a subject matter expert. Subject matter experts are still important contributors to biomarker discovery; however, the proliferation of bioinformatics databases has made investigating these components much more accessible. In this process, a researcher is looking for common threads between the genes (or genotypes or metabolites, etc.) that have been suggested. This requires determining the functions of these genes, what they interact with, and which pathways they participate in. Gene Ontology is a primary source of information on a gene's function, what biological processes it plays a role in, and where in the cell it is located. Using tools like ArrayTrack™'s Gene Ontology for Functional Analysis (GOFFA), researchers can identify molecular function or biological processes that are overrepresented in a set. Similar enrichment analyses can also be performed using pathway data from KEGG, GeneGo MetaCore, and Ingenuity Pathway Analysis. These tools do not work at the level of individual genes but at the pathway level. Using a database of pathway composition, these tools can identify pathways that are overrepresented in the set of genes being studied, offering insight into what pathways are affected in the samples. By using these bioinformatics tools, researchers are more quickly able to screen and identify potential biomarkers and better understand the particular space the biomarker occupies in the biological process being investigated.

Performance Evaluation for Risk Assessment Applications

Once a candidate biomarker has been identified, the final step of the discovery process is performance estimation and statistical validation. Before a biomarker can be approved for real-world use, researchers must have a thorough understanding of the performance characteristics of the biomarker and its final application. Typically, the biomarker will form part of a risk prediction model, where it is assessed along with other known predictors to make a final determination about the status of an unknown sample. The biomarker performance is usually measured by sensitivity and specificity and predictive power for a training data set as well as for an external data set. Especially when a predictive model has been developed from omics markers, it should be validated from two different perspectives: (1) internal validation, the performance on training data or samples; and (2) external validation, the performance on real-world or unknown data or samples. A model with high predictive performance in training and lower predictive performance in external validation is a typical failure case in model development.

Once an omics biomarker has been validated as thoroughly as possible, its predictive capability needs to be tested in a risk assessment context against real data which link chemical exposures to a health outcome. This has been done on a limited basis by correlating data from the National Health and Nutrition Examination Survey (NHANES) on SNPs' polymorphisms for the enzyme ALA dehydratase, which exists as ALAD 1, ALAD 2, and ALAD 1-2, and is the main lead binding protein in blood, with blood lead values and prevalence of hypertension. Overall, the data showed that carriers of the ALAD-2 allele had a greater prevalence of hypertension linked to lead exposure relative to ALAD-1 carriers in the 3.85 μg/dl quartile. These data are useful in helping to identify human subpopulations at special risk for lead-induced hypertension,[45] which is an important public health outcome.

DECISION FOREST FOR OMICS BIOMARKERS

Machine learning techniques have shown great potential for biomarker identification and qualification.[42–44] However, there are many ways to build a classifier, and even within a specific type of algorithm, many parameters can be tuned. The choice of method depends on the familiarities and resources of the researcher and the particular application of the biomarker. We believe that ensemble or consensus prediction methods have tremendous potential in biomarker research and can play an important role in reducing the overfitting that is often seen in omics biomarker development.

Omics data are usually noisy and have many more predictor variables (e.g., genes of DNA microarray data and m/z peaks of mass spectrometry data) than samples available for developing and testing the classifiers. These characteristics magnify the importance of assessing potential random correlation and overfitting for a classification model based on omics data. We developed a classification method, decision forest (DF), for class prediction using omics data.[30,46,47] DF is an ensemble methodology that combines multiple heterogeneous, but comparably predictive, decision tree models to produce a consensus prediction. The method is less prone to overfitting noise and chance correlation than other classification methodologies.

DF combines *heterogeneous* yet *comparable* trees in order to better encode the different functional dependencies of a disease or toxic endpoint with molecular data (Figure 11-4). The heterogeneity requirement assures that each of the decision trees uniquely contributes to the consensus prediction, while the quality requirement ensures that all trees contribute comparably to the consensus prediction. Heterogeneity is attained by excluding independent variables (such as genes, proteins, metabolites) already used for tree construction from the pool of variables from which subsequent trees are constructed. Since a certain degree of noise is always present in biological data, optimizing a tree inherently risks overfitting the noise. DF mitigates overfitting by maximizing the difference among individual trees, which could result in cancellation of some random noise when trees are combined.

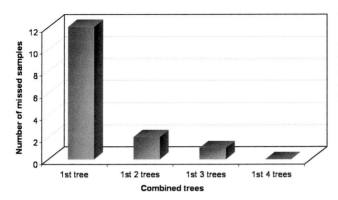

FIGURE 11-4 Overview of the method of decision forest for omics. (1) Data preprocessing; (2) tree development; (3) multiple tree development; (4) tree combination; and (5) statistical test. DF not only produces a classifier but also identifies the biomarkers (genes, proteins, or metabolites).

There are three benefits associated with DF compared with other similar consensus modeling methods: (1) since the difference in individual trees is maximized, the best consensus is usually realized by combining only a few trees (i.e., 4 or 5), which consequentially also results in reduced computational expense; (2) since DF is entirely reproducible, the disease-patterns associations are constant in their interpretability for biological relevance; and (3) since all samples are included in the individual tree development, the information in the original data set is fully utilized in the consensus modeling process. We present here three specific examples of using DF for omics biomarkers.

Application to SNP Genotype Data

Esophageal squamous cell carcinoma is a high-mortality cancer with complex etiology and progression involving both genetic and environmental factors. We examined the association between esophageal cancer risk and the genotypic patterns of 61 SNPs in a case-control study of a population from Shanxi Province in North Central China that has the highest rates of esophageal squamous cell carcinoma in the world. High-throughput Masscode mass spectrometry genotyping was done on the genomic DNA from 574 individuals (394 cases and 180 age-frequency matched controls). SNPs were chosen from the genes involving DNA repair enzymes, and Phase I and Phase II enzymes.

We developed a novel pattern recognition method called Decision Forest for SNPs (DF-SNPs), specifically for discovery of genetic biomarkers based on SNPs' genotypic data.[48] The classifier for separating the cases with esophageal squamous cell carcinoma from the matched controls was developed with DF-SNPs. The evaluation of the classifier by cross-validations was conducted to estimate its performance, resulting in 94.7% concordance, 99.0% sensitivity, and 85.1% specificity, respectively. The results demonstrate the usefulness of DF-SNPs for identifying genetic biomarkers (SNPs or combinations of SNPs) that can be used to assess the susceptibility to esophageal cancer. Importantly, the DF-SNPs algorithm incorporated a randomization test for assessing the relevance (or importance) of genotypes of individual SNPs (homozygous common, heterozygous, and homozygous variant) and the genetic patterns (haplotypes) of SNPs that differentiate cases with esophageal squamous cell carcinoma from the matched controls. For example, we found that the different genotypes of SNP GADD45B E1122 are associated with esophageal cancer risk.

Cross-validations of DF models inherently contain the information for rank-ordering the independent variables according to their ability to differentiate classes and, thus, according to their biomarker potential. The rationale behind this approach is that the number of classifiers selecting a particular variable should tend to increase in direct proportion to the biological relevance of the variable. Thus, the independent variables that are most frequently used by the multiple classifiers generated from the cross-validations have the most potential as valid biomarkers. In the DF-SNPs modeling of esophageal cancer study, we performed 2,000 runs of 10-fold cross-validations that generated 20,000 classifiers. Each classifier was constructed based on a set of different samples. Examining the frequencies of the SNPs used in the 20,000 classifiers provided a more robust and unbiased way for identifying the potential genetic biomarkers. Analyzing the distribution of the 61 SNPs in the case-control study of esophageal cancer successfully identified a set of SNPs, genotypes of the SNP types (homozygous common, heterozygous, and homozygous variant), and the genetic patterns (haplotypes) of the SNPs as the potential genetic biomarkers to differentiate the esophageal squamous cell carcinoma cases from the matched controls.[48]

Application to Proteomics Data

DF was applied to the proteomics data set for distinguishing cancer from noncancer. The model was developed to predict the presence of prostate cancer using a proteomics data set generated from surface-enhanced laser deposition/ionization time-of-flight mass spectrometry (SELDI-TOF MS).[49] The data set contained 326 samples, of which 167 were prostate cancer patients and 159 were either healthy or had benign prostatic hyperplasia. The fitted DF model for the data set consisted of four DT models, each of them having the comparable misclassifications ranging from 12 to 14 (3.7%–4.3% error rate). The misclassification was significantly reduced as the number of DT models to be combined increased to form the DF model (Figure 11-5). The four-tree DF model gave 100% classification accuracy. However, it is important to note that a statistically sound fitted model provides a

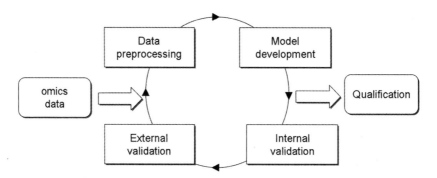

FIGURE 11-5   Plot of misclassifications versus the number of tree classifiers that were combined in DF.

limited indication of whether the identified pattern is biologically relevant or is solely due to chance. Moreover, a fitting result cannot be used as the validation of the model's capability for predicting unknown samples that were not included in the training set used for model development. It is important to carry out a rigorous validation procedure to determine the fitted model with respect to the degree of chance correlation and the level of confidence for predicting unknown samples.

The degree of chance correlation and prediction confidence of the DF model developed based on the proteomics data were rigorously assessed using extensive cross-validations and randomization testing. Comparison of the model predictions with the randomization testing results demonstrated the biological relevance of the DF model and the capability of the DF algorithm to reduce overfitting. Furthermore, the analysis of prediction accuracy with different prediction confidence levels (high and low confidences) revealed that most misclassifications were associated with the low-confidence predictions. For the high-confidence predictions, the DF model achieved a sensitivity of 99.2% and a specificity of 98.2%. The DF model also identified a list of significant peaks that could be useful for biomarker identification.

In addition to development of a predictive model for proteomic diagnostics, identification of potential biomarkers is another important use of the SELDI-TOF MS technology.[50] Each DT model in DF determines a sample's classification through a series of rules based on the selection of predictor variables. Thus, it is expected that the DF-selected variables could be useful as a starting point for biomarker identification.

Application to Microarray Gene Expression

DF was primarily developed for two-class classification modeling.[30] It was extended to multiclass classification modeling and was applied to prediction of disease stages based on microarray gene expression data for prognosis. The gene expression data used in this study contained 83 small round blue-cell tumor (SRBCT) samples that were obtained from cancer patients at four different stages.[47] Based on the 500 runs of 10-fold cross-validation of the DF models, tumor prediction accuracy was ~97%, sensitivity was ~95%, diagnostic sensitivity was ~91%, and diagnostic accuracy was ~99.5%. Among the 25 genes selected by the DF models to distinguish tumor class, 12 genes have functional information in the literature that confirms their involvement in cancer. The four types of SRBCT samples are also distinguishable in the clustering analysis based on the expression profiles of the 25 genes. The results demonstrated that the multiclass DF is an effective classification method for analysis of gene expression data for the purpose of prognosis (Table 11-2).

Any classifiers developed based on omics data usually have many embedded independent variables, some of which may be covariate, that potentially qualify as biomarkers. For example, in order to verify the anticipated biological relevance of the 25 genes used in the DF classification models for SRBCTs, the hierarchical cluster analysis was applied to the 83 tumor samples based on the 25 genes. The samples grouped into four distinct clusters corresponding to the four types of SRBCTs with no misclassification, indicating that the DF model-selected 25 genes can accurately differentiate the four tumor types.

TABLE 11-2 Diagnostic Sensitivity and Accuracy Derived from 500 Runs of 10-Fold Cross-Validation*

Sample Category	Diagnostic Sensitivity (%)				
	NT**	EWS	BL	NB	RMS
EWS	1.61	90.50	0.11	0.00	0.01
BL	0.00	0.20	94.49	0.00	0.00
NB	0.03	0.00	0.00	92.47	0.00
RMS	0.12	1.37	0.06	0.00	89.07
Diagnostic Accuracy %		98.63	99.56	100	99.96

*The average diagnostic sensitivity is the percentage of samples of a tumor type correctly predicted by the model. The bottom row lists the diagnostic accuracy; that is, the model's ability to predict a tumor to be in the correct class.
**NT: Not-Determinable.

CONCLUSION

Omics biomarker discovery has been receiving increased attention, and emerging molecular technologies like microarrays, mass spectrometry-based proteome assay, SNP genotyping, and NGS offer the potential to develop more comprehensive and sophisticated biomarkers. In order to realize this potential, however, researchers must understand the bioinformatics tools that make dealing with omics data tractable. The use of biomarker-based monitoring approaches as a tool for environmental risk assessment is often critically limited by a lack of integrated bioinformatics approaches, statistical analyses, and predictive models. The application of bioinformatics solutions to these problems will allow researchers to improve the quality and interpretability of new environmental biomarkers while limiting the time required for development.

In this chapter we discussed the key steps for omics biomarker discovery. Understanding the available technologies and choosing the correct one for the application are essential, as are understanding and choosing the bioinformatics tools best suited to the problem. Careful application of statistical and bioinformatics tools allows researchers to identify candidate biomarkers, and rigorous statistical techniques improve the performance of a possible biomarker while reducing the risk of overfitting. This chapter illustrated the power of the combination of omics technologies with bioinformatics approaches for biomarker discovery and identification. An integrated bioinformatics approach with the correct choice of samples, omics technologies, and statistical techniques will allow the development of powerful new biomarkers for safety assessment.

DISCLAIMER

The views presented in this article do not necessarily reflect those of the U.S. Food and Drug Administration.

References

[1] Ulrich R, Friend SH. Toxicogenomics and drug discovery: will new technologies help us produce better drugs? Nature Rev Drug Discov 2002;1:84—8.

[2] van de Vijver MJ, He YD, van't Veer LJ, Dai H, Hart AAM, Voskuil DW, et al. A gene-expression signature as a predictor of survival in breast cancer. N Engl J Med 2002;347:1999—2009.

[3] Frank R, Hargreaves R. Clinical biomarkers in drug discovery and development. Nat Rev Drug Discov 2003;2:566—80.

[4] Ghosh D, Poisson LM. "Omics" data and levels of evidence for biomarker discovery. Genomics 2009;93: 13—6.

[5] Frueh FW, Amur S, Mummaneni P, Epstein RS, Aubert RE, DeLuca TM, et al. Pharmacogenomic biomarker information in drug labels approved by the United States Food and Drug Administration: prevalence of related drug use. Pharmacotherapy 2008;28:992—8.

[6] Shi Q, Hong H, Senior J, Tong W. Biomarkers for drug-induced liver injury. Expert Rev Gastroenterol Hepatol 2010;4:225—34.

[7] Aardema MJ, MacGregor JT. Toxicology and genetic toxicology in the new era of "toxicogenomics": impact of "-omics" technologies. Mutat Res—Fundam Mol Mech Mutagen 2002;499:13—25.

[8] Beliaeff B, Burgeot T. Integrated biomarker response: a useful tool for ecological risk assessment. Environ Toxicol Chem 2002;21:1316—22.

[9] Bjornstad A, Larsen BK, Skadsheim A, Jones MB, Andersen OK. The potential of ecotoxicoproteomics in environmental monitoring: biomarker profiling in mussel plasma using protein chip array technology. J Toxicol Environ Health A Curr Issues 2006;69:77—96.

[10] Van Aggelen G, Ankley GT, Baldwin WS, Bearden DW, Benson WH, Chipman JK, et al. Integrating omic technologies into aquatic ecological risk assessment and environmental monitoring: hurdles, achievements, and future outlook. Environ Health Perspect 2010;118:1—5.

[11] Robbens J, van der Ven K, Maras M, Blust R, De Coen W. Ecotoxicological risk assessment using DNA chips and cellular reporters. Trends Biotechnol 2007;25:460—6.

[12] Tong WD, Cao XX, Harris S, Sun HM, Fang H, Fuscoe J, et al. Arraytrack—supporting toxicogenomic research at the US Food and Drug Administration National Center for Toxicological Research. Environ Health Perspect 2003;111:1819—26.

[13] Shi LM, Reid LH, Jones WD, Shippy R, Warrington JA, Baker SC, et al. The MicroArray Quality Control (MAQC) project shows inter- and intraplatform reproducibility of gene expression measurements. Nat Biotechnol 2006;24:1151—61.

[14] Shi LM, Campbell G, Jones WD, Campagne F, Wen ZN, Walker SJ, et al. The MicroArray Quality Control (MAQC)-III study of common practices for the development and validation of microarray-based predictive models. Nat Biotechnol 2010;28:827—38.

[15] Atkinson AJ, Colburn WA, DeGruttola VG, DeMets DL, Downing GJ, Hoth DF, et al. Biomarkers and surrogate endpoints: preferred definitions and conceptual framework. Clin Pharmacol Ther 2001;69: 89—95.

[16] Dhanasekaran SM, Barrette TR, Ghosh D, Shah R, Varambally S, Kurachi K, et al. Delineation of prognostic biomarkers in prostate cancer. Nature 2001;412:822—6.

[17] Colburn WA. Biomarkers in drug discovery and development: from target identification through drug marketing. J Clin Pharmacol 2003;43:329—41.

[18] Marrer E, Dieterle F. Impact of biomarker development on drug safety assessment. Toxicol Appl Pharmacol 2010;243:167—79.

[19] Swenberg JA, Fryar-Tita E, Jeong YC, Boysen G, Starr T, Walker VE, et al. Biomarkers in toxicology and risk assessment: informing critical dose-response relationships. Chem Res Toxicol 2008;21:253—65.

[20] Mayyeux R. Biomarkers: potential uses and limitations. J Am Soc Exp Neurother 2004;1(2):182—8.

[21] Galloway TS. Biomarkers in environmental and human health risk assessment. Mar Pollut Bull 2006;53:606—13.

[22] Scinicariello F, Abadin HG, Edward Murray H. Association of low-level blood lead and blood pressure in NHANES 1999—2006. Environ Res 2011;111:1249—57.

[23] Mendrick D, Tong WD. Biomarkers. In: Dubitzky W, Wolkenhauer O, Kwang-Hyun C, Hiroki Y, editors. Encyclopedia of systems biology. Heidelberg/New York: Springer; 2012 [in press].

[24] Simon R. Diagnostic and prognostic prediction using gene expression profiles in high-dimensional microarray data. Br J Cancer 2003;89:1599–604.
[25] Huang J, Shi W, Zhang J, Chou JW, Paules RS, Gerrish K, et al. Genomic indicators in the blood predict drug-induced liver injury. Pharmacogenomics J 2010:S37–47.
[26] Bock C. Epigenetic biomarker development. Epigenomics 2009;1:99–110.
[27] Xu J, Kelly R, Zhou G, Turner S, Ding D, Harris S, et al. SNPTrackTM: an integrated bioinformatics system for genetic association studies. Hum Genomics 2012;6:5.
[28] Ding Y, Chen M, Liu Z, Ding D, Ye Y, Zhang M, et al. atBioNet—an integrated network analysis tool for genomics and biomarker discovery. BMC Genomics 2012 [in press].
[29] Arasappan D, Tong W, Mummaneni P, Fang H, Amur S. Meta-analysis of microarray data using a pathway-based approach identifies a 37-gene expression signature for systemic lupus erythematosus in human peripheral blood mononuclear cells. BMC Medicine 2011;9:65.
[30] Tong WD, Hong HX, Fang H, Xie Q, Perkins R. Decision forest: combining the predictions of multiple independent decision tree models. J Chem Inf Comput Sci 2003;43:525–31.
[31] Tong WD, Fang H, Xie Q, Hong HX, Shi LM, Perkins R, et al. Gaining confidence on molecular classification through consensus modeling and validation. Toxicol Mech Method 2006;16:59–68.
[32] Somorjai RL, Dolenko B, Baumgartner R. Class prediction and discovery using gene microarray and proteomics mass spectroscopy data: curses, caveats, cautions. Bioinformatics 2003;19:1484–91.
[33] Tan Y, Shi L, Tong W, Wang C. Multi-class cancer classification by total principal component regression (TPCR) using microarray gene expression data. Nucleic Acids Res 2005;33:56–65.
[34] Tan Y, Shi L, Tong W, Gene Hwang GT, Wang C. Multi-class tumor classification by discriminant partial least squares using microarray gene expression data and assessment of classification models. Comput Biol Chem 2004;28:235–43.
[35] Khan J, Wei JS, Ringner M, Saal LH, Ladanyi M, Westermann F, et al. Classification and diagnostic prediction of cancers using gene expression profiling and artificial neural networks. Nat Med 2001;7:673–9.
[36] Olshen AB, Jain AN. Deriving quantitative conclusions from microarray expression data. Bioinformatics 2002;18:961–70.
[37] Zhang HP, Yu CY, Singer B, Xiong MM. Recursive partitioning for tumor classification with gene expression microarray data. Proc Natl Acad Sci U S A 2001;98:6730–5.
[38] Brown MPS, Grundy WN, Lin D, Cristianini N, Sugnet CW, Furey TS, et al. Knowledge-based analysis of microarray gene expression data by using support vector machines. Proc Natl Acad Sci 2000;97:262–7.
[39] Zhou XB, Liu KY, Wong STC. Cancer classification and prediction using logistic regression with Bayesian gene selection. J Biomed Inform 2004;37:249–59.
[40] Dudoit S, Fridlyand J, Speed T. Comparison of discrimination methods for the classification of tumors using gene expression data. J Am Stat Assoc 2002;97:77–87.
[41] Fang H, Tong WD, Shi LM, Jakab RL, Bowyer JF. Classification of cDNA array genes that have a highly significant discriminative power due to their unique distribution in four brain regions. DNA Cell Biol 2004;23:661–74.
[42] Golub TR, Slonim DK, Tamayo P, Huard C, Gaasenbeek M, Mesirov JP, et al. Molecular classification of cancer: class discovery and class prediction by gene expression monitoring. Science 1999;286:531–7.
[43] Yeoh EJ, Ross ME, Shurtleff SA, Williams WK, Patel D, Mahfouz R, et al. Classification, subtype discovery, and prediction of outcome in pediatric acute lymphoblastic leukemia by gene expression profiling. Cancer Cell 2002;1:133–43.
[44] Petricoin EF, Ardekani AM, Hitt BA, Levine PJ, Fusaro VA, Steinberg SM, et al. Use of proteomic patterns in serum to identify ovarian cancer. Lancet 2002;359:572–7.
[45] Scinicariello F, Yesupriya A, Chang MH, Fowler BA. Modification by ALAD of the association between blood lead and blood pressure in the US population: results from the Third National Health and Nutrition Examination Survey. Environ Health Perspect 2010;118:259–64.
[46] Tong WD, Hong HX, Fang H, Xie Q, Perkins R, Walker JA. From decision tree to heterogeneous decision forest: a novel chemometrics approach for structure-activity relationship modeling. In: Lavine BK, editor. Chemometrics and chemoinformatics, vol. 894. 2005. p. 173–85.
[47] Hong HX, Tong WD, Perkins R, Fang H, Xie Q, Shi LM. Multiclass decision forest—a novel pattern recognition method for multiclass classification in microarray data analysis. DNA Cell Biol 2004;23:685–94.

[48] Xie Q, Ratnasinghe L, Hong H, Perkins R, Tang Z-Z, Hu N, et al. Decision forest analysis of 61 single nucleotide polymorphisms in a case-control study of esophageal cancer; a novel method. BMC Bioinformatics 2005;6:S4.

[49] Tong WD, Xie W, Hong HX, Fang H, Shi LM, Perkins R, et al. Using decision forest to classify prostate cancer samples on the basis of SELDI-TOF MS data: assessing chance correlation and prediction confidence. Environ Health Perspect 2004;112:1622–7.

[50] Diamandis EP. Point—proteomic patterns in biological fluids: do they represent the future of cancer diagnostics? Clin Chem 2003;49:1272–5.

CHAPTER 12

Translation of Computational Model Results for Risk Decisions

William Mendez and Bruce A. Fowler
ICF International, Fairfax, VA, USA

This chapter discusses the impacts of computational toxicology on risk assessment—that is, the extent to which the methods discussed in the other chapters have been applied in evaluating human health risks associated with exposures to chemical agents. Given the broad scope of the technologies that constitute "CompTox," the chapter begins with a brief historical and definitional overview, which is followed by discussions of major ongoing efforts in the areas of CompTox risk assessment. Note that the focus is on potential and actual CompTox applications to risk assessment; a full review of computational toxicology methods is far beyond the scope of this chapter.

ORIGINS AND NATURE OF THE COMPUTATIONAL TOXICOLOGY APPLICATIONS IN RISK ASSESSMENT

The extent to which particular technologies have been accepted as part of standard components of risk assessment varies greatly. Some of these methods (for example, quantitative structure-activity relationships, QSARs) have rather deep historical roots, evolving in parallel with the earliest techniques for quantitative health risk assessment, while others are more recent in origin. Similarly, some have arisen primarily in response to economic forces (the need to efficiently target resources in pharmaceutical development, for example), whereas other techniques were developed and are being applied in the context of changing regulatory requirements. Table 12-1 provides a list of some important historical milestones in computational toxicology applications in risk assessment. Included are both major technological and scientific breakthroughs that enabled applications to occur and key statutory and regulatory developments that facilitated or dictated risk assessment

TABLE 12-1 Timeline of Computational Toxicology Development

Date	Authors	Accomplishment/Requirement
1959	Russell and Burch (1959)[1]	Publication of *The Principles of Humane Experimental Technique*; start of "3Rs" (reduce, replace, refine) movement in Europe, which strongly encouraged the search for alternatives to conventional toxicity testing.
1962	Hansch et al. (1962)[2]	Successful application of QSAR model incorporating parameters related to chemical configuration and partitioning behavior to estimate the activity of plant growth activators (building on the work of many earlier investigators).
1976	U.S. Congress	Passage of Toxic Substances Control Act (TSCA), requiring EPA to assess and regulate risks to health and the environment from all existing and newly introduced chemicals.
1976	Guess and Crump (1976)[3]	Proposed linear multistage model for cancer dose-response; adapted version of Armitage and Doll (1954)[4] model
1980	OSHA (1978)[5]	Applied simple biokinetic model to forecast proportions of workers in auto, smelting, and pigments industry exceeding blood lead targets over time.
1984	Crump et al. (1984)[5]	Proposed benchmark dose methodology for dose-response assessment; adopted by EPA as preferred method for noncancer dose response in 1995.
1986	Periera and Williams (2007)[6]	First industrial application of high-throughput screening technologies used for therapeutic agents. Pfizer Inc. began screening 800 (eventually expanded to 7,200) natural product compounds/week for activity against two potential therapeutic targets, later for ADME properties.
1987	Anderson et al. (1987)[7]	Published first multicompartment PBPK model (for methylene chloride) to be applied in rulemaking by OSHA (1997)[9]
1990	Goodsell and Olson (1990)[10]	Development of AutoDock, first publicly available computational tool to model binding of small molecules to proteins.
1994	U.S. EPA (1994)[11]	Agency issues *Methods for Derivation of Inhalation Reference Concentrations and Application of Inhalation Dosimetry*. (Endorses PBPK modeling as the preferred method for inhalation dosimetry.)
1994	USEPA-TRW	Release of DOS version of IEUBK biokinetic blood lead model; Windows© version released in 2002.
1994	Affymetrix	First commercially available DNA microarray system, greatly facilitated analyses of gene expression patterns.
2000	Human (Mouse, Rat) Genome Project	Draft human genome released, 2000; "complete" genome, 2002; genome of C57BL/6J mouse completed in 2002; Norway rat completed in 2004.
2003–4	U.S. EPA[12]	Agency issues "Framework for a Computational Toxicology Research Program" (2003) formalizing CompTox research and

(Continued)

TABLE 12-1 (Continued)

Date	Authors	Accomplishment/Requirement
		budgetary priorities; National Center for Computational Toxicology begins operation (2004).
2003	European Union Seventh Amendment to Council Directive 76/768[13]	Directive prohibits use of animal toxicity testing in support of any cosmetics ingredient marketed in the European Union after 2013.
2007	National Research Council (2007)[14]	Publication of *Toxicity Testing in the 21st Century*; strongly endorsed move from *in vivo* to *in vitro* testing, apical to basal tox endpoints, increased use of systems toxicology.
2007	European Community (2007)[15]	REACH (Registration, Evaluation, Authorisation, and Restriction of Chemicals) legislation requires submission of physical, chemical, and toxicological data on chemicals manufactured in or imported into the European Union. Annex XI specifies data from appropriately validated QSAR models may be used to indicate "presence or absence of a certain dangerous property."
2007	OECD (2007)[16]	Organization issues *Guidance on the Validation and Use of QSAR Models*.
2007	USEPA-ORD	ToxCast program initiated to conduct a broad range of HTS screening tests on chemicals of potential regulatory interest to assist in testing, regulatory priority setting for EPA programs (approximately 2,000 chemicals, to be tested for about 500 endpoints through 2012); validated data entered into ToxCast database to support predictive models.
2008	U.S. Food and Drug Administration (2008)[17]	*Guidance for Industry—Genotoxic and Carcinogenic Impurities in Drug Substances and Products: Recommended Approaches* (defines role of QSAR modeling).
2008	USEPA-ORD	ACToR—Aggregated Computational Toxicology Resource (combines toxicological, physical-chemical, HTS data for ~500,000 chemicals).
2008	OECD/ECHA (2008)[18]	OECD QSAR Toolbox (version 1.0) released to public; Windows©-based system supports "read-across" and structure-activity regressions for ecological, *in vivo*, and *in vitro* endpoints.
2008	NTP, DHHS, EPA	Memorandum of Understanding forming Tox21 program for testing and validation of *in vitro* high-throughput methods, HTS screening of 8,100 first generation chemicals.
2009	European Commission and Colipa (European Cosmetics Association)[19]	Launch of SEURAT (Safety Evaluation Ultimately Replacing Animal Testing) program. SEURAT-1 program allocates 50 million euros to research on replacements to animal testing for 2011–2015.
2009	National Research Council (2009)[20]	Publication of *Risk and Decision Making*; describes framework for evolution of decision making to include advanced technologies, provision for public consultation, and improved transparency.

applications of computational toxicology. Technical milestones include the development of modern quantitative structural activity relationships (based on chemical structure and partitioning variables) by Hansch and coworkers in the early 1960s; the development of the multistage model for carcinogenesis by Guess and Crump in 1976; the earliest implementation of high-throughput screening methods for drug screening by Pfizer and Co. starting in 1986; the release of the first commercial DNA microarray technology in 1994; and the completion of the human, rat, and mouse genome sequences between 2000 and 2004.

Also included are specific regulatory initiatives that require or imply acceptability of computational toxicological methods, including what is apparently the first use of a PBPK model in human health standard setting[8] along with statutes that made the use of such methods necessary or obligatory (U.S. Toxic Substances Control Act, European Community REACH).[15,21]

Very few methods used in modern risk assessment are not in some way "computational." Kavlock and Dix[22] provide a good general definition, which narrows the scope somewhat:

> Computational toxicology is the application of mathematical and computer models and molecular biological and chemical approaches to explore both qualitative and quantitative relationships between chemical exposure and adverse health outcomes.

Because a comprehensive review of the entire field would be impractical, the scope of this chapter is limited to recent developments related to the following technologies, recognizing that there is considerable overlap in some cases:

- **(Q) Structure-activity relationships.** As noted previously, *in silico* structure-activity relationships (QSARS) have achieved a high degree of technical sophistication over the last decades, and the range of agents and structural/activity parameters has increased greatly. SAR and QSAR methods have also achieved a high degree of regulatory acceptance where large numbers of chemical agents have to be prioritized for additional toxicity testing or with regard to regulatory concern. QSAR models are widely integrated into product development processes in the pharmaceutical, pesticide, and chemical industries.
- **Biokinetic and PBPK modeling.** Biokinetic and PBPK models are an important element in modern quantitative risk assessment. Such models are used to improve dosimetry in dose-response analysis of *in vivo* bioassay data, in interspecies extrapolation, in low-dose extrapolation, and, more recently, to support the evaluation of alternative metabolic pathways and pharmacodynamic mechanisms and characterize variability in individual response.
- **Characterization of adverse outcome pathways.** Since the publication of "Toxicity Testing in the 21st Century,"(11) attention has been focused on a new generation of toxicological models, based on the detailed characterization of biochemical modes of action and disturbances in key control pathways that result in "apical" adverse effects. Characterization of adverse outcome pathways (AOPs) often involves a combination of *in vivo* and *in vitro* data and can require substantial computational resources.

Proponents argue that focusing on AOPs can greatly reduce the need for expensive animal testing, and shed light on commonalities of toxicological mechanisms among chemicals affecting the same pathways.

- **Systems toxicology.** As defined in this chapter, systems toxicology broadens the focus of toxicological analysis from single AOPs to include complex biological control networks whose interactions contribute multifactorially to the development of adverse effects. The ultimate objective of "systems toxicology" is to derive a comprehensive description of all of the events and processes leading to toxic outcomes, at all levels of biochemical and biological organization. Systems toxicology includes the experimental methods required to elucidate biological networks (often involving large amounts of data from *in vitro* assays) and development of *in silico* models of biological networks, and "virtual tissues."
- **Incorporation of high-throughput screening and "omics" data.** Advances in molecular biology and in the automation of *in vitro* biochemical assays have resulted in dramatic expansions in the types and amounts of data that have the potential to be "translated" for use in risk assessment. Knowledge of DNA sequences and genetic makeup of humans and common laboratory animals has been applied to develop assays of gene expression in response to toxic chemicals and other agents. In addition, advances in laboratory automation have been applied to assays of cytotoxicity in numerous cell types for numerous organisms, gene expression, enzyme activity and inhibition, and chemical-protein interactions. Obviously, advances in "omics" and high-throughput screening methods support the development of systems toxicological approaches to risk assessment noted earlier.
- **Data management and information technologies.** Implementation of many of the approaches listed previously requires novel techniques for the management and analysis of large volumes of data generated through conventional toxicity testing, high-throughput screening (HTS) results, and omics data. Even before the advent of "omics" and HTS screening methods, managing physical, chemical, and toxicological data for large numbers of chemicals posed a significant challenge to available information technologies. These challenges range from the population and maintenance of relatively conventional (although very large) relational databases to the development of sophisticated methods for identifying patterns and dependencies in *in vitro* test results. The advent of HTS data has posed new challenges associated with data management and curation, and requires entirely new methods of analysis.

The other issue to be addressed is what constitutes "risk assessment." As the term is commonly used, it refers to the calculation of numerical estimates of the risk of specific outcomes associated with a given set of exposures (like a cancer slope factor), or alternatively, derivation of health-based criteria that represent regulatory estimates of "safe doses" for the exposed population (like a reference dose, or RfD.) In the discussion that follows, risk assessment is also interpreted to include methods for hazard- and risk-based screening, as well as quantitative risk assessment per se. The first reason for doing so is that by far the most widespread use of computational technologies described previously is currently in the hazard- and risk-based screening and prioritization of chemicals. For example, SAR and QSAR are integral to the implementation of major regulatory initiatives to control chemical hazards (REACH and TSCA, and regulation of pesticide uses under the U.S. FIFRA legislation being important examples). In contrast, regulatory acceptance of nontraditional methods and data in quantitative risk assessment for high-saliency chemicals and the development of health-protective criteria have been very slow to come.

Another reason for addressing hazard and risk screening methods is that many computational toxicology applications proposed for use of quantitative risk assessment have grown out of and incorporate methods originally used in hazard-based screening. In particular, the application of systems biology approaches supported by more "classical" CompTox methods (such as QSAR and read-across) is contributing more and more to quantitative risk assessment through their use in identification of adverse outcome pathways and mode-of-action and weight-of-evidence documentation that underlies regulatory risk assessment.

As will be seen in the following section, the applications of computational toxicology in risk assessment have been driven over the last five years to a large extent by the new paradigm for toxicological analysis presented in the NRC's 2007 "Toxicity Testing in the 21st Century."[14] Goals of this new program include the movement away from empirical modeling of dose response for "apical" endpoints measured in animal bioassays to mechanism-based estimation of human health risks using apical and pathway-specific endpoints, with a much reduced reliance on *in vivo* bioassay data. The prospects for meeting these objectives are discussed in more detail in the Summary.

DRIVERS FOR THE APPLICATION OF COMPUTATIONAL TOXICOLOGY TO RISK ASSESSMENT

In its 2007 report, the NRC[14] identified four basic objectives of an effective testing strategy for chemical hazards, which were not currently being met:

> ... *depth*, providing the most accurate, relevant information possible for hazard identification and dose-response assessment; *breadth*, providing data on the broadest possible universe of chemicals, end points, and life stages; *animal welfare*, causing the least animal suffering possible and using the fewest animals possible; and *conservation*, minimizing the expenditure of money and time on testing and regulatory review.

To take the last issue first, decreasing costs are often cited as a primary driver for adoption of CompTox methods in risk assessment. While reduced costs of *in vivo* (whole animal) studies are often identified as being most important, it has been pointed out that actual animal testing costs account for only a very small fraction of the total R&D budgets of the corporations most responsible for developing and manufacturing chemical products subject to European health and safety regulation,[24] and this situation is likely to be similar in the United States. Conducting a battery of animal tests is not likely to be a major cost component in the development of high-value-added products, such as drugs or very high-production-volume chemicals. Arguably, where cost issues arise as a major concern is for the much larger number of moderate- and low-volume chemical products, and a large component of the societal "costs" imposed by regulatory requirements for animal testing actually take the form of impediments to innovation that keep useful products from reaching the market in a timely manner.[24] Animal testing, even where it is cost-effective, is time-consuming, and one of the most important potential benefits of adopting CompTox methods will be to increase the overall capacity of testing strategies (increasing "throughput").

With regard to reduction in animal usage, a fully implemented CompTox-based risk assessment strategy would clearly reduce the requirement for animal testing of "apical" endpoints that were not deemed relevant to the critical toxic effects. The use of *in vitro* data and adverse outcome pathway analysis in a tiered assessment scheme could theoretically limit animal testing to a bare minimum of key endpoints and obviate the need entirely in some cases.

How effectively the increased application of CompTox methods would be in reducing the demand for animal testing would depend on the spectrum of agents being evaluated, and on the "depth" and "breadth" of the universe of *in vitro* and mechanistic data available. As will be seen below, controversy exists among experts as to how realistic it is to expect that (1) testing methods for predicting impacts on the most important adverse outcome pathways have been, or will shortly be, elucidated and (2) how large the universe of agents tested needs to be before an adequate database is assembled to support prediction of toxic impacts in untested agents. To the extent that these assumptions are not fulfilled, the potential for reduction in animal usage is unlikely to be realized in the short term.

In addition, factors other than the regulation of environmental chemicals are known to strongly affect the use of animal testing. For example, in Europe, where "3Rs" (reduce, refine, replace) programs have long been in place, it is estimated that overall animal usage has been reduced approximately 60% over the last 30 years[24] and approximately 50% in the United States.[25] The potential for the reduction in animal usage through the application of CompTox methods is further limited by the fact that only about 10% of total animal usage in Europe is associated with regulatory toxicology.

Perhaps the greatest "cost" that the application of CompTox can reduce is the "cost of ignorance" regarding risks associated with the large bulk of chemicals to which humans currently are exposed.[14,24] Being able to more accurately assess risks for a much larger number of agents, and being better able to allocate resources to address these risks, has the potential to provide significant improvements in public health. This concept of effective risk management is an express element of some proposed strategies for incorporating CompTox methods into risk assessment. The idea that "adaptive stress responses" can compensate for disturbances in biological pathways (NRC 2007, Figure 2-2)[14] naturally leads to a focus on tests that determine when disturbances become potentially pathological. This (in theory) allows resources to be focused on agents and exposures that exceed the thresholds, rather than on determining the numerical level of risk associated with specific exposures.

TRANSLATIONAL RESEARCH

Much remains to be done to develop and perfect computational toxicological models so that they can be used in support of quantitative risk assessment. While the NRC estimated that full implementation of the "Tox21" agenda would take two decades, research in risk assessment applications of computational toxicology is ongoing in a number of institutions under a number of programs.

U.S. Environmental Protection Agency

As discussed in the following section, although specific CompTox methods have been in use in EPA program offices for many years, the field was first formally recognized as a research priority in the U.S. EPA through its 2003 "Framework for a Computational Toxicology Research Program."[12] The Framework laid out a broad strategy for allocating CompTox research resources across the various disciplines and organizations within the agency. In the fall of 2004, the EPA announced the formation of the National Center for Computational Toxicology (NCCT) in the Office of Research and Development that would provide a central home for CompTox research within the agency, collaborating with the National Health and Environmental Effects Research Laboratory (NHEERL) and the National Exposure Research Laboratory (NERL), the National Center for Environmental Research (NCER), and other federal agencies. While the NCCT managed a wide range of research efforts, the first five-year charter included a clear focus on developing CompTox resources to support risk assessment applications.[22] The EPA's 2009 "Strategic Plan for Evaluating the Toxicity of Chemicals"[26] further emphasized the drive to develop risk assessment supporting technologies through setting goals for the identification and screening of toxicity pathways, developing approaches for incorporating toxicity pathway modeling into risk assessment, and helping program offices find ways to incorporate CompTox methods into their decision-making processes.

While the EPA, through the NCCT and other programs, has supported a wide range of translation research in toxicology, its most significant achievements have been in the development of data repositories for conventional toxicity and *in vitro* and high-throughput screening (HTS) data, support for the expansion of chemical space and range of *in vitro* assay data available to support identification of toxicity pathways, and systems toxicology research, including research into "virtual tissues." Results from all of these efforts are accessible through the overarching ACToR database management system, as discussed next.

Aggregated Computational Toxicological Resource, or **ACToR,** is a collection of databases developed and managed by the U.S. EPA Computational Toxicology Research Program. It serves as the primary repository supporting the EPA's computational toxicology program. The goals of ACToR are

> to aggregate all publicly available information on chemicals in the environment by (1) making information on the health effects and exposure potential for environmental chemicals readily accessible; (2) characterizing gaps in knowledge of the toxicology of environmental chemicals; and (3) providing a resource for model-building to fill data gaps in environmental health risk information.[27]

ACToR provides centralized access to four databases: DSSTox, ToxRefDB, ToxCastDB, and ExpoCastDB. Between them, these sources provide conventional toxicity data, HTS and genomics test results, chemical and physical data, structural parameters, rigorous

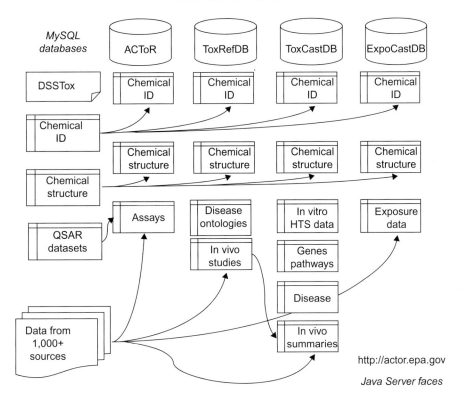

FIGURE 12-1 Structure of ACToR Data System.[27]

chemical identifiers, and exposure information. The overall structure of ACToR is summarized in Figure 12-1. ACToR aggregates data from over 500 public sources on over 500,000 environmental chemicals searchable by chemical name, other identifiers, and by chemical structure.[28] The top level of the data hierarchy in ACToR is organized around seven information categories: capture level, data level, inherent chemical property, exposure, hazard, risk management, and use category. A range of "assays" (data elements) identified within each top-level category serve as search filters that can be used to narrow and refine searches. All data within the constituent data bases can be accessed independently or using the ACToR framework. The search capabilities of ACToR are currently limited but continue to be upgraded.[27]

Distributed Substructure Searchable Toxicity, or DSSTox, is a major source of chemical structure and conventional toxicity data within the ACToR system. DSSTox provides access to 16 databases of *in vivo* toxicity and *in vitro* toxicity, along with uniformly curated files containing chemical structure data (Table 12-2.) Chemical structure data and toxicological data in DSSTox are standardized, searchable, and fully documented, and a chemical structure browser is also provided, making DSSTox a powerful tool for structure-activity modeling.[22]

TABLE 12-2 DSSTox Databases

Database ID	Database Name (Number of Records)
ARYEXP	European Bioinformatics Institute (EBI) ArrayExpress Repository for Gene Expression Experiments (958)
CDBPAS	Carcinogenic Potency Database Summary Tables—All Species (1,547)
DBPCAN	EPA Water Disinfection By-Products with Carcinogenicity Estimates (209)
EPAFHM	EPA Fathead Minnow Acute Toxicity Database (617)
FDAMDD	FDA Center for Drug Evaluation & Research—Maximum (Recommended) Daily Dose Database (1,216)
GEOGSE	National Center for Biotechnology Information (NCBI) Gene Expression Omnibus (GEO) Series Experiments (1,179)
HPVCSI	EPA High Production Volume Challenge Program Structure-Index File (3,548)
HPVISD	EPA High Production Volume Information System (HPV-IS) Data Structure-Index Locator File (1,006)
IRISTR	EPA Integrated Risk Information System (IRIS) Toxicity Review Database & Structure-Index Locator File (544)
KIERBL	EPA Estrogen Receptor Ki Binding Study [23] Database File (278)
NCTRER	FDA National Center for Toxicological Research (NCTR)—Estrogen Receptor Binding Database (232)
NTPBSI	National Toxicology Program (NTP) On-line Chemical Bioassay Database Structure-Index Locator File (2,330)
NTPHTS	National Toxicology Program (NTP) High-Throughput Screening Project Structure-Index File (1,408)
TOX21S	Tox21 Chemical Inventory for High-Throughput Screening Structure-Index File (8,193)
TOXCST	Research Chemical Inventory for EPA's ToxCast TM Program Structure-Index File (1,892)
NAMEID	External (non-DSSTox) SDF Data File Title & Description
ISSCAN	ISSCAN: Istituto Superiore di Sanita, "Chemical Carcinogens: Structures and Experimental Data"

ToxRefDB, another database that can be accessed through ACToR, is a repository of conventional toxicity data (acute, chronic, carcinogenesis, developmental, and reproductive toxicity), drawn from high-quality studies, primarily of pesticide active ingredients.[27] As of mid-2012, ToxRefDB included data on 474 chemicals (Table 12-3.) Files include details of study design, dosing regimen, and dose-response data points for each exposed group. File formats are standardized to support searching and data extraction (directly from ToxRefDb, or through ACToR.) Data from ToxRefDB represent a valuable resource for the development of structure-activity relationships, and in the validation and "phenotypic anchoring" of *in vitro* high-throughput study results.

Data generated as part of EPA-ORD's **ToxCast** project are made available to ACToR through the **TOXCastDB** database. ToxCast is a "large-scale experiment using a battery of

TABLE 12-3 Toxicity Data in ToxRefDB[25]

Studies/Chemicals	Number of Studies
Total Chemicals	474
Total Studies	1,978
Cancer/Chronic—Rat	324
Cancer/Chronic—Mouse	324
Multigeneration Reproduction—Rat	352
Prenatal Development—Rat	365
Prenatal Development—Rabbit	331
Subchronic Rodent	302

in vitro, high-throughput screening (HTS) assays, applied to a relatively large and diverse chemical space, to develop methods to predict potential toxicity of environmental chemicals at a fraction of the cost of full-scale animal testing."[29] Begun in 2007, the ToxCast project is a broad-based initiative to improve the utility of high-throughput screening methods and *in vitro* toxicogenomics in characterizing chemical toxicity, and to expand the availability of HTS data for a broadening spectrum of chemicals.[30] A specific goal of the project is to support the development of toxicity pathway-based human and ecological risk assessment. Under Phase I, 309 chemicals, mostly pesticide active ingredients, were tested in 467 different screening assays, including cell-free and cell-based (human and rat) assays.[29] The Phase I chemicals were chosen in part because they had been subject to a wide range of conventional (*in vivo*) toxicity testing.

An additional 676 chemicals were selected for ToxCast Phase II screening. In addition to pesticides, chemicals in Phase II included those that had been evaluated in the National Toxicity Program (NTP) and NCEA's IRIS program, as well as selected high-production-volume chemicals, and chemicals of regulatory concern in water, air, cosmetics, foods, and consumer products. The EPA also screened 111 pharmaceutical chemicals that were never marketed due to toxicity concerns.[31] This last group of chemicals is of particular interest because (1) they were generally designed to affect a single pharmacological pathway or a small number of pathways, and (2) they were provided with relatively rich databases of animal and human toxicity.

The data set for combined ToxCast Phases I and II, as this chapter was written, includes data on 1,101 chemicals (960 excluding replicates), characterized in approximately 650 bioassays.[28] Approximately 2,000 additional chemicals are being screened under the ongoing Phase II, including 880 potential endocrine-disrupting chemicals.

ToxCastDB serves as the repository for data generated in the ToxCast Project. ToxCastDB is intended to support the primary aim of the ToxCast program, which is "to build predictive models of *in vivo* toxicity using *in vitro* HTS data and associated chemical and biological data."[27] ToxCastDB contains data from all of the *in vitro* tests and accompanying *in vivo* data concerning the ToxCast chemicals (with a time lag owing to data validation needs). Information is accessible using the "assay table" structure of ACToR (see below). ToxCastDB includes individual dose-response data points for all HTS tests, and assays are linked to specific genes, control pathways, and diseases to facilitate pathway-related dose-response analyses. Data in ToxCastDB can be accessed directly or through ACToR.

The **ExpoCast** program supports and coordinates a wide range of exposure science activities conducted by the NCCT, NERL, and stakeholders; the goal of the program is "to advance characterization of exposure required to translate findings in computational toxicology to information that can be directly used to support exposure and risk assessment for decision making and improved public health."[32] To date, ExpoCast activities have focused on developing methods and identifying data sources that can be used, along with other computational toxicology tools, such as pathway analysis, to support screening and prioritization of chemicals based on potential human health hazards. The **ExpoCastDB** database serves as the repository of exposure data and tools assembled under the ExpoCast program. ExpoCastDB includes compilation of data on the levels of chemicals found in environmental media (air, soil, dust, food, water) and biomarkers of exposure (primarily urinary excretion levels). Exposure data obtained from several recent surveys are currently available for 99 chemicals, mostly pesticide active ingredients. Both the numbers and types of chemicals, and the types of information provided are expected to expand (e.g., to include nonpesticides and nanomaterials, as well as sampling and laboratory analytical methods). ExpoCastDB data can also be accessed through ACToR.

Under the EPA's Chemical Safety for Sustainability (CSS), the ORD is developing a series of **CSS Dashboards** that are designed to assist nonexpert users to access and analyze the data in ACToR and other databases to support particular applications.[33] In the 2012–2013 time frame, ORD is developing prototype dashboards tailored to support the needs of the endocrine disruptor screening program, the EPA Office of Water in setting priorities for its Candidate Compound List (CCL), prioritization and screening decisions in the regulation of pesticides, and to support risk analyses for high-priority "workplan" chemicals under the Toxic Substances Control Act.[34] Each dashboard is being crafted to provide easy and seamless access to complex data concerning chemical "inherent properties," occurrence, use, toxicity, and high-throughput screening data relevant to each decision setting, as well as clear graphical summaries of the information (Figure 12-2).

Tox21 Collaboration

The Tox21 Collaboration is a cooperative effort between the U.S. National Toxicology Program (NTP) of the U.S. Department of Health and Human Services, the NIH Chemical Genomics Center, and the EPA National Center for Computational Toxicology (NCCT). Formed through a Memorandum of Understanding in 2008, the Tox21 Collaboration was designed to build infrastructure to support *in vitro* high-throughput screening of chemicals and to begin the process of gathering data on a broad range of *in vitro* endpoints on high-priority chemicals.[35] Under the collaboration, robotic methods are being used to test and validate specific test methods, and to conduct high-throughput screening of a primary chemical inventory of approximately 8,100 distinct compounds. Tests in the initial round of

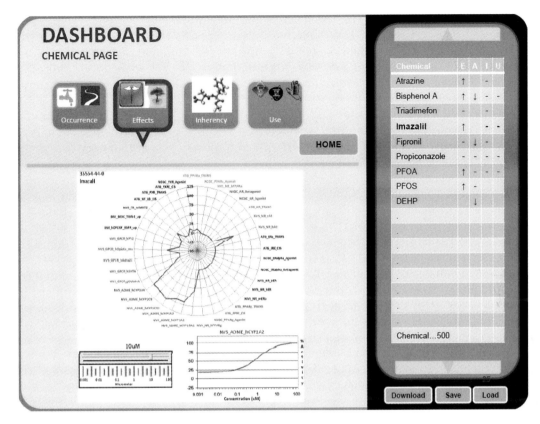

FIGURE 12-2 Sample CSS Dashboard output screen.

screening include assays of cell viability, DNA damage, apoptosis, epigenetic DNA modification, mitochondrial toxicity, and a range of nuclear receptors and indicators of other cellular control pathways.[36] Compounds are tested at between 7 and 15 concentrations, across an approximate 4-log range,[35] and output data are managed through the ToxCast program, making them available to support the evaluation of adverse outcome pathways, the development of structure-activity relationships, or in quantitative risk assessment.

OECD and European Union Activities

The Organisation for Economic Cooperation and Development (OECD) also conducts an active research program aimed at developing computational toxicological methods into chemical safety evaluation and risk assessment. These efforts have been partly motivated by changes in the regulatory climate in Europe that discourages the use of animals in testing to the extent practical (the "three-R" principles of "reduce, replace, refine" have been widely adopted as official policy). As discussed in the following section, the OECD has been a leader in the development and application of quantitative structure-activity relationships to chemical categorization and hazard- and risk-based screening.[37,38] In addition to QSAR modeling, the OECD has been supporting research related to the development of adverse outcome pathways, a key element in the implementation of pathway-driven risk assessment. The OECD's "Proposal for a Template, and Guidance for Developing and Assessing the Completeness of Adverse Outcome Pathways"[39] describes the applications of AOPs in development of testing and assessment methodologies, chemical categories, and risk characterization. The proposal also describes the main impediment to the implementation of AOP methodologies and provides guidance as to the minimal level of data needed to support an AOP description. A key concept in the proposal is that an AOP be firmly "anchored" at one end by a molecular-initiating event and at the other by a clearly defined apical adverse outcome, and that causal relationships linking the initiation event, intermediate events, and apical outcome be fully documented. Stress is given to identifying experimental support for both the qualitative and quantitative nature of the interactions in a pathway. Examples of applications of AOPs are provided to the analysis of selected adverse outcomes.[40]

The European Union also supports a range of CompTox-related activities. Prominent among these is the Safety Evaluation Ultimately Replacing Animal Testing (SEURAT) program, initiated in 2009. The goals of the project are "the development of a concept and corresponding long-term research strategy for future research and development work leading to pathway-based human safety assessments in the field of repeated dose systemic toxicity testing of chemicals..." and "to establish animal-free Innovative Toxicity Testing (ITT) methods, enabling robust safety assessment that will be more predictive than existing testing procedures."[19]

The SEURAT 1 project, funded jointly by the European Union (EU) and Colipa (the European Cosmetics Association), has allocated 50 million euros to seven projects over the years 2011 to 2015. The projects include research on stem cell differentiation to provide better models for assessing target organ-specific endpoints *in vitro*, development of a "hepatic microfluidic bioreactor" for evaluating liver toxicity, evaluation of selected cellular biomarkers for predictive value in repeated-dose testing, development of organotypic cell cultures for long-term toxicity testing, and identification of adverse outcome pathways. In addition to these laboratory projects, SEURAT 1 supports a project to develop computational methods for the use of *in vitro* data to predict long-term health endpoints and a coordination function to manage data and cell lines, and to support the selection of agents for testing.[40]

COMPUTATIONAL TOXICOLOGY APPLICATIONS IN RISK- AND HAZARD-BASED SCREENING

U.S. Environmental Protection Agency: Industrial Chemicals and Pesticides

With the passage of the Toxic Substances Control Act of 1976, the U.S. EPA became responsible for managing risks to human health and the environment from chemicals already "in commerce" and all chemicals that might be "introduced into commerce."[42] The need to manage risks from both new and existing chemicals necessarily involves the applications of a range of computational toxicology tools for chemical property estimation and for risk- and hazard-based testing and regulatory priority setting. As of 2006, the EPA had identified approximately 2,750 high-production-volume (HPV) chemicals (annual production greater than one million pounds) and approximately 4,000 medium-production-volume (MPV) chemicals, with annual production volumes between 25,000 and one million pounds.[41] In addition, the EPA processes more than 1,000 premanufacturing notifications (PMNs) for new chemical substances or new uses of existing chemicals each year, and toxicity data are submitted for fewer than half of these chemicals.[28] In response to these challenges, the EPA has developed a range of tools and techniques to facilitate the analysis of potential health risks associated with industrial chemicals. As discussed later, the EPA and OECD began a partnership in 1998 to manage risks from HPV chemicals. This collaboration involved cooperation in the development of databases and decision tools. The EPA's Office of Pollution Prevention and Toxics (OPPT) took the lead in the development of several such tools.

EPI Suite is the "workhorse" chemical/physical property estimation software suite developed by OPPT, starting in 2000.[42] EPI Suite includes modules for estimating a wide range of physical and environmental fate properties including solubility, K_{OW}, biodegradation and aqueous hydrolysis constants, Henry law constants, soil absorption constants, and bioaccumulation factors. Higher-level models are available to assess environmental partitioning based on fugacity calculations and for predicting removal efficiency by activated sludge treatment. A graphical user interface (GUI) is provided, and the user inputs SMILES notation, a CAS number, or MDB file for the substance being evaluated. The property estimation modules include implementations of well-documented structure-activity models (for which full documentation is provided), and the models are supported by a physical property database of over 40,000 chemicals. EPI Suite is routinely applied as a screening tool to fill in missing physical property data for new PMN chemicals[43] as part of an integrated mode-of-action and in-category-based approach for evaluation of chemical hazards.[42] The software is publicly available and widely used by industry and academics, as well as by the EPA.

The **ECOSAR** program was developed by the OPPT to encapsulate and make available a wide range of structure-activity-based aquatic toxicity models that had been used in chemical prioritization starting in the early 1980s.[44] ECOSAR is a Windows®-based software package primarily designed to support QSAR-based estimation of toxicity to fish, aquatic invertebrates, and algae of unknown chemicals. ECOSAR includes over 700 QSAR models derived for 111 defined chemical classes. In addition to SMILES notation and CAS numbers, chemical identity can be specified through a drawing module. ECOSAR also requires K_{OW}, solubility, and melting point information; missing values estimated by EPI Suite (see preceding description) can be accepted as inputs, and ECOSAR can be invoked through the EPI Suite after properties are estimated. Like EPI Suite, ECOSAR is made available to the public on the OPPT website (http://www.epa.gov/oppt/newchems/tools/21ecosar.htm).

OncoLogic is an expert system, developed by the EPA with contractor support, to screen chemicals for carcinogenic potential.[45] The publicly available software incorporates toxicological information (bioassay results) from a range of published sources (the *Chemical Induction of Cancer Series*, *IARC Monographs*, *NTP Bioassay Reports*, the U.S. PHS *Survey of Compounds Which Have Tested for Carcinogenic Activity*, as well as nonconfidential data submitted to the EPA). OncoLogic includes modules for estimating the carcinogenic potential of fibers, polymers, metals/metalloids, and organic chemicals. Carcinogenic potential is estimated based on expert-derived "rules packages" and structure-activity relationships within chemical categories. Depending on the nature of the substance being evaluated, the main drivers may be structural and physical properties (fibers, polymers, metals), the presence of reactive function groups (RFGs), or a combination of the two. Within the organic module, 48 distinct chemical categories are defined based on the presence of specific RFGs (aldehyde, epoxide, aromatic amine, etc.). Chemicals with more than one RFG must be evaluated iteratively for each group that is present. The outputs of the OncoLogic program consist of a summary level of concern (low, marginal, low-moderate, moderate, moderate-high, or high), a listing of structural features and/or RFGs of concern, and the expert rule(s) used to assess the carcinogenic potential. The model does not predict carcinogenic potency.

EPA-OPPT also uses two additional CompTox tools in the risk- and hazard-based screening of new chemicals. The **Exposure Fate Assessment Screening Tool**, or **E-FAST**, accepts inputs related to chemical and physical characteristics, production volumes, and process flows and generates screening-level estimates of aquatic, terrestrial, and ambient air exposures, as well as nonoccupational human exposures.[46] **Chem-Steer** is a suite of expert tools for estimating occupational exposures and environmental releases based on physical/chemical properties, production volume, batch size, and industry-specific profiles and generic exposure scenarios. As with ECOSAR and EPI Suite, E-FAST and Chem-Steer are interoperable, so E-FAST can accept release estimates directly from the Chem-Steer model. Both E-FAST and Chem-Steer are clearly designated as screening-level models, not to be used as substitutes for actual exposure or release data.[46]

Based on their experience with the preceding models, and more than 30 years of experience assessing hazards from chemicals with little toxicological data, EPA-OPPT has developed a set of hazard-based chemical categories for the New Chemicals Program (NCP).[47] These 58 categories, mostly for families of organic chemicals, are based on chemical structure and most probable mode of action. Intended to aid both industry and the EPA in anticipating chemical hazards and the need for additional testing, information for each category includes

- Basic category definition (structural features)
- Human health and environmental toxicity concerns (types of toxic endpoints, sensitive species)
- Defined boundaries of the chemical category (molecular weight, solubility, vapor pressure, presence of functional group variations or spatial configurations, etc.)
- Suggested testing strategies for filling in important missing data gaps related to ecological and human health hazards.

Environmental CompTox Applications in the OECD and European Union

As noted previously, the Organisation for Economic Co-operation and Development (OECD) has supported a wide range of activities related to the use of computational toxicology in two major areas: (1) the use of SAR and QSAR methods to evaluate risks from categories of chemicals and predict toxic risks associated with individual chemicals within categories, and (2) the development of systems biological approaches to risk assessment including approaches for identifying adverse outcome pathways (AOPs). As discussed next, the need for computational toxicology applications in Europe has been driven to a large extent by OECD projects related to harmonization, changes to EU regulations governing cosmetic ingredient testing, and passage and implementation of the REACH legislation.

SAR and QSAR Methods

The OECD has been promoting the development and application of SAR and QSAR techniques since the early 1990s.[37] Early efforts included the development of guidelines for the use of QSAR data in Minimum Premarketing Datasets, developed in cooperation with the U.S. EPA. The OECD's "Guidance for Aquatic Effects Assessment" (1995)[48] also described approaches for the use of QSAR models in estimating no-effects levels for ecotoxicological endpoints. In 1998, as part of a collaboration with the U.S. EPA to address risks from high-production-volume (HPV) chemicals, the OECD incorporated the EPA's guidance document on the application derivation of chemical categories developed by the U.S. EPA into its own OECD "Guidance document on the validation of (quantitative) structure-activity relationships (Q)SAR models."[16]

OECD-sponsored work on the development and application of SAR and QSAR models continued in the current century, spurred on not only by the need to fill data gaps related to the HPV chemicals, but also by regulatory and statutory developments. Reduce and Replace (R and R) initiatives were well under way in the pharmaceutical and cosmetics industries when the European Community Seventh Amendment to Council Directive 76/768 in 2003[13] was promulgated, prohibiting outright the use of animal safety testing of cosmetic ingredients marketed in Europe. The passage of the REACH legislation in 2006 required chemical manufacturers and distributors to develop "dossiers" of chemical, physical, and toxicological data on all chemicals to be produced or imported into Europe. In addition to protection of public health, a key objective of the REACH statute was also to reduce reliance on animal toxicity testing, and the law specifically allowed manufacturers to apply structure-activity data to demonstrate "the presence or absence of a toxic effect."

In 2004, an expert panel sponsored by the OECD issued a brief set of principles for the validation of QSAR models for regulatory applications.[38] The principles set forth in the document stated that, in order to be considered for regulatory application, QSAR models should have (1) a defined endpoint; (2) an unambiguous algorithm; (3) a defined domain of applicability; (4) appropriate measures of goodness-of-fit, robustness, and predictivity, and a mechanistic interpretation, if possible. While these very general guidelines left much up to the judgment of submitters of data and regulatory bodies, they helped to further establish the legitimacy of applying QSAR methods in support of regulatory decisions.

As implementation of the REACH legislation began, the OECD issued further, more specific guidance on the use of QSAR and related methods. In early 2007, a report was issued discussing the widespread application of QSAR methods within the organization,[49] and later that year, a detailed guidance document was issued on the validation of QSAR models for regulatory use.[16] This document expanded on the principles annunciated in 2004, stressing the need for adequate data quality to support model development, clear definitions of endpoints and algorithms, and careful specification of applicability domains. Frequently used model algorithms were described, along with general validation approaches (internal and external), and appropriate goodness-of-fit and predictivity measures were discussed. The mechanistic bases of some common QSAR models were reviewed, useful molecular descriptors were presented, and methods for drawing mechanistic inferences from QSAR models were described. Model transparency was stressed, and the provision of training set data was specified as a key element of documentation for any model. Finally, a checklist for QSAR model validation was presented; however, the guidance document stressed that the validation needs, even for identical models, could vary substantially, depending on the intended regulatory application.

The OECD's 2007 "Guidance on the Grouping of Chemicals"[50] describes approaches for defining categories of chemicals based on chemical, physical, or toxicological properties. The primary rationale for grouping was to facilitate risk management for high-production volume and other chemicals for which toxicological data may be incomplete. The rationale for the categorization approach was that it could "provide significant efficiencies and benefits when identifying data gaps and filling data needs that are ultimately deemed necessary." The context for the categorization guidance was that if chemical categories were sufficiently well defined (that is, based on appropriate analyses of as much data on as many member chemicals as possible), important toxicological properties of the category members could be defined, quantified, and, most importantly, some toxicological effects could be ruled out with a reasonable degree of certainty. This would implicitly open the door for regulatory treatment of certain categories of chemicals as a class rather than as individual substances, and would greatly reduce the need for conventional toxicity (*in vivo*) testing and safety analyses for large groups of chemicals.

The guidance recognizes the close relationship between grouping and QSAR modeling activities, noting that category definitions are much stronger if they are based on a presumed common mode of action. Also, it is noted that finding commonalities in toxic activity among groups of chemicals with similar specific physical properties or functional groups can shed light on the plausibility of specific modes of action. In addition to prescribing a detailed approach to category definition (Figure 12-3), methods for predicting properties of new chemicals belonging to specific categories are also described. The OECD guidance also provides very clear descriptions of the "analogue," "read-across," and QSAR model approaches to filling data gaps within chemical categories and predicting toxicological properties of untested chemicals. Examples are provided regarding how categories might be defined for a range of situations (structural analogues, chemicals metabolized by similar pathways, complex natural mixtures, metals and inorganic compounds) and recommended reporting formats are provided, although, again, the guidance was not linked to any specific regulatory requirement.

One of the most important products of the OECD QSAR project is the **OECD QSAR Toolbox**.[18] The toolbox is a PC- or server-based expert system that incorporates the OECD guidance related to categorization, read-across, and QSAR models. It also incorporates a large number of data sets containing physical and chemical property data, molecular descriptors, mammalian and nonmammalian toxicity test data, *in vitro* and high-throughput data, and categorical and endpoint/mechanistic descriptors derived by the U.S. EPA, OECD, and other organizations for thousands of chemicals. A GUI allows the user to enter or retrieve data on individual chemicals on a point-and-click basis; define category criteria; and conduct read-across, trend analyses or run QSAR models to fill data gaps for untested chemicals. The workflow of a typical analysis includes the following:[18]

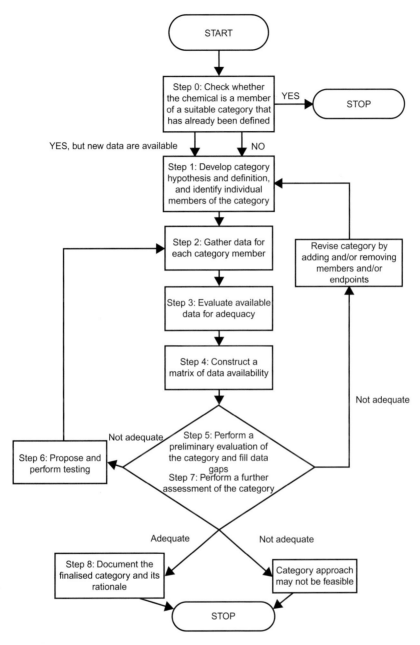

FIGURE 12-3 OECD approach to chemical category development.[16]

- **Chemical input.** The user enters the identity of the chemical for which properties are to be estimated, either by CAS number, chemical name, SMILES notation, or through a structure drawing interface. The toolbox provides data that allow the user to confirm

that the correct (intended) chemical has been entered and provides identification data (CAS numbers, SMILES notations, etc.) for missing fields.
- **Profiling.** The user chooses from a list of over 50 predefined categorization schemes ("profilers") defined according to a regulatory framework (for example, OECD HPV chemicals), general mode of action (DNA binding by OASIS, etc.), or endpoint-specific mode of action (Aquatic Toxicity by ECOSAR, etc.). The toolbox then displays information as to whether the target compound is classified under each profiler and by what criteria it is classified (regulatory program, occurrence in specified databases, structural parameters, reactivity, and/or biochemical mechanisms). Data from the profiling step are used to determine the best ways to fill data gaps for the target compound.
- **Endpoints.** The user next specifies endpoints that are of interest, and the toolbox displays data on the endpoints from one of over 30 data sets included in the toolbox. Searchable data sets include physical-chemical properties (EPISUITE); bioaccumulation bioconcentration; and biodegradation, aquatic, and terrestrial toxicity data (U.S. EPA ECOTOX, ECETOX Aquatic); chronic, reproductive, and developmental toxicity (U.S. EPA ToxRefDB); and databases on mutagenicity, genotoxicity, dermal and ocular irritation, skin sensitization, and single- and repeated-dose toxicity.
- **Category definition.** If the user wishes to estimate toxicity (or some other property) based on similarity to related chemicals, the Category Definition function of the toolbox is then invoked. The toolbox offers a range of grouping methods, based on structural similarity, mechanism, and endpoint. The selection of the most appropriate grouping method(s) is informed by the results of the previous profiling step. Data on selected endpoints and relevant properties are then retrieved for chemicals in the various databases, according to the selected grouping method. Selections can be limited only to chemicals with complete data sets, or chemicals with missing data elements can be included, and the missing elements can be filled in by read-across methods. The user has the opportunity to review the structure and endpoint data on all the candidate chemicals, edit the group, change grouping methods, or develop more narrowly defined subcategories of chemicals.
- **Filling data gaps.** At this point, the user can estimate attributes for the target chemical based on similarity to other chemicals in the defined group. Options for filling data gaps include read-across, trend analysis, and QSAR modeling. Read-across analyses are conducted by averaging the numeric values (or positive/negative results for dichotomous endpoints) for a user-selected numbers of "nearest neighbor" (most similar) chemicals. Selecting trend analysis allows the user to conduct a simple regression, wherein selected property values (for example, aquatic LC50 value) are plotted against predictor values [for example, $\log(K_{ow})$]. A sample trend analysis plot is shown in Figure 12-4. The toolbox also supports more complex QSAR modeling as fill-in methods, in which third-party models are applied to the selected chemical groups (it is up to the user to assure that the model structure is appropriate, and the model applicability space encompasses the target compound).
- **Reports.** The toolbox provides a range of documentation and reporting options for property estimation results. Customized print and electronic summaries can be

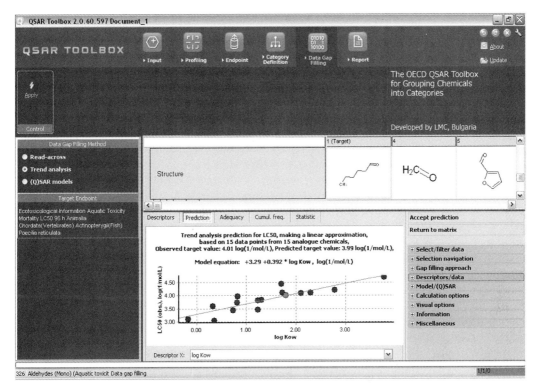

FIGURE 12-4   OECD QSAR Toolbox trend analysis screen.[51]

generated, and the toolbox also includes an interface for directly exporting the results to the IUCLID database in the appropriate format. Thus, the toolbox is well tuned to REACH reporting requirements.

In contrast to the use of QSAR methods, it appears that the REACH program has to date had only limited impact on the use of new *in vitro* test data in support of hazard characterization. A recent study[52] of dossiers submitted for REACH preregistration found that a large proportion of the dossiers relied on older conventional toxicity studies, that few proposed new testing (*in vivo* or *in vitro*), and that for some endpoints (notably, reproductive toxicity), where chemical-specific data were not available, submitters relied solely on read-across methods from similar compounds. These findings are perhaps not surprising, given that submitters have been encouraged to use QSAR methods, and few wish to suggest additional testing using methods whose predictive power has not been validated. As noted in Drivers for the Application of Computational Toxicology to Risk Assessment, the strongest incentives to reduce animal usage and find alternative test methods to date have come from regulations placing direct limitations on animal studies (as embodied in the EU prohibition of animal testing of cosmetics), rather than changes in the regime of chemical risk regulation.

CURRENT STATUS OF COMPUTATIONAL TOXICOLOGY IN QUANTITATIVE RISK ASSESSMENT

As discussed to this point, various CompTox methods (QSAR, read-across, *in vitro* testing, and "omics" analyses) have been widely adopted as tools for the prioritization of chemicals for testing and for generating indices of regulatory concern. QSAR methods, high-throughput screening programs, and expert systems are also routinely used in pharmaceutical development in lead generation and in the evaluation of leads for potential human toxicity. A different, and smaller, set of computational methods has found acceptance in quantitative human health risk assessment.

The U.S. NRC 2007 report "Toxicity Testing in the 21st Century"[14] has been very influential in defining a new risk assessment paradigm, although precursors can be found in the EPA's 2003 "Framework for Computational Toxicology Research"[12] and the "3Rs" (reduce, replace, and refine) framework developed in Europe, supported by the OECD and European Union.[24] Table 12-4 provides a useful summary of the ingredients in a computational toxicology paradigm for risk assessment as defined by Andersen and Krewski.[53]

Of these techniques, PBPK modeling is by far the most widely accepted in quantitative risk assessment. The EPA's "Guidelines for Cancer Risk Assessment"[54] identifies "toxicokinetic" modeling as the preferred dosimetric approach when such models are available, as does the International Program for Chemical Safety's "Guidance on the Development of Chemical-Specific Adjustment Factors" for dose-response assessment.[55] The U.S. EPA's IRIS program (which conducts detailed quantitative health risk assessments that are used to support regulatory measures) has issued guidelines for the use of PBPK methods,[56] and the majority of IRIS assessments under review as this chapter was being written incorporate PBPK models in dose-response assessment, or as an aid in interpreting and comparing *in vivo* toxicological results. Most of the models include major metabolic pathways as well as absorption, intercompartment transport, and excretion.

In contrast, few of the other computational toxicology tools shown in Table 12-4 have found wide application in quantitative risk assessment. As discussed later, while high-throughput screening and omics methods have been used for hazard- and risk-based prioritizing, their role in risk assessment has remained limited. "Bioinformatics," broadly defined, has found a place in risk assessment, however. As noted in previous sections, risk assessors now have at their fingertips a range of large, publicly available databases of validated *in vivo* and *in vitro* data, greatly reducing the costs and effort to obtain information. Assessors start with more information related to observed apical effects, biochemical modes of action, and adverse outcome pathways. Thus, even if the assessment is performed using standard approaches (derivation of health-protective criteria based on animal bioassay or epidemiological studies) it is likely to be more complete (and will certainly be less resource-intensive) than before the development of high-quality informatics resources. In addition to the databases specifically designed to support computational toxicology, specialized databases are being implemented to facilitate the performance and documentation of high-saliency assessments. The U.S. EPA Office of Research and Developments' HERO database (http://hero.epa.gov/), for example, contains over 300,000

TABLE 12-4 Toxicity Testing Tools and Their Application in Risk Assessment[53]

Tool	Application
High-throughput screens	Efficiently identify critical toxicity pathway perturbations across a range of doses and molecular and cellular targets
Stem cell biology	Develop *in vitro* toxicity pathway assays using human cells produced from directed stem cell differentiation
Functional genomics	Identify the structure of cellular circuits involved in toxicity pathway responses to assist computational dose-response modeling
Bioinformatics	Interpret complex multivariable data ftom HTS and genomic assays in relation to target identification and effects of sustained perturbations on organs and tissues
Systems biology	Organize information from multiple cellular response pathways to understand integrated cellular and tissue responses
Computational systems biology	Describe dose-response relationships based on perturbations of cell circuitry underlying toxicity pathway responses giving rise to thresholds, dose-dependent transitions, and other dose-related biological behaviors
PBPK models	Identify human exposure situations likely to provide tissue concentrations equivalent to *in vitro* activation of toxicity pathways
Structure-activity relationships	Predict toxicological responses and metabolic pathways based on the chemical properties of environmental agents and comparison to other active structures
Biomarkers	Establish biomarkers of biological change representing critical toxicity pathway perturbations

references related to agents that are the subject of Integrated Science Assessments (ISAs) and IRIS assessments. It is designed to be an "evergreen" toxicological information resource, and is continuously updated with new articles as they are published. When HERO is fully implemented, EPA personnel will be able to access the full text of every article cited in ISA and IRIS documents, either by keyword searching or through hyperlinks in the documents themselves. Stakeholders will be able to access publicly available documents in the same fashion.

Under the fully realized "Tox21" paradigm,[57] information from *in vitro* tests and high-throughput screening methods would be used to identify adverse outcome pathways and critical concentrations (points of departure) at critical receptors below which no pathway disturbance would occur. Subsequently, systems-biology-based models, including PBPK models *in silico* modeling of critical control networks,[58] and biomarkers of human exposure,[32] would be used to estimate human exposure concentrations corresponding to the mechanistic points of departure to serve as health criteria.

However, to date, such an approach has not been fully implemented for chemicals of direct regulatory interest. The major reason for this is that the causal path from biochemical insult to effects at the whole-organism levels are poorly characterized, even for well-studied chemicals. There is currently a gap between the identification of individual AOPs through *in vitro* testing and the identification of which AOP(s) is (are) critical at the organism or population level. In addition, many high-saliency chemicals have large databases of conventional *in vivo* toxicological data and even epidemiological study results, which are seen as being sufficient to support quantitation for risks and the derivation of health criteria.

Several authors[59,60] have argued that a productive way forward to the full integration of computational toxicological methods into risk assessment would start with the use of cases studies illustrating how currently available data can be incorporated and elucidating where conceptual and information gaps remain. Notable attempts include two case studies of pesticide active ingredients conducted by the U.S. EPA's Office of Pesticide Programs[61] to illustrate the implementation of the proposed Integrated Testing and Assessment (IATA) framework. In the first of these studies, toxicogenomic data were evaluated for elucidating the mode of action for propiconizole (a fungicide) in the causation of mouse liver tumors. Objectives of the study were to examine how to integrate conventional toxicological data (propiconazole is well characterized in *in vivo* studies) with omics information to examine the significance of various apical endpoints (liver weight increase/hypertrophy, increased CYP 2b activity, cell proliferation) and to place them in the context of a detailed mode of action and adverse outcome pathway for tumor induction. Omics data were used to identify key events (oxidative stress, increase in atRA metabolism, dysregulation of cholesterol biosynthesis and metabolism, increased mevalonic acid formation, increased cell proliferation, and decreases in apoptosis and specific gene expression pathway disturbances) associated with propiconazole toxicity. The combination of omics and apical events data strengthened and clarified the proposed mitogenic mode of action. Gene expression analyses also identified a number of key events that were not part of the main toxicity pathway, but nonetheless modified or modulated apical outcomes. The case study did not include a dose-response or human risk assessment for propiconazole exposure.

The second case study involved the use of conventional and *in vitro* and *in silico* toxicological data to elucidate the adverse outcome pathway and quantify dose response for the adverse effects of triclosan on thyroid function. Toxicological data on apical events (changes in circulating thyroid hormone levels) were analyzed to evaluate relative sensitivity of weanling and adult rats to triclosan. Data on gene expression in liver (some from the ToxCast program) confirmed the key role of pregnane-X-receptor and constitutive androstane receptor activation in the proposed mode of action, followed by upregulation of enzymes responsible for the peripheral catabolism of thyroid hormones, resulting in the apical effect of reduced hormone levels. Benchmark dose (BMD) modeling was used to estimate triclosan doses associated with 20% reduction in T4 levels in weanling rats, which were identified as the most sensitive animal model. While providing strong confirmatory evidence for the proposed mode of action, the case study did not address the complex issue of how (or if) the dose-response modeling in animals could be extrapolated to humans.

SUMMARY

The preceding sections discussed how a wide range of computational toxicology methods have been developed over the past decades with potential applications in risk assessment. Within the generally accepted scope of computational toxicology, the techniques that have found the widest application and acceptance are (quantitative) structure-activity relationships. Although this technique was initially developed based primarily on chemical structure analyses and classical *in vivo* toxicity data, recent research has broadened the scope of QSAR modeling to include a wider range of endpoints, including omics and *in vitro* test results. The "chemical space" covered by QSAR models and their supporting databases has also expanded, so that they now are used routinely to set testing priorities and support chemical category definitions in major regulatory programs, including the U.S. Toxic Substances Control Act and the EU REACH program. QSAR and related methods also have found wide use in pharmaceutical and commercial product development for many years.

Applications of computational toxicology for quantitative risk assessment are slowly being developed but have yet found limited application. As discussed in Computational Toxicology Applications in Risk- and Hazard-Based Screening, biokinetic and PBPK modeling have become the preferred methods for dosimetric analysis in health risk assessment, where data are available to support them. Other advanced computational technologies have found only limited application in high-saliency risk assessments. To date (to the authors' knowledge), no regulatory human risk assessment has been conducted solely based on quantitative analysis of adverse outcome pathways identified through omics and high-throughput screening data. However, the use of pathway analyses based on either *in vivo* or *in vitro* data has been demonstrated in a number of case studies (Computational Toxicology Applications in Risk- and Hazard-Based Screening) and used as supporting evidence in risk assessments for high-saliency chemicals.

As discussed in Translational Research, much translational research is under way to validate and refine high-throughput screening data, to expand the universe of HTS data, and to make these data available for use in the elucidation of adverse outcome pathways. Large data sets generated by systematic high-throughput screening programs (such as the ToxCast program and Tox21 Collaboration) will help to elucidate adverse outcome pathways for known chemicals, and provide additional confidence in predictions of mode of action for untested agents.

The use of HTS screening and omics data, coupled with human exposure biomarkers or biokinetic modeling, has been suggested to be a useful tool in chemical safety assessment.[57] Applications of such approaches to subsets of the ToxCast chemicals[62,63] have been used to identify agents (mostly pesticide active ingredients) for which current estimates of human exposure and dose exceed levels that correspond to internal concentrations that show significant impacts in one or more *in vitro* screening tests. The results of such analyses have not yet been accepted by regulatory agencies (at the time this chapter was written). However, there does not appear to be any statutory bar to the regulatory application of such data, at least in the context of the Toxic Substances Control Act.[64]

It is clear that computational toxicology will play an increasing role in risk assessment; the important questions are: What roles and how soon? The strongest advocates of an expanding role of CompTox in risk assessment suggest that the state of science is very nearly that which is needed to support a major role for pathway-based risk assessment for high-saliency assessment (that is, in establishing regulatory guidance or standards for chemical exposures). For example, Bhattacharya et al.[57] suggest that many adverse outcome pathways for important endpoints and pathways are close to being sufficiently well characterized to support risk assessment. They propose as an example that the role of DNA damage response pathways in mutagenic carcinogenicity is well enough understood that a manageable program of appropriately targeted *in vitro* studies, along with carefully designed computational pathway models, could prove sufficient to support the pathway analysis in quantitative risk assessment. In contrast, other commentators[65,66] cite the historical lack of success of pathway-based (BBDR) models and point out that the statistical uncertainties that are inherent in the interpretation of *in vitro* screening test results are as likely to be as large, or larger, than those associated with the use of conventional (*in vivo*) toxicity data for risk assessment. Thus, even the determination of "safe" levels becomes a statistical exercise, where conventional *in vivo* test data are more likely to be influential to decision makers than even well-documented pathway analysis. Thomas et al.[67] evaluated the power of 600 *in vitro* test results on approximately 300 of the ToxCast Phase 1 chemicals in predicting the 60 *in vivo* endpoints. They found that, with very few exceptions, individual test results and combinations of *in vitro* results failed to predict *in vivo* toxicity beyond levels that could be achieved based on chemical properties alone. These results remain controversial, in that they appear to be inconsistent with the findings of other researchers that *in vitro* test results from the ToxCast database can be used to successfully predict specific *in vivo* endpoints.[68–70] Additional research is needed to clarify the best approaches for selecting and applying *in vitro* test data for the identification of adverse outcome pathways and for the prediction of potential human risks.

There is broad agreement that one area where computational toxicity has potential for improving the quality of risk assessment in the short term is in the assessment of potential population variability, both with regard to exposures[32] and genetic susceptibility.[71] In addition, the large amounts of HTS data being generated will increase generalizability of QSAR models, inform hazard-based screening, and inform *in vivo* and *in vitro* testing strategies. The drive to reduce the use of experimental animals will be given added impetus by the continued growth of CompTox methods.

For high-importance chemicals, it is hard to see how, in the near future, the application of advanced computational toxicity methods will either obviate dependence on *in vivo* toxicological testing or substantially reduce resources required to conduct defensible risk analyses. The 20-year time frame suggested for implementation of the "Tox21" vision continues to look like a reasonable schedule for implementation of fully CompTox-based risk assessments. Something like the phased strategy proposed by Chiu et al.[59] would seem to provide a logical path forward (Figure 12-5). Under "Phase I" of the process (more or less the current state of science), the primary role of high-throughput and molecular data is to augment human biomarker and traditional toxicological data and strengthen the weight of evidence supporting specific mechanistic and dose-response hypotheses. Under Phase II, HTS data, "anchored" to *in vivo* endpoints, could be used to provide dose-response surrogates for some animal toxicity endpoints, and virtual tissue and control network models could provide additional insights regarding the interpretation of human biomarkers of exposure and effect. The final, "reorientation" phase (Phase III) postulates the use of anchored *in vitro* data as an independent third line of evidence providing direct input into dose-response and risk assessment, playing a prominent role, along with findings from human and animal toxicity studies. The outstanding question is at what point regulatory

SUMMARY

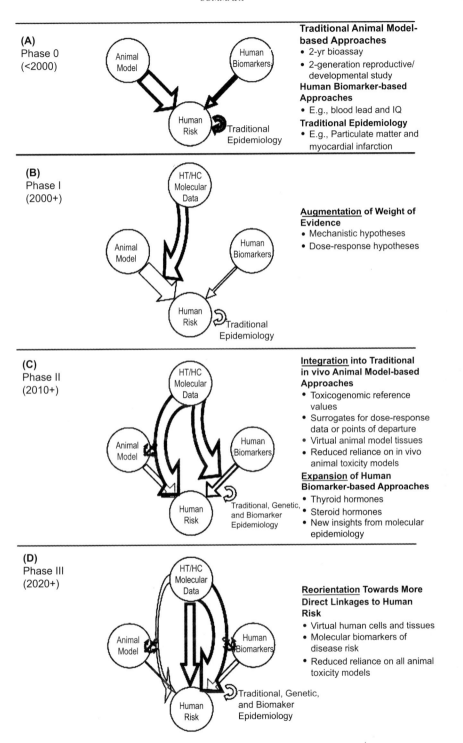

FIGURE 12-5 General schema for expanding the role of CompTox chemicals in quantitative risk assessment.[59]

agencies and stakeholders will seriously entertain the use of CompTox methods as the primary basis for decision making. It is likely that the contribution of CompTox to risk assessment will be very agent-specific; where modes of action are seen to be simple, and the risk and cost implications of specific decisions do not involve large costs or risks to public health, acceptance is likely to be easier than where risk management decisions have major public health or cost implications and where modes of action are less clearly established.

Important technical obstacles remain to the adaptation of CompTox methods for risk assessment. Doubts that *in vitro* methods (or even virtual tissues) can adequately predict organ, organism-, or population-level effects; difficulties in adapting *in vitro* methodologies; to predicting developmental, endocrine, or neurotoxicologic endpoints seem particularly troublesome. To date, relatively little has been done with regard to responses of *in vitro* assay systems to mixtures of toxicants. While considerable progress has been made in modeling endocrine control pathways, (HPT and HPG axes, estrogen and androgen synthesis and metabolism), it is still not clear when these efforts will bear fruit in terms of usable models of human adverse effects.

References

[1] Russell WMS, Burch RL. The principles of humane experimental technique. London: Methuen; 1959. Available at <http://altweb.jhsph.edu/pubs/books/humane_exp/het-toc>.
[2] Hansch C, Leo A. Exploring QSAR 1. Fundamentals and applications in chemistry and biology. Washington, DC: American Chemical Society; 1995.
[3] Guess HA, Crump KS. Low-dose-rate extrapolation of data from animal carcinogenicity experiments—analysis of a new statistical technique. Math Biosci 1976;32:15−36.
[4] Armitage P, Doll R. The age distribution of cancer and a multi-stage theory of carcinogenesis. Br J Cancer 1954;8:1−12.
[5] Occupational Safety and Health Administration, Department of Labor. Final Standard for Occupational Exposure to Lead.43 Federal Register 52952, Nov. 14, 1978.
[6] Crump KS. A new method for determining allowable daily intakes. Fundam Appl Toxicol 1984;4:854−71.
[7] Pereira DA, Williams JA. Review: historical perspectives pharmacology origin and evolution of high throughput screening. Br J Pharm 2007;152:53−61.
[8] Andersen ME, Clewell HJ, Gargas ML, Smith FA, Reitz RH. Physiologically based pharmacokinetics and the risk assessment process for methylene chloride. Toxicol Appl Pharmacol 1987;87:185−205.
[9] Occupational Safety and Health Administration (OSHA). Occupational exposure to methylene chloride. Fed Regist 1997;62(7):1493−619.
[10] Goodsell DS, Olson AJ. Automated docking of substrates to proteins by simulated annealing proteins: structure. Funct Genet 1990;8:195−202.
[11] U.S. Environmental Protection Agency. Methods for derivation of inhalation reference concentrations and application of inhalation dosimetry. EPA/600/8-90/066F. <http://www.epa.gov/raf/publications/methods-derivation-inhalation-ref.htm>; 1994.
[12] U.S. Environmental Protection Agency. 2003 A framework for a computational toxicology research program. EPA/600/R-03/065.
[13] European Community. Directive 2003/15/EC of the European Parliament and of the Council of 27 February 2003 amending Council Directive 76/768/EEC on the approximation of the laws of the Member States relating to cosmetic products. 2003.
[14] U.S. National Research Council. Toxicity testing in the 21st century: a vision and strategy. Committee on Toxicity Testing and Assessment of Environmental Agents. Board on Environmental Studies and Toxicology. National Academy Press. Washington, DC; 2007. <http://www.nap.edu/catalog/11970.html>.

[15] European Community. 2003 Regulation (EC) No 1907/2006 of the European Parliament and of the Council of 18 December 2006 concerning the Registration, Evaluation, Authorisation and Restriction of Chemicals (REACH).
[16] OECD. Guidance document on the validation of (quantitative) structure-activity relationships [(Q)SAR] models. OECD Environment Health and Safety Publications. Series on Testing and Assessment. No. 69; 2007.
[17] U.S. Food and Drug Administration. Guidance for industry. Genotoxic and carcinogenic impurities in drug substances and products: recommended approaches. <http://www.fda.gov/downloads/Drugs/.../Guidances/ucm079235.pdf>; 2008.
[18] OECD. Guidance document for using the OECD (Q)SAR Toolbox to develop chemical categories according to the OECD guidance on the grouping of chemicals. Series on Testing and Assessment No. 102. <http://www.oecd.org/officialdocuments/displaydocumentpdf/?cote=env/jm/mono(2009)5&doclanguage=en>; 2009.
[19] SEURAT (Safety Evaluation Ultimately Replacing Animal Testing). Towards the replacement of *in vivo* repeated dose systemic toxicity testing. Safety Evaluations Ultimately Replacing Animal Testing (SEURAT) Program Annual Report. vol. 1. <http://www.seurat-1.eu/pages/library/seurat-1-annual-report.php>; 2011.
[20] NRC (National Research Council). Science and decisions. Advancing risk decisions. Committee on Improving Risk Analysis Approaches Used by the U.S. EPA. Board on Environmental Studies and Toxicology. National Academy Press, Washington, DC; 2009. <www.nap.edu>.
[21] Toxic Substance Control Act. 1976 15 USC (C. 53) 2601—92.
[22] Kavlock R, Dix D. Computational toxicology as implemented by the U.S. EPA: providing high throughput decision support tools for screening and assessing chemical exposure, hazard and risk. J Toxicol Env Health 2010;13:197—217.
[23] Laws SC, Yavanhxay S, Copper RL, Eldridge JC. Nature of the binding interaction for 50 structurally diverse chemicals with rat estrogen receptors. Tox Sci 2006;9(1):46—56. doi:10.1093/toxsci/kfl092.
[24] Hartung T. Lessons learned from alternative methods and their validation for a new toxicology in the 21st century. J Toxicol Environ Health B 2010;13:277—90. <http://search.ebscohost.com/login.aspx?direct=true&db=mnh&AN=20574902&site=ehost-live>.
[25] Kulp-Eddy J, Snyder M, Stokes W. A review of trends in animal use in the United States (1972—2006). ALTEX 2007;14:163—5.
[26] U.S. Environmental Protection Agency. The US Environmental Protection Agency's strategic plan for evaluating the toxicity of chemicals. EPA/100/K-09/001. <http://www.epa.gov/spc/toxicitytesting/docs/toxtest_strategy_032309.pdf>; 2009.
[27] Judson RS, Martin MT, Egeghy P, Gangwal S, Reif DM, Kothiva P, et al. Aggregating data for computational toxicology applications: the U.S. Environmental Protection Agency (EPA) Aggregated Computational Toxicology Resource (ACToR) system. Int J Mol Sci 2012;13(2):1805—31.
[28] U.S. Environmental Protection Agency. New chemical program: Assessing risk. <http://www.epa.gov/oppt/newchems/pubs/assess.htm>; 2012.
[29] Judson RS, Houck KA, Kavlock RJ, Knudsen TB, Martin MT, Mortensen HM, et al. *In vitro* screening of environmental chemicals for targeted testing prioritization: the ToxCast project. Env Health Pers 2010;118:485—92.
[30] Dix DJ, Houck KA, Martin MT, Richard AM, Setzer RW, Kavlock RJ. The ToxCast program for prioritizing toxicity testing of environmental chemicals. Toxicol Sci 2007;95:5—12.
[31] U.S. Environmental Protection Agency. ToxCast Phase I and II chemicals. Office of Research and Development/National Center for Computational Toxicology. <http://www.epa.gov/ncct/toxcast/files/ToxCast%20Chemical%20Summary%2014Dec2010.pdf>; 2010.
[32] Cohen Hubal EA, Richard A, Aylward L, Edwards S, Gallagher J, Goldsmith MR, et al. Advancing exposure characterization for chemical evaluation and risk assessment. J Toxicol Env Health 2010;13:299—313.
[33] U.S. Environmental Protection Agency. Chemical safety for sustainability: Strategic Research Action Plan 2012—2016. <http://www.epa.gov/research/docs/css-strap.pdf>; 2012.
[34] Dix DJ. EPA's Comp Tox program—current and future research. Presentation to the EPA Science Advisory Board. <http://yosemite.epa.gov/sab/sabproduct.nsf/8F23F800B73DE667852579F20055CCEE/$File/NCCT+SAB+may2012.pdf>; 2012.

[35] Shukla SJ, Huang R, Austin CP, Xia M. The future of toxicity testing: a focus on *in vitro* methods using a quantitative high-throughput screening platform. Drug Discov Today 2010;15:997−1007.

[36] Tice R. Overview of the Tox21 Collaboration. Presentation at the NTP Workshop: Role of Environmental Chemicals in the Development of Diabetes and Obesity. <http://ntp.niehs.nih.gov/ntp/ohat/diabetesobesity/presentations/Tox21OverviewTiceRevised_508.pdf>; 2011.

[37] OECD. History of the QSAR project. <http://www.oecd.org/document/11/0,3746,en_2649_34379_42920331_1_1_1_1,00.html>; 2012.

[38] OECD. OECD Principles for the Validation, for Regulatory Purposes, of (Quantitative) Structure-Activity Relationship Models. <[http://www.oecd.org/document/23/0,2340,en_2649_34379_33957015_1_1_1_1,00.html]>; 2004.

[39] OECD. Proposal for a template, and guidance for developing and assessing the completeness of adverse outcome pathways. <http://www.oecd.org/dataoecd/50/39/49963554.pdf>; 2012.

[40] OECD. 2011 Report of the workshop on using mechanistic information in forming chemical categories. OECD Environment, Health and Safety Publications Series on Testing and Assessment No. 138. ENV/JM/MONO(2011)8.

[41] Henry T.U.S. EPA-OPPT QSAR and Expert Systems Tools. Presentation to the PPDC meeting,October 8. <http://www.epa.gov/pesticides/ppdc/testing/oct08/presentation-10-08.pdf>; 2008.

[42] Dellarco V, Henry T, Sayre P, Seed J, Bradbury S. Meeting the needs of a more effective and efficient testing and assessment paradigm for chemical risk management. J Toxicol Environ Health B 2010;13:347−60.

[43] Patel N, Boethling B. Overview of EPI Suite™: Software for chemical property and fate estimation. U.S. EPA Office of Pollution Prevention and Toxics presentation to the Science Advisory Board. <http://www.epa.gov/sab/pdf/oppt_slide_presentation_feb-22-2006.pdf>; 2006.

[44] Mayo-Bean K, Nabholz JV, Meylan WM, Howard PH, Moran-Bruce K. Estimating toxicity of industrial chemicals to aquatic organisms using the ECOSAR (Ecological Structure Activity Relationship Class Program). <http://www.epa.gov/oppt/newchems/tools/ecosarusersguide.pdf>; 2011.

[45] U.S. Environmental Protection Agency. OncoLogic user's manual; An expert system for prediction of the carcinogenic potential of chemicals. <http://www.epa.gov/oppt/sf/pubs/onco-user-man.pdf>; 2009.

[46] U.S. Environmental Protection Agency. Exposure Assessment Fate and Screening Tool (E-FAST) version 2.0 documentation manual. <http://www.epa.gov/opptintr/exposure/pubs/efast2man.pdf>; 2007.

[47] U.S. Environmental Protection Agency. TSCA New Chemical Program (NCP) chemical categories. Office of Pollution Prevention and Toxics. <http://www.epa.gov/oppt/newchems/pubs/npcchemicalcategories.pdf>; 2010.

[48] OECD 1995 Guidance for Aquatic Effects Assessment. OECD Environment Monographs No. 92. <http://www.oecd.org/dataoecd/5/42/34290206.pdf>; 1995.

[49] OECD. 2007 Report on regulatory uses and applications in OECD countries of (quantitative) structure-activity relationships [(Q)SAR] models in the assessment of new and existing chemicals. Series on Testing and Assessment No. 58.<http://www.oecd.org/env/ehs/risk-assessment/38131728.pdf>.

[50] OECD. Guidance on the grouping of chemicals. Series on Testing and Assessment No. 80. <http://www.oecd.org/officialdocuments/displaydocumentpdf/?cote=env/jm/mono(2007) 28&doclanguage=en>; 2007.

[51] OECD. QSAR user manual tips and tricks. <http://www.oecd.org/dataoecd/6/43/47089540.pdf>; 2011.

[52] Gilbert N. Data gaps threaten chemical safety law. Nature 2011;475:150−1.

[53] Andersen ME, Krewski D. Toxicity testing in the 21st century: bringing the vision to life. Tox Sci 2009;107(2):324−30.

[54] U.S. Environmental Protection Agency. Guidelines for carcinogen risk assessment. Risk Assessment Forum. EPA/630/P-03/001F. <http://www.epa.gov/raf/publications/pdfs/CANCER_GUIDELINES_FINAL_3-25-05.pdf>; 2005.

[55] IPCS. Chemical-specific adjustment factors for interspecies differences and human variability: guidance document for use of data in dose/concentration-response assessment. World Health Organization. <http://whqlibdoc.who.int/publications/2005/9241546786_eng.pdf>; 2005.

[56] U.S. Environmental Protection Agency. Approaches for the application of physiologically based pharmacokinetic (PBPK) models and supporting data in risk assessment. <http://cfpub.epa.gov/ncea/cfm/recordisplay.cfm?deid = 157668#Download>; 2006.

[57] Bhattacharya S, Zhang Q, Carmichael PL, Boekelheide K, Andersen ME. Toxicity testing in the 21 century: defining new risk assessment approaches based on perturbation of intracellular toxicity pathways. PLoS ONE 2011;6(**6**):e20887. doi:10.1371/journal.pone.0020887.

[58] Zhang Q, Bhattacharya S, Andersen ME, Conolly RB. Computational systems biology and dose-response modeling in relation to new directions in toxicity testing. J Toxicol Environ Health B 2010;13:253−76.

[59] Chiu WA, Euling SY, Siegel Scott C, Subramanian RP. Approaches to advancing quantitative human health risk assessment of environmental chemicals in the post-genomic era. Toxicol Appl Pharmacol 2010;10.1016/j.taap.2010.03.019.

[60] Andersen ME, Clewell HJ, Carmichael PL, Boekelhyde K. Can a case-study approach speed implementation of the NRC report: "Toxicity testing in the 21st century: a vision and strategy?" ALTEX 2011;28:175−82.

[61] U.S. Environmental Protection Agency. Integrated approaches to testing and assessment strategy: Use of new computational and molecular tools. Office of Pesticide Programs. <http://yosemite.epa.gov/sab/sabproduct.nsf/373C1DB0E0591296852579F2005BECB3/$File/OPP+SAP+document-May2011.pdf>; 2011.

[62] Rotroff DM, Wetmore BA, Dix DJ, Ferguson SS, Clewell HJ, Houck KA, et al. Incorporating human dosimetry and exposure into high-throughput *in vitro* toxicity screening. Tox Sci 2010;117(**2**):348−58.

[63] Wetmore BB, Wambaugh JF, Ferguson SS, Sochaski MA, Rotroff DM, Freeman K, et al. Integration of dosimetry, exposure, and high-throughput screening data in chemical toxicity assessment. Tox Sci 2012;125(**1**):157−74.

[64] Locke PA, Bruce Myers Jr. D. Implementing the national academy's vision and strategy for toxicity testing: opportunities and challenges under the U.S. Toxic Substances Control Act. J Toxicol Environ Health B 2010;13:376−84.

[65] Crump KS, Chen C, Lewis TA. The future use of *in vitro* data in risk assessment to set human exposure standards: challenging problems and familiar solutions. Env Health Pers 2010;118(10):1351−4.

[66] Crump KS, Chen C, Chiu W, Louis TA, Portier CJ, Subramanian RP, et al. What role for biologically based dose-response models in estimating low-dose risk? Env Health Pers 2010;118(5):585−8.

[67] Thomas RS, Black MB, Li L, Healy E, Chu T-M, Bao W, et al. A comprehensive statistical analysis of predicting in vivo hazard using high-throughput in vitro screening. Tox Sci 2012;128(**2**):398−417.

[68] Dix DJ, Houck K, Judson RS, Knudson T, Martin M, Reif D, et al. Incorporating biological, chemical and toxicological knowledge into predictive models of toxicity. Toxicol Sci 2012. [Advanced Access September 14, 2012. Downloaded from <http://toxsci.oxfordjournals.org/> at Society of Toxicology on October 16, 2012.]

[69] Thomas RS, Black MB, Li L, Healy E, Chu T-M, Bao W, et al. Response to "Incorporating biological, chemical and toxicological knowledge into predictive models of toxicity." Toxicol. Sci 2012. [Advanced Access September 14, 2012. Downloaded from <http://toxsci.oxfordjournals.org/> at Society of Toxicology on October 16, 2012.]

[70] Rotroff DM, Dix DJ, Houck KA, Knudson TB, Martin MT, McLaurin KW, et al. Using *in vitro* high throughput screening assays to identify potential endocrine-disrupting chemicals. Env Health Persp 2012. [Online publication September 28, 2012. <http://dx.doi.org/10.1289/ehp.1205065>.]

[71] O'Shea SH, Schwartz J, Kosyk O, Ross P, Jin Ha M, Wright FA, et al. *In vitro* screening for population variability in chemical toxicity. Tox Sci 2011;119(2):398−407.

CHAPTER 13

Future Directions for Computational Toxicology for Risk Assessment

Bruce A. Fowler, Editor
ICF International, Fairfax, VA, USA

This volume has presented specific examples of a variety of computational toxicology methods that were used to inform risk assessment practice for the protection of the public's health. The computational tools in use and under development are indeed impressive today and even more so for their promise of future achievements. Employing these methods can sharpen the clarity and precision of answers to risk issues. In order for this rapidly growing area of scientific endeavor to realize its full potential as a valuable set of tools for risk assessment, there are a number of concrete steps which must be implemented to reach these goals. This will require hard work, patience, and commitment of essential resources. The following is a list of needed elements that must be included in helping computational toxicology to continue to move forward as it evolves. As the fields of computational toxicology and risk assessment continue to develop, the specific needs articulated here will also change. These suggestions should be considered only those of the editor, based upon the material included in this book and his personal experience with computational toxicology and in dealing with practical chemical risk assessment issues on a national and international basis. It is hoped that this information will be of use to others in guiding the future evolution of computational toxicology and risk assessment in the 21st century.

NEEDED ESSENTIAL ELEMENTS

As with any other field of science, computational toxicology requires robust primary data to model. This is particularly important when the output data are used for risk assessment purposes.

The key elements here are an established QA/QC program for the data modeled and documentation of analytical fidelity if translation of chemical exposure data is to be interpreted for risk assessment. QA/QC, cellular viability for *in vitro*–*in vivo* extrapolations, standard operating procedures (SOPs) for data mining projects, and finally a review step to be sure the final output of the modeling meets the Bradford Hill criteria for causality are other essential elements. These kinds of protections are usually not difficult or expensive to execute but will help the results of the computational toxicology gain acceptance and confidence in the larger scientific community and with the risk assessment community in particular.

SPECIFIC ELEMENTS IN COMPUTATIONAL TOXICOLOGY NEEDED FOR THE FIELD TO MOVE FORWARD

In my view, the following elements represent an essential but not inclusive list of key elements needed in order for the field of computational toxicology to move ahead:

1. Standardize terms and make a concerted effort to support development of a common vocabulary among computational modelers, risk assessors, and societal decision makers. The Society of Toxicology is well positioned to undertake such an initiative as a component of its continuing educational activities.
2. Further integrate diverse data sets from other disciplines, such as molecular biology/toxicology, which are also rapidly evolving, to yield a more 360-degree perspective on the findings for mode of action risk assessment. A common vocabulary will promote integration of computational modeling into risk assessment practice.
3. Develop interdisciplinary graduate education programs to train the next generation of scientists who will be applying both computational toxicology and molecular toxicology tools to mode of action risk assessments. It must be remembered that the evolving intertwined fields of computational toxicology and molecular toxicology are competing with other modern disciplines for the very best students who need to be challenged by stimulating problems and state of the art curricula.
4. In particular, such graduate education programs should also include a section on translation of computational results into "Plain English." The information is from analyses of ever-more complex data sets and is needed by those who are engaged in rendering regulatory or societal decisions on chemicals and may not have extensive technical backgrounds. This need for clear communication of analytical results is essential for the field to move forward and to secure the support of societal decision makers. Only with clearer communication can society realize the benefits.
5. There is a need for an ongoing continuing education (CE) program to keep practitioners in the fields of computational toxicology and risk assessment up to date on linkages between these rapidly evolving fields. This could be accomplished by providing CE course opportunities focused on application of computational toxicology methods to risk assessment practice. The Society of Toxicology is again well positioned to take a leadership role in this area. Application of artificial intelligence (AI) and

cloud computing approaches to risk assessment practice problems, for example, might be topics of broad interest.

6. Currently, computational modeling and, in the near future, AI can be used to delineate previously less obvious and thus unappreciated relationships between complex biological pathways and networks. When these tools are used, a more precise mode of action for new classes of chemicals and chemical mixtures can be discerned. Further, the increased risk to sensitive subpopulations can be more confidently pinpointed. Such analyses can be based on individual genetic inheritance as a function of age, gender, race, and ethnicity. An integrative systems biology approach should greatly improve the precision of risk assessments and hence chemical regulation by addressing the central issues of relating dose or exposure of a chemical or mixture of chemicals to whom.
7. Utilize data mining of the numerous existing public databases for information to generate hypotheses, guide laboratory studies, and link disparate but relevant data sets. Such an approach would clearly produce great economies of time and fiscal resources and provide valuable information for risk assessment purposes. One specific example would be to overlay the high-quality analytical data from the USGS database on chemicals in groundwater with the high-quality analytical and clinical biomonitoring data from the CDC-NHANES database and identify correlates. This is possible since both data sets are organized on a county-by-county basis but could be done on a regional basis if necessary for administrative reasons.
8. Include causality criteria, such as those of *Bradford Hill,* in *all* studies to the maximum degree possible. This time-tested approach is a sound general checklist for assuring that the results of computer modeling efforts meet reasonable standards of good science and that they are based on the integration of the best available data and rigorous analyses.
9. Apply computational modeling tools to testing previously articulated hypotheses and *hypothesis-driven risk assessment research projects,* which advance the field of science and address clear issues. This approach should lead to generating refined data that are of direct value to societal decision makers, since such information would be targeted to addressing specific risk assessment problems. Computational tools are too precious a resource to be wasted in attempting to dress up more mundane issues that could be handled by more conventional methods or existing regulatory instruments.
10. Utilize computational modeling tools to promote "green chemistry" in the development of safer chemicals *prior* to industrial production or widespread usage. The "green chemistry" approach based on up-front computer modeling is already being used but needs to be expanded into a common and accepted practice of industry.

In summary, the preceding suggestions should hopefully stimulate spirited discussions of how the field of computational toxicology can best meet its full potential in providing valuable, timely, and cost-effective information to inform modern risk assessment practice for the 21st century. It is hoped that these ideas will be incorporated into both developing academic curricula for training the next generation of scientists in this promising field and increasingly be embraced in the near term by the risk assessment community to address significant and pressing problems.

Index

Note: Page numbers followed by "*f*" and "*t*" refers to figures and tables, respectively.

A

Acetylcholinesterase (AChE), 141
AChE. *see* Acetylcholinesterase (AChE)
ACToR. *see* Aggregated Computational Toxicological Resource (ACToR) data system
Acute lymphocytic leukemia (ALL), 204–205
Acute myeloid leukemia (AML), 204–205
Adenosine diphosphate (ADP), 69
Adenosine monophosphate (AMP), 69
Adenosine triphosphate (ATP), 69
ADP. *see* Adenosine diphosphate (ADP)
Adverse Event Reporting System (AERS) data, 175
Adverse outcome pathway (AOP), 3, 116–117, 116*f*
  characterization of, 218–219
AERS data. *see* Adverse Event Reporting System (AERS) data
Affymetrix GeneChip, 204–205
Affymetrix microarrays, 182–183
Agency of Toxic Substances and Disease Registry (ATSDR) programs
  SAR/QSAR modeling in, 6
Aggregated Computational Toxicological Resource (ACToR) data system, 222
  structure of, 223*f*
ALAD gene. *see* Aminolevulinic acid dehydratase (ALAD) gene
ALL. *see* Acute lymphocytic leukemia (ALL)
Ames mutagenicity assay, 172–173
Aminolevulinic acid dehydratase (ALAD) gene, 158–159
AML. *see* Acute myeloid leukemia (AML)
AMP. *see* Adenosine monophosphate (AMP)
Animal PBPK models, for sensitive subpopulations, 58–61, 59*f*, 60*f*
ANNs. *see* Artificial neural networks (ANNs)
Anotation
  gene/protein, 120–123
  of large gene sets, 120–123
AOP. *see* Adverse outcome pathway (AOP)
ArrayTrack™, 196–197, 203, 205
Artificial neural networks (ANNs), 204–205
Assay Depot, 178
atBioNet, 203
ATP. *see* Adenosine triphosphate (ATP)
ATSDR programs. *see* Agency of Toxic Substances and Disease Registry (ATSDR) programs
ATSDR toolkit, 8, 11

B

Bayesian approach
  for DILI prediction, 175
  in PBPK modeling for sensitive subpopulations, 56–58, 56*f*, 57*f*
BBDR models. *see* Biologically based dose response (BBDR) models
BEI. *see* Biological Exposure Index (BEI)
Berkeley Madonna, 10
BEs. *see* Biomonitoring equivalents (BEs)
Bile salts
  metabolism and transport, 73
Bing, 180
Biocompare, 178
Bioconductor project, 184
BioEpisteme, 175
Bioinformatics tools and databases
  omics biomarkers, 202–203
Biokinetics, 218
Biological Exposure Index (BEI), 145
Biologically based dose response (BBDR) models, 5
  of HPT axis, 14–16
Biomarkers
  identification, 204–205
  omics. *see* Omics biomarkers
  overview, 196, 198
Biomonitoring equivalents (BEs)
  defined, 138
Bisphenol A (BPA)
  physiological model for, 12–14
BPA. *see* Bisphenol A (BPA)

C

CA. *see* Cholic acid (CA)
Carbofuran (2,3-dihydro-2,2-dimethyl-7-benzofuranyl-N-methylcarbamate), 141
  biological exposure guidance value estimation, 143–145
  biological monitoring in pesticide production, 141–145, 142*t*

Carbofuran (2,3-dihydro-2,2-dimethyl-7-benzofuranyl-N-methylcarbamate) (*Continued*)
  IndusChemFate PBTK model application, 143–145
    metabolism pathways, 143–145, 143*f*, 144*t*
    physical-chemical and biological properties for, 144*t*
  metabolism pathways in PBTK model simulation, 143–145, 143*f*, 144*t*
  provisional biomonitoring equivalent, 145
  TLV-TWA of, 142
Carcinogenicity
  in silico prediction of, 173–174
CASE, 6
CASRNs. *see* Chemical Abstract Services Registry Numbers (CASRNs)
Causal reasoning (CR) analysis, 186
CDC. *see* Centers for Disease Control and Prevention (CDC)
CDC-NHANES database, 249
Centers for Disease Control and Prevention (CDC), 152
Cerep, 178
Charles River Laboratories, 178
Chemical Abstract Services Registry Numbers (CASRNs), 159–161
Chemical Safety for Sustainability (CSS), EPA's, 226, 227*f*
ChemSpider.com, 159
Chem-Steer, 230
Cholestasis, 76–78
Cholic acid (CA), 73
Clustering, 186–187
Co-clustering analysis, 187
CompTox methods, in risk assessment
  in animal testing, 221
  costs, 220–221
  current status in quantitative risk assessment, 236–238, 237*t*
  drivers for application of, 220–221
  origins and nature, 215–220
    timeline of, 216*t*
  in risk- and hazard-based screening, 228–235
    environmental applications in OECD and European Union, 230–235
    industrial chemicals and pesticides (U.S. EPA), 228–230
    SAR and QSAR methods, 231–235
  translational research, 221–228
    OECD and European Union activities, 227–228
    Tox21 Collaboration, 226–227
    U.S. EPA, 222–226
Computational toxicology *see also* CompTox methods
  defined, 218
  future directions, 247
  elements, 247–249
  QA/QC program, 247–248
Computer modeling, of adverse side effects
  bioinformatics aspect of, 3
  general approach, 1, 2*f*
  need for, 2–3
Cost(s)
  CompTox methods in risk assessment, 220–221
  estimates in pharmaceutical industry, 171
CR analysis. *see* Causal reasoning (CR) analysis
Critical Path Initiative (FDA), 171
Cross-species analysis, using networks, 128–130
CSS Dashboards, 226, 227*f*

D

Data management
  information technologies and, 219
DAVID tool, 185
Decision forest (DF), for omics biomarkers, 206–209, 207*f*
  benefits, 207
  cross-validations of, 208
  microarray gene expression, 209, 210*t*
  proteomics data, 208–209, 208*f*
  SNP genotype data, 207–208
Decision trees (DT), 186–187, 204–205
Deepwater Horizon Gulf oil spill, 1–2
Derek, 175
Derek Nexus, 172–173
Derived no effect levels (DNELs), 138
DF. *see* Decision forest (DF)
DF-SNPs modeling, 207–208
2,3-dihydro-2,2-dimethyl-7-benzofuranyl-N-methylcarbamate. *see* Carbofuran
DILI. *see* Drug-induced liver injury (DILI)
Dimethyl arsenic (DMA), 10–11
Discriminant partial least squares, 204–205
"Distinct subpopulation" approach, 55
Distributed Substructure Searchable Toxicity (DSSTox), 223, 224*t*
DMA. *see* Dimethyl arsenic (DMA)
DNELs. *see* Derived no effect levels (DNELs)
Drug-induced liver injury (DILI), 67–68
  integrated systems approach, 67
    bile salt metabolism and transport, 73
    energy homeostasis, 69–70
    fatty acid metabolism, 71–72
    general principles, 68
    glutathione homeostasis, 70–71
    mitochondrial impairment, 75–76, 75*f*
    model building, 68–69
    model validation and predictions, 74–79, 74*t*, 75*f*, 76*f*, 79*f*
    overview, 67–68

solving equation-set, 74
prediction of, 174–176
DSSTox. *see* Distributed Substructure Searchable Toxicity (DSSTox)
DT. *see* Decision trees (DT)

E

ECOSAR program, 229
E-FAST. *see* Exposure Fate Assessment Screening Tool (E-FAST)
Elderly
  PBPK modeling of physiological changes, 47–48
  see also Sensitive subpopulations
Energy homeostasis, 69–70
EPA-OPPT, 230
EPI Suite, 229
Esophageal squamous cell carcinoma
  DF-SNPs in, 207–208
European Union
  activities, OECD and, 227–228
  environmental CompTox applications in, 230–235
  SAR and QSAR methods, 231–235
  REACH legislation (*see* Registration, Evaluation, and Authorization of Chemicals (REACH))
ExpoCast program, 226
Exposure Fate Assessment Screening Tool (E-FAST), 230

F

Fatty acid metabolism, 71–72
Finite sample model (FSM), 46
Fisher's linear discriminant (FLD), 204–205
FLD. *see* Fisher's linear discriminant (FLD)
Forbes (magazine), 171
Fourth National Report on Human Exposure to Environmental Chemicals, 10, 10t
"Framework for a Computational Toxicology Research Program," 222
FSM. *see* Finite sample model (FSM)

G

Gamma glutamyl cysteine (γ-GC), 70–71
Gene function
  and adverse biological phenotype, relationship between
    analysis approaches, 184–188
    causal reasoning, 186
    classical statistical approaches, 186–187
    gene sets, pathways, and gene signatures, 185–186, 185t
    genomic-analysis methods, 182–184
    literature mining, 178–182
    microarrays, 182–183
    next-generation sequencing, 183–184, 183t
    pharmaceutical industry, risk assessment in, 177–188, 179t
    proteomics/metabolomics, 184
    systems biology/virtual liver or heart, 187–188
Gene network enrichment analysis (GNEA), 128–129
Gene Ontologies (GOs), 185–186, 203, 205
Gene Ontology for Functional Analysis (GOFFA), 205
Gene set, 185–186
Gene set enrichment analysis (GSEA), 128–129
Gene signatures, 185–186
GeneGo, 203, 205
Gene/protein annotation, 120–123
  mammalian ortholog identification and, 120–122
Genetic algorithm (GA)-based SVM, 204–205
Genomic-analysis methods, 182–184
  microarrays, 182–183
  next-generation sequencing, 183–184, 183t
  proteomics/metabolomics, 184
Genotoxic impurities (GTIs), 173
Genotoxicity
  computational prediction of, 172–173
Glutathione (GSH), 70–71
  homeostasis, 70–71
Google, 180
Google Scholar, 180
GOs. *see* Gene Ontologies (GOs)
GSH. *see* Glutathione (GSH)
GTIs. *see* Genotoxic impurities (GTIs)

H

Hazard and risk assessment
  industrial chemicals and pesticides (EPA), 228–230
  pathway-based, 115–118, 116f
HCA. *see* Hierarchical clustering analysis (HCA)
Health risk assessment
  sensitive subpopulations modeling and interindividual variability in pharmacokinetics for, 45
  see also Sensitive subpopulations
  UFH use in, 45–47
Hepatotoxicity
  prediction of, 174–176
hERG. *see* Human Ether-a-go-go Related Gene (hERG)
Herper, Matthew, 171
HGP. *see* Human Genome Project (HGP)
Hierarchical clustering analysis (HCA), 204
High Production Volume (HPV) Challenge program, 114
High-throughput screening (HTS) data, 156–159
  future efforts in use of, 130–131
  and "omics" data, incorporation of, 219
  ToxCast™ program, 156–159

High-throughput screening (HTS) data (*Continued*)
   in vitro effects, data semantics and limitations, 165–166
HPT axis. *see* Hypothalamic-pituitary-thyroid (HPT) axis
HPV program. *see* High Production Volume (HPV) Challenge program
HTS data. *see* High-throughput screening (HTS) data
Human biological monitoring (HBM) techniques, 137
   IndusChemFate PBTK-model, 139–140
      biological monitoring of carbofuran in pesticide production (example), 141–145
      differentiation of biomonitoring equivalents of methyl ethyl ketone for men and women (example), 145–147
      discussion, 148–149
      graph of sigmoide relation of tubular resorption, 139, 140*f*
      standardized subjects, 141*t*
      structure, 139, 140*f*
      supplementary information, 149
      version 2.00 of, 139–140
   overview, 137–139
   in silico modeling, 138
Human Ether-a-go-go Related Gene (hERG), 178
Human Genome Project (HGP), 163–164
Human health risk assessment
   described, 5–6
Human interindividual uncertainty factor (UF_H), 45
   in health risk assessment, 45–47
Human PBPK toolkit, 8, 11
Hypothalamic-pituitary-thyroid (HPT) axis, 14
   BBDR models of, 14–16

I

In vivo toxicity
   and off-target promiscuity, relationship between, 177
   and physico-chemical properties of compounds, relationship of, 176–177
InChI. *see* International Chemical Identifier (InChI)
Individual-based PBPK modeling, for sensitive subpopulations, 48–50, 48*f*, 49*f*, 51*f*, 51*t*
IndusChemFate, 139–140
   biological monitoring of carbofuran in pesticide production (example), 141–145
   differentiation of biomonitoring equivalents of methyl ethyl ketone for men and women (example), 145–147
   discussion, 148–149
   graph of sigmoide relation of tubular resorption, 139, 140*f*
   standardized subjects, 141*t*
   structure, 139, 140*f*
   supplementary information, 149
   version 2.00 of, 139–140
Information technologies
   data management and, 219
Ingenuity Pathway Analysis, 203, 205
International Chemical Identifier (InChI), 159–161

K

KEGG, 203, 205
K-nearest neighbor (KNN), 204–205
KNN. *see* K-nearest neighbor (KNN)

L

LC-MS/MS method, 13
Leadscope Inc., 172–173
Leadscope Predictive Data Miner (LPDM), 172–173, 175
Lhasa Limited, 172–173
Liver toxicant, 201–202
LOAEL. *see* Lowest observed adverse effect level (LOAEL)
Logistic regression (LR), 204–205
Lowest observed adverse effect level (LOAEL), 126
   defined, 176–177
LR. *see* Logistic regression (LR)

M

Mammalian animal testing
   nonanimal alternatives of, 115
Mammalian ortholog identification
   gene/protein annotation and, 120–123
Manganese superoxide dismutase (MnSOD), 78
MAPE. *see* Median absolute performance error (MAPE)
Mapping, 128
MAQC project. *see* MicroArray Quality Control (MAQC) project
Margin of exposure (MOE) approach, 164
Markov Chain Monte Carlo (MCMC) simulation method
   for Bayesian analysis of PBPK models, 56–58, 56*f*, 57*f*
Maximum recommended daily dose (MRDD), 175
Maximum tolerated dose (MTD) values, 172
MCMC simulation method. *see* Markov Chain Monte Carlo (MCMC) simulation method
MC4PC, 172–173, 175
Median absolute performance error (MAPE), 10–11
Median performance error (MPE), 10–11
Medical Subject Headings (MESH), 180–182
MEK. *see* Methyl ethyl ketone (MEK)
Metals PBPK models, 10–16, 10*t*
   BBDR HPT axis models, 14–16
   BPA, 12–14, 13*f*

Methyl ethyl ketone (MEK)
  biomonitoring equivalents differentiation of, for men and women, 145–147, 146t, 147f
Methylmercury model, 11
Michaelis–Menten saturable metabolism, 139
Microarray gene expression data
  DF application to, 209, 210t
MicroArray Quality Control (MAQC) project, 196–197
Microarray technology, 182–183, 201–202
  background, 182–183
Microsoft Academic Search, 180
Minimal risk levels (MRLs), 138
Mitochondrial impairment, in liver, 75–76, 75f
MMA. see Monomethyl arsenic (MMA)
MnSOD. see Manganese superoxide dismutase (MnSOD)
MOE approach. see Margin of exposure (MOE) approach
Monomethyl arsenic (MMA), 10–11
Monte Carlo simulation method
  and PBPK modeling for sensitive subpopulations, 53–56, 54f, 55t
MPE. see Median performance error (MPE)
MRDD. see Maximum recommended daily dose (MRDD)
MRLs. see Minimal risk levels (MRLs)
MTD values. see Maximum tolerated dose (MTD) values
MultiCASE, 6
MultiCASE Inc., 172–173

N

National Center for Computational Toxicology (NCCT), 152, 222
National Center for Environmental Research (NCER), 222
National Center for Health Statistics (NCHS), 152
National Center for Toxicological Research (NCTR), 12, 196–197
National Exposure Research Laboratory (NERL), 222
National Health and Environmental Effects Research Laboratory (NHEERL), 222
National Health and Nutrition Examination Survey (NHANES), 8–9, 152, 154–155, 206
  NHANES IV vs. ToxCast™ data sets, 159, 160t
National Morbidity, Mortality, and Air Pollution Study (NMMAPS), 155–156
National Research Council (NRC), 220
National Toxicology Program (NTP), 173–174, 225
Natural-language processing (NLP), 181–182
NCCT. see National Center for Computational Toxicology (NCCT)
NCER. see National Center for Environmental Research (NCER)
NCHS. see National Center for Health Statistics (NCHS)
NCTR. see National Center for Toxicological Research (NCTR)
Neonates
  PBPK modeling of physiological changes, 47
  see also Sensitive subpopulations
NERL. see National Exposure Research Laboratory (NERL)
Network analysis, of gene expression, 128
Network inference methods, 128
Next-generation sequencing (NGS) technologies, 183–184, 183t
NGS. see Next-generation sequencing (NGS) technologies
NHANES. see National Health and Nutrition Examination Survey (NHANES)
NHANES IV vs. ToxCast™ data sets, 159, 160t
NHEERL. see National Health and Environmental Effects Research Laboratory (NHEERL)
NLP. see Natural-language processing (NLP)
NMMAPS. see National Morbidity, Mortality, and Air Pollution Study (NMMAPS)
No observed adverse effect level (NOAEL), 46, 126
NOAEL. see No observed adverse effect level (NOAEL)
Nonmammalian species data to mammalian species data, computational translation of
  alternatives of, 115
  annotation of large gene sets, 122–123
  AOP concept, 118–120
  cross-species analysis using networks, 128–130
  effects on species, 118–120, 119f
  future perspectives, 130–131
  gene/protein annotation and mammalian ortholog identification, 120–122
  network inference and mapping, 128
  pathway-based dose-response relationships, 125–128
  pathway-based extrapolation to mammals, 124–125
  pathway-based hazard and risk assessment, 115–118, 116f
  pathway-level comparison/translation, 123–124
  at systems level, 130
  tools for extrapolation of, 121t
Nonsteroidal anti-inflammatory drugs (NSAIDs), 75–76
NRC. see National Research Council (NRC)
NSAIDs. see Nonsteroidal anti-inflammatory drugs (NSAIDs)

O

Occupational exposure limits (OELs), 138
OCR. see Optical character recognition (OCR)

Ocular toxicity
    prediction of, 175
OECD. *see* Organisation for Economic Cooperation and Development (OECD)
OECD QSAR Toolbox, 232–235, 233f
    category definition, 234
    chemical input, 233–234
    endpoints, 234
    filling data gaps, 234
    profiling, 234
    reports, 234–235
OELs. *see* Occupational exposure limits (OELs)
Off-target promiscuity
    in vivo toxicity and, relationship between, 177
Omics biomarkers, 195
    decision forest for, 206–209, 207f
        microarray gene expression, 209, 210t
        proteomics data, 208–209, 208f
        SNP genotype data, 207–208
    discovery, 196–199, 197f, 199f
        bioinformatics tools and databases, 202–203
        biomarker identification, 204–205
        challenges and solutions in, 201–206
        ensuring data quality, 203–204
        performance estimation and statistical validation, 205–206
        samples and technologies selection, 201–202
    future efforts in use of, 130–131
    overview, 195–197
    predictive models development and, 200–201, 200f
    risk assessment and, 198–199, 198t
    types and representative molecular technologies, 198t
    workflow, 196–197, 197f
Omics data
    and HTS data, incorporation of, 219
OncoLogic, 229–230
Optical character recognition (OCR), 181
Organisation for Economic Cooperation and Development (OECD), 162
    environmental CompTox applications in, 230–235
        SAR and QSAR methods, 231–235
    European Union activities and, 227–228
    SAR and QSAR models, 231–235
Oxidative stress, in liver, 76, 76f

P

Pathway analysis, 185–186
    commonly used collections, 185t
Pathway-based dose-response effects, 125–128
Pathway-based hazard and risk assessment, 115–118, 116f
P-bounds modeling. *see* Probability-bounds (P-bounds) modeling

PBPK models. *see* Physiologically based pharmacokinetic (PBPK) models
PE. *see* Performance error (PE)
Performance error (PE), 10–11
Pharmaceutical industry, risk assessment in
    background, 171–172
    compound's molecular properties and adverse biological phenotype, relationship between, 172–177
        carcinogenicity prediction, 173–174
        computational prediction of genotoxicity, 172–173
        physical properties to *in vivo* toxicity, relationship of, 176–177
        polypharmacology approaches for adverse effects prediction, 177
        specific organ toxicities, prediction of, 174–176
    cost estimates, 171
    gene function and adverse biological phenotype, relationship between, 177–188, 179t
        analysis approaches, 184–188
        causal reasoning, 186
        classical statistical approaches, 186–187
        gene sets, pathways, and gene signatures, 185–186, 185t
        genomic-analysis methods, 182–184
        literature mining, 178–182
        microarrays, 182–183
        next-generation sequencing, 183–184, 183t
        proteomics/metabolomics, 184
        systems biology/virtual liver or heart, 187–188
    SAR/QSAR models in, 7
Physiologically based pharmacokinetic (PBPK) models, 5, 218
    case studies, 7–8
    human PBPK toolkit, 8
    IndusChemFate. *see* IndusChemFate
    metals models, 10–16, 10t
        BBDR HPT axis models, 14–16
        BPA, 12–14, 13f
    sensitive animal subpopulations, 58–61, 59f, 60f
    sensitive human subpopulations, 47–58
        Bayesian approach in, 56–58, 56f, 57f
        individual-based PBPK modeling, 48–50, 48f, 49f, 51f, 51t
        Monte Carlo simulations and, 53–56, 54f, 55t
        probability-bounds modeling, 50–53, 52f, 53f
    VOCs models, 8–9, 9t
Pregnancy
    PBPK modeling of physiological changes, 47
    see also Sensitive subpopulations
    in rat, 58–61, 59f, 60f
Principal component analysis, 186–187, 204
Probability-bounds (P-bounds) modeling

sensitive subpopulations, 50–53, 52f, 53f
Proteomic data set
  DF application to, 208–209, 208f
Proteomics/metabolomics, 184
PubChem, 159, 162
Publicly available data sets, 151
  chemical domain and limitations to data analysis, 164–165
  data compiling methods from multiple sources, 159–161
  designing, 162
  HTS DATA and *in vivo* effects, data semantics and limitations, 165–166
  human biomonitoring chemical exposure data, 154–156
    HTS data, 156–159
    NHANES data set, 154–155
    NHANES IV *vs.* ToxCast™ data sets, 159, 160t
    NMMAPS database, 155–156
    ToxCast™ program, 156–159
  Human Genome Project (HGP), 163–164
  overview, 151–152
  with uses in risk assessment, 152–159, 153t
PubMed, 180–181

Q

QA/QC program, 247–248
QSAR models. *see* Quantitative structure-activity relationship (QSAR) models
QSAR Toolbox, 162
QSPR algorithms. *see* Quantitative structure-property relationship (QSPR) algorithms
Quantitative structure-activity relationship (QSAR) models, 5, 172, 215–218, 231–235
  application of, 6–7
  CompTox methods, current status of, 236–238
  examples, 6–7
  role in pharmaceutical industry, 7
Quantitative structure-property relationship (QSPR) algorithms, 139

R

RAMAS Risk Calcs software, 52
REACH. *see* Registration, Evaluation, and Authorization of Chemicals (REACH)
Reference concentrations (RfCs), 138
Reference doses (RfD), 138
Registration, Evaluation, and Authorization of Chemicals (REACH), 113
  nonanimal testing models alternatives for mammalian testing, 115
  see also Nonmammalian species data to mammalian species data, computational translation of
  overview, 113–114

Regression analysis, 187
RfCs. *see* Reference concentrations (RfCs)
RfD. *see* Reference doses (RfD)
Risk assessment
  defined, 219–220
RMSPE. *see* Root median square performance error (RMSPE)
Root median square performance error (RMSPE), 10–11

S

Safety Evaluation Ultimately Replacing Animal Testing (SEURAT) program, 228
SAR models. *see* Structure-activity relationship (SAR) models
SCIRUS, 180
SDF. *see* Structured data file (SDF)
SELDI-TOF MS. *see* Surface-enhanced laser deposition/ionization time-of-flight mass spectrometry (SELDI-TOF MS)
Sensitive population model (SPM), 46
Sensitive subpopulations
  animal PBPK models, 58–61, 59f, 60f
  defined, 45
  human, physiological differences and PBPK modeling of, 47–58
    Bayesian approach in, 56–58, 56f, 57f
    individual-based PBPK modeling, 48–50, 48f, 49f, 51f, 51t
    Monte Carlo simulations and, 53–56, 54f, 55t
    probability-bounds modeling, 50–53, 52f, 53f
SEURAT program. *see* Safety Evaluation Ultimately Replacing Animal Testing (SEURAT) program
SIMCA. *see* Soft independent modeling of class analogy (SIMCA)
Simplified molecular-input lineentry system (SMILES), 159–161
Single nucleotide polymorphisms (SNPs), 177
  DF-SNPs, 207–208
SMILES. *see* Simplified molecular-input lineentry system (SMILES)
SNP genotyping array, 177, 203–204
SNPs. *see* Single nucleotide polymorphisms (SNPs)
SNPTrack, 203
Soft independent modeling of class analogy (SIMCA), 204–205
Specific organ toxicities, prediction of, 174–176
SPM. *see* Sensitive population model (SPM)
Standard operating procedures (SOPs), 247–248
Steady-state volume of distribution (VD_{ss}), 177
Structure-activity relationship (SAR) models, 172, 218, 231–235
  see also Quantitative structure-activity relationship (QSAR) models

Structure-activity relationship (SAR) models (*Continued*)
  application of, 6–7
  examples, 6–7
  role in pharmaceutical industry, 7
Structure-based computational modeling, 175
Structured data file (SDF), 159–161
Supervised learning techniques, 204
Supervised/class prediction approaches, 186–187
Support vector analysis, 187
Support vector machines (SVMs), 204–205
Surface-enhanced laser deposition/ionization time-of-flight mass spectrometry (SELDI-TOF MS), 208–209
SVMs. *see* Support vector machines (SVMs)
Systems biology, defined, 187–188
Systems biology model
  to predict compound toxicity, 187–188
Systems toxicology, 219

T

Taurocholic acid (TCA), 73
TCA. *see* Taurocholic acid (TCA)
TDIs. *see* Tolerable daily intakes (TDIs)
"3Rs" (reduce, refine, replace) programs, 198–199
Threshold limit values (TLVs), 138
  of carbofuran, 142
TLVs. *see* Threshold limit values (TLVs)
Tolerable daily intakes (TDIs), 138
TOPKAT, 6–7
Total principal component regression (TPCR), 204–205
Tox21 Collaboration, 226–227
TOXCastDB database, 224–226
ToxCast™ program, 152, 156–159
  NHANES IV *vs.*, 159, 160*t*
Toxic Substances Control Act of 1976, 228–229
Toxicity Reference Database (ToxRefDB), 157, 224, 225*t*
"Toxicology in the 21st Century" initiative, 5–6
ToxRefDB. *see* Toxicity Reference Database (ToxRefDB)
TPCR. *see* Total principal component regression (TPCR)
Translational research
  CompTox methods in risk assessment, 221–228
    OECD and European Union activities, 227–228
    Tox21 Collaboration, 226–227
    U.S. EPA, 222–226
Troglitazone, 185
T-test analysis, 187

U

UF_H. *see* Human interindividual uncertainty factor (UF_H)
UMLS. *see* Unified Medical Language System (UMLS)
Unified Medical Language System (UMLS), 182
Unsupervised clustering methods, 186–187
Unsupervised learning techniques, 204
U.S. Environmental Protection Agency (U.S. EPA), 45, 58, 113–114, 152, 222–226
  ACToR, 222, 223*f*
  CSS Dashboards, 226, 227*f*
  DSSTox, 223, 224*t*
  ECOSAR program, 229
  EPI Suite, 229
  ExpoCast program, 226
  HPV program, 114
  OncoLogic, 229–230
  risk- and hazard-based screening, 228–235
    industrial chemicals and pesticides, 228–230
  TOXCastDB database, 224–226
  ToxCast™ program, 156–159
  ToxRefDB, 224, 225*t*
U.S. EPA. *see* U.S. Environmental Protection Agency (U.S. EPA)
U.S. FDA. *see* U.S. Food and Drug Administration (U.S. FDA)
U.S. Food and Drug Administration (U.S. FDA), 12, 171, 175, 196–197, 203

V

VD_{ss}. *see* Steady-state volume of distribution (VD_{ss})
VGDS. *see* Voluntary genomic data submissions (VGDS)
VOC PBPK models, 8–9, 9*t*, 11
VOCs. *see* Volatile organic compounds (VOCs)
Volatile organic compounds (VOCs), 8
Voluntary genomic data submissions (VGDS), 182

W

Weather forecasting models, 1
WuXi, 178

Y

Yahoo!, 180